AF443667

MOLECULE-BASED MATERIALS
The Structural Network Approach

LARS ÖHRSTRÖM

Department of Chemical
and Biological Engineering,
Chalmers Tekniska Högskola,
Göteborg, Sweden

KRISTER LARSSON

National Electron Accelerator Laboratory
for Nuclear Physics and Synchrotron
Radiation Research (MAX-lab),
Lunds Universitet,
Lund, Sweden

ELSEVIER

Amsterdam – Boston – Heidelberg – London – New York – Oxford
Paris – San Diego – San Francisco – Singapore – Sydney – Tokyo

ELSEVIER B.V.   ELSEVIER Inc.   ELSEVIER Ltd.   ELSEVIER Ltd.
Radarweg 29   525 B Street   The Boulevard   84 Theobalds Road
P.O. Box 211, 1000 AE   Suite 1900, San Diego   Langford Lane, Kidlington,   London WC1X 8RR
Amsterdam, The Netherlands   CA 92101-4495, USA   Oxford OX5 1GB, UK   UK

First edition 2005

Library of Congress Cataloging in Publication Data
A catalog record is available from the Library of Congress.

British Library Cataloguing in Publication Data
A catalogue record is available from the British Library.

ISBN-13: 978-0-444-52168-2
ISBN-10: 0-444-52168-2

♾ The paper used in this publication meets the requirements of ANSI/NISO Z39.48-1992 (Permanence of Paper).
Printed in The Netherlands.

Preface

Once upon a time one of the authors, who came from a solution chemistry laboratory where people were happy to think about one or perhaps two molecules at the time, had some very naïve ideas about three-dimensional nets: Take a ligand with enough binding sites, mix with a metal complex with some vacant coordination sites and you should get a three-dimensional coordination polymer. Not so surprising perhaps that this guy did not make a single three-dimensional net during his entire post-doc!

Things improved, however, and when the second author came along as a PhD-student to the first, *he* made a number of 3D-nets. None of these corresponded to the nice designs of his supervisor…

Obviously, these two chemists needed to improve their understanding of three-dimensional nets, and what better way than to write a book? So here we are; we have learnt a lot in the process and hope that you will do so too.

Naturally, we could not have made it to this point without a number of people whose efforts and encouragement were absolutely crucial.

First we want to thank our editor Alexandra Migchielsen and her staff at Elsevier in Amsterdam that came to us with this idea in the first place. Without her enthusiastic support this book would never have been written.

Secondly, a number of colleagues have read various parts of the manuscript during different stages of its progress and we are in great dept to them for their corrections, suggestions and fruitful discussions concerning this work. We thus thank: Jörgen Albertsson, Stuart Batten, Neil Champness, Nina Kann, Joel Miller, and Guy Orpen for devoting some of their precious time to this work.

Moreover, the manuscript of this book formed the framework for a graduate course in "Molecule-Based 3D-nets Nets" at the Department of Chemical and Biological Engineering at Chalmers Tekniska Högskola, and the ideas, improvements and input from Cédric Borel, Mohamed Ghazali and John Tumpane is gratefully acknowledged.

We also thank Michael O'Keeffe for answering questions about the "Reticular Chemistry Structural Resource" database and related issues, Davide Proserpio for help with "ring-counting" and the TOPOS program, Oleg Dolomanov for assistance and discussions concerning the OLEX program, Helene Antonsson at Chalmers Library for VIP treatment, and John Tanaka for providing details about the life of Alexander "Jumbo" Wells.

Remaining errors, misunderstandings and linguistic or logical somersaults are entirely due to our own inadequacies and stubbornness and should not in any way be associated with the above-mentioned persons.

L. Öhrström wants to thank Susan Jagner, Göran Svensson, Joacim Gustafsson, Roger Forsberg and Britt Bergström at the former department of Inorganic Chemistry at Chalmers for their generous "cover-up" for prolonged periods of absence of both body and mind (mostly the latter) and a special thanks is due to the colleagues teaching the courses "Chemistry and Biochemistry" and "Inorganic and Organic Chemistry" whose good natured tolerance of unreliability and absentmindedness during the progress of this work is beyond belief.

L. Öhrström is also grateful to a number of persons who has hosted him (and various parts of his family) during several weeks of writing sessions: The staff at MAX-lab in Lund, Elsie and Claes-Rune Öhrström in Stockholm, Lena Öhrström and Lars Navéus in Malmö, Ulla Kann in Windhoek and Emeric and Sabine Malevergne in Lans-en-Vercors.

Above all, L. Öhrström is thankful to his family for their keen interest and support of this project. K. Larsson would like to thank his fiancé for all her support and for not minding the lost weekends.

Table of Contents

Chapter 1

Introduction and a short dictionary of network terminology

Molecule-based materials are "bottom-up" designed materials where the starting point is one or several molecules with some specific properties that will go through a synthetic process to give a bulk material with desired specifications as to magnetism, porosity, catalytic potential, luminescence, solubility, colour, etc. In order to achieve these properties, there may be a specific three-dimensional arrangement of the molecular building blocks that is required.

Providing that the building blocks are able to connect to each other in specific ways (that is forming directional intermolecular or coordination bonds) the molecules in the crystal may form infinite patterns, see Figure 1.1. If these patterns extend in all three dimensions we may regard the final structure as a three-dimensional or 3D net. We should also be able to carry out the reverse act, specify what net we want, adjust our molecules accordingly and then achieve the specific three-dimensional arrangement (net) that is needed.

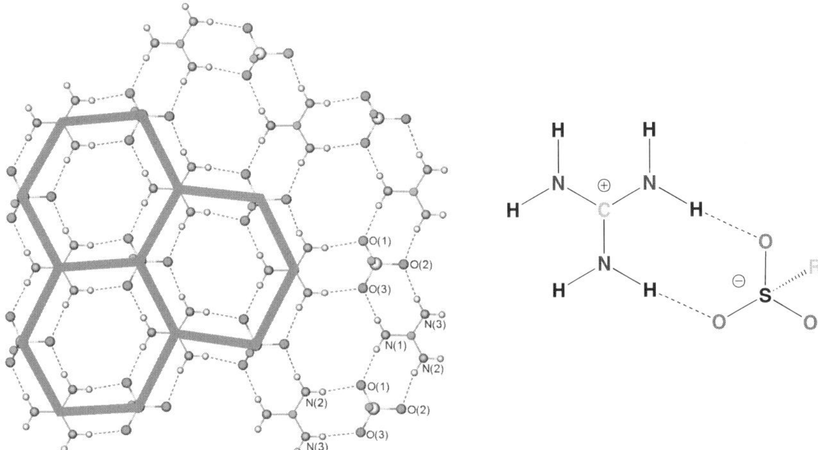

Figure 1.1 A two dimensional net (emphasised in grey) formed by the hydrogen bonds between the cations and the anions in $[C(NH_2)_3][O_3SC_6H_4N=NC_6H_4N(CH_3)_2]$. Only part of the anions are shown for clarity [1,2]. The intersections of the links we call nodes or sometimes vertices and in this case they are centred on the sulphur and carbon atoms. Adapted from ref.1. Reproduced by permission of The Royal Society of Chemistry.

The analysis, preparation, properties and nomenclature of such molecule-based 3D-nets are the subject of this book. We find that this is a subject deserving broader attention than it has hitherto received, and thus this book is aimed at a wider audience.

We should also add that although "molecule-based materials containing 3D nets" may sound like a fancy new term, such compounds found their way into chemical industry and everyday life very early on as the "coordination polymers" Berlin Green, $[Fe(III)Fe(III)(CN)_6]$, Prussian Blue, $K[Fe(II)Fe(III)(CN)_6]$, and Turnbull's Blue, $K_2[Fe(II)Fe(II)(CN)_6]$, synthesised and used as pigments since the early 18th century [3].

1.1. Who is this book written for?

- Our main goal with this book is to give a practical rather than mathematcial introduction for chemists starting in, or curious about, this field; be they faculty, new graduate students or scientists in industry.
- This book is also intended for crystallographers who do not routinely deal with nets, to give you motivation and ideas why to look for these things when solving a structure. A word of warning, however, the presentation is not based on crystallography and you will not find extensive use of space groups, symmetry operators and other such elements.
- Another possible audience is synthetically minded chemists, both "organic" and "inorganic", that wish to grasp this subject without going through a complete crystallographic education.
- We also want to inspire all scientists in the fields of "supramolecular chemistry", "crystal engineering", "self assembly" and "nanoscience" to explore and use these concepts and ideas in their work.
- On a different note, the concept of nets in the analysis of molecular crystals may give us insight to unfold some of the complex relations between intermolecular forces in a crystal leading to a specific structure. Thus we hope that people interested in polymorphism can also profit from this book.
- Lastly, if a few non-scientists should dare to venture beyond the attractive (we hope) cover we trust there are a few items digestible for them as well. All we really demand from you is some memory of LEGO-building or tile laying.

1.2. Nomenclature used in this book

Before starting our journey through the nets, we would like to say something about the nomenclature of nets used in this book.

Numbers appear in the description of nets, and these usually refer to the shortest rings found in the net, and in Figure 1.2 you can see a net based on 10-rings, or 10-gons. If you examine the connection points in this net, and these may represent atoms or molecules, you will see that each of them connects to only three neighbours. Thus, the connectivity of this net is three, and we can call it a (10,3)-net or alternatively a 10^3-net.

Throughout we will try to use the common names that chemists and structural scientist in different areas may be likely to recognize, for example (10,3)-a or $SrSi_2$, but we will also adopt the lower-case, three letter codes proposed by O'Keeffe et al. [4] and print these in bold face, i.e. **srs** (short for $SrSi_2$). These codes, and a lot of useful additional information, including coordinates and space groups and examples, can be found in an electronic database created by O'Keeffe and Yaghi, available free of charge on the world wide web [5]. The reader will find more on nomenclature in Chapters 4 and 10. Our principle will be to use all common names for the nets the first time they appear in each chapter but for brevity reasons we will use the three letter codes when we continue to discuss them.

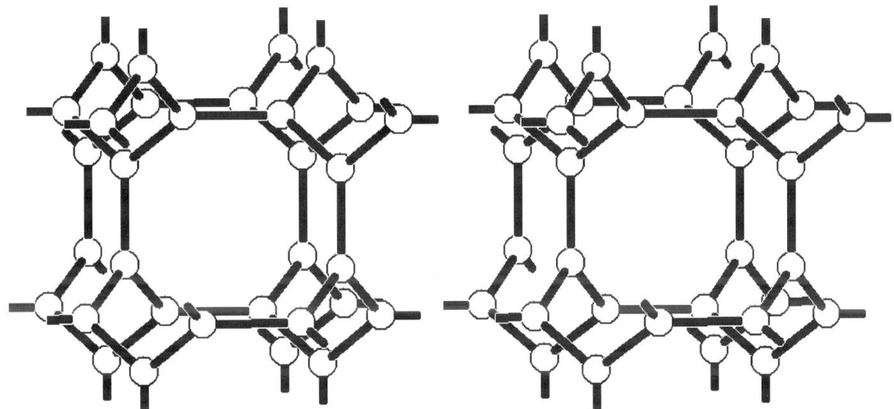

Figure 1.2 In this stereo-drawing[1] you see a net where the shortest circuit that gets you back to where you began contains 10 nodes (in white), and all nodes are three-connected. This is the (10,3)-a, $SrSi_2$ or **srs** (short for $SrSi_2$) net that is found in solid materials as diverse as $[Zn_2(1,3,5-$ benzenetricarboxylate)$(NO_3)]\cdot(H_2O)(C_2H_5OH)_5$ [7] (in which case the nodes are placed in the centres of the benzene rings) and hydrogen peroxide [8] (where the nodes are oxygen atoms and the links are one O-O bond, one O-H bond and one hydrogen bond between O and H).

---

[1] A number of stereo-drawings will appear throughout the book. The stereo effect depends on an angular separation between the left and right images, corresponding to the different view directions for our left and right eyes. How to do it: "Gaze at the stereo pair, keeping your eyes level and cross your eyes slightly. As you know, crossing your eyes makes you see double, so you will see four images. Try to cross your eyes slowly, so that the two images in the center come together. When they converge or fuse, you will see them as a single 3D image. The fused image will appear to lie between two flat images, which you should ignore. When you are viewing correctly, you see three images instead of four." This and other helpful hints can be found on a web page by Dr. Gale Rhodes [6]. The stereo-images in this book are convergent (cross-eye viewing).

1.3. Why did we write this book?

1.3.1. Some history

About 50 years ago Alexander F. Wells, then at ICI[2] in England, published the first paper in what was to become a remarkable series of articles laying the foundations of the concept of nets[3] in solid-state chemistry. The paper was entitled "The geometrical basis of crystal chemistry I", [9] and during the next 20 years he continued this work, by shrewd insight, intuition and keen model-building. The series culminated in the book "Three dimensional nets and polyhedra" published in 1977 and a shorter sequel in 1979 [10,11]. A short version was also included in later editions of his classic textbook "Structural Inorganic Chemistry" [12].

Although the nets in question were mostly classical "inorganic" ionic solids, building nets through for example shared oxygen atoms, or strong ion-ion interactions, Wells did a service to chemistry by not excluding "organic" molecular based nets. In certain areas he was also well before his time, deriving many types of nets before they were actually found in real structures.

1.3.2. A rapid expansion

The rising interest in this area during the last 10 years seems to be the consequence of several factors converging:

- "Supramolecular chemistry" and "self-assembly" have evolved from mainly taking place in solution to include the solid state. This is sometimes referred to as "crystal engineering".
- Many applications of new molecule-based materials envisaged by scientists demand the organisation of the building blocks in an ordered three-dimensional architecture, often well described as a net.
- Finally, the evolution of computers and programs for crystallographers, including access to the Cambridge Structural Database (CSD), and molecular graphics in general, has given us tools to discover, analyse and visualise (i.e. communicate) three-dimensional nets in an unprecedented way.

But of course, science progresses also largely because of individual achievements, and without diminishing the importance of other contributions we would like to mention the work of M.C. Etter [13] and G. Desiraju [14] in the

---

[2] Imperial Chemical Industries, Former UK based chemical conglomerate, split up in the mid 1990's.

[3] We normally prefer the shorter "net" to the longer, and as far as we can see synonymous word "network".

context of hydrogen bonded solid state compounds and R. Robson for coordination polymer nets [15]. Indeed, Hoskins and Robson in their 1991 landmark paper on the "Design and construction of a new class of scaffolding-like materials…" prophesise that "…it may be possible to devise…solids with relatively huge empty cavities. Materials combining good or even high thermal, chemical, and mechanical stability with unusually low density may therefore be afforded." [15]. As we shall see, 15 years later this had all come true!

Although Wells' work is still widely read and cited, see Figure 1.3, it has been long out of print and it is not obvious that a copy can be located by your university library.[4] This is also the case for a related book by Michael O'Keeffe and Bruce Hyde, "Crystal Structures I: Patterns and Symmetry" that is also difficult to obtain [16].

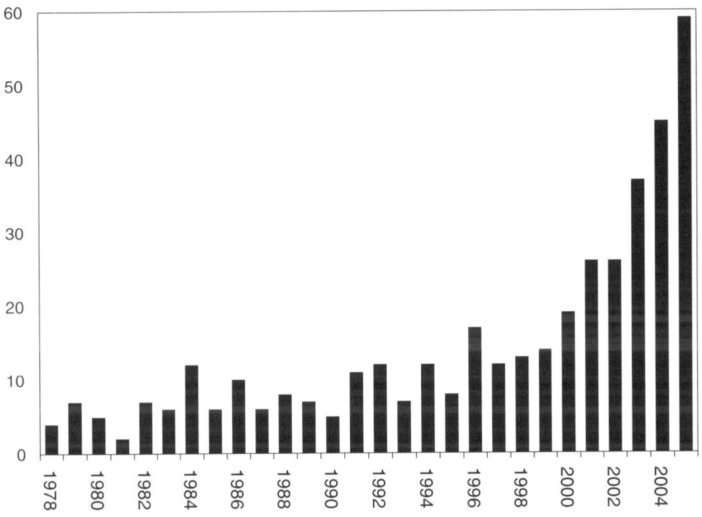

Figure 1.3 Number of citations per year to A. F. Wells, 'Three-dimensional nets and polyhedra', John Wiley & Sons, 1977 [10] according to Science Citation Index. Projected data for 2005.

1.3.3. Problems with names and nomenclature

"Three-dimensional nets and polyhedra" contains virtually the only (and the original) definitions of the still widely used and understood "Wellsian nomenclature" of 3D-nets. This simple designation is based on connectivity, number of nodes in the shortest circuit, and, when needed, a letter designating symmetry in a descending scale as in the net called (10,3)-a (-a being the most

---

[4] We once in desperation had a copy flown in from Berkely, CA.

symmetric), see Figure 1.2 (one of our favourite nets and now designated as the **srs** net).

The trouble is that without access to this book, errors will be made, propagated and soon the "Wellsian nomenclature" may become meaningless. We actually admit to making such an error for this very reason [17][5]. This embarrassing fact, however, prompted us to go through and review all (10,3)-nets possible, including a couple not discovered by Wells, [18] their geometric requirements, space groups etc so that an easily available resource could be created [19]. This initial effort was the starting point of the present work.

Another problem with names of the type (10,3)-x and even $SrSi_2$ is that they are notoriously difficult to track in Scifinder (CA), Science Citation Index and other databases (a problem we will briefly touch upon in section 4.4). Thus we should be grateful to O'Keeffe, Yaghi, Delgado-Friedrichs, and their groups, for their classification work and for the comprehensive web site dedicated to nets, the source of the three letter codes appearing throughout the text [4,5].

1.3.4. What this book is and is not

However, this book is more than "Wells for beginners" or "Nets for Nuts", since there are many issues that concern molecular 3D-nets that were not dealt with by Wells, or in subsequent related books [16]. Moreover, many topics are of a character not customarily dealt with in a normal research paper, review, highlight or perspective article.

We also want to emphasis that, unlike the books by Wells, O'Keeffe & Hyde, and others, this publication is not an original contribution to science, and not the result of long and fruitful labour in this field. It is rather the work of two curious minds that started in this field not long ago. We have been writing the book that we ourselves are missing, making a comprehensive introduction to this field, from application and synthesis to mathematics. If the book lacks in depth in certain areas we hope that it will make up for this in breadth.

Our intention is also to present the most important work in the field, but we make no promises of completeness, nor any excuses if a reader finds his or her own work missing, our selection is subjective. A number of reviews have been very helpful in this context, [4,20-33] although our goal has been to include a fair number of examples from very recent research as well.

1.3.5. The contents of the book

First, however, we have to convince you that the "net perspective" of molecular crystals is something useful. This we do by discussing how nets can be an organising principle for solid state compounds when there are strong

---

[5] The 3D net made out of hydrogen bonded [Co(2,2'-biimidazolato)$_3$] molecules described in this article is a **utp** or (10,3)-d net not a $ThSi_2$ net (**ths** net) as stated.

directional intermolecular forces in play. We also deal with how this organising principle can be utilised to make useful materials for the future science and industry.

Hopefully being convinced at this point, we move on to some definitions. This is important since *all* molecular solids contain three-dimensional arrangements of molecules, and one can, with a little imagination, analyse any such solid in terms of a 3D-net. We thus need to answer the questions: What is a node? How are nodes connected? What forces can be used to connect them? When is it useful to analyse a material in terms of 3D-nets, 2D-nets or chains?

Nomenclature needs special attention, and is one of the reasons for writing this book. Not only because of the risks of possible confusions arising from difficulties in accessing the "old" literature, but also since there are currently at least four different ways of naming nets: From an indifferent note of "a 3D-net" to descriptions based on similarities with "inorganic" crystal structures (as in "$SrSi_2$"), the Wellsian description(s), and the more exhaustive Extended Schläfli, or Vertex Symbols (as in 10_5 10_5 10_5 for the net in Figure 1.1 where the subscript refers to the number of different 10-rings formed by the net) to the zeolite type three letter codes (**srs**). This is certainly far from an ideal situation, especially when one wants to retrieve information through a text-and-molecules only database, containing no searchable structural information, as the Chemical Abstracts.

Searching databases for nets, interpenetration and related problems also needs special treatment, and a section at the end of Chapter 4 deals with this. We also briefly comment on the use of the Cambridge Structural Database, and in this context we should mention two very useful papers from the groups of Proserpio [31] and O'Keeffe & Yaghi [30].

We are then ready to present an overview and some real examples of the most common 3D-nets derived by Wells and others. The purpose being to provide a good visual reference and an understanding of how the nets are connected, and how they differ and relate to each other. It will also provide a spectrum of different net-forming compounds and here and there we will briefly comment on synthesis and properties of selected examples.

Next we move on to some less common molecule-based nets. This part is divided into several chapters, starting with the three-connected nets and subsequently dealing with the other nets in order of the connectivity of the nodes. Doing so, we will not discriminate by atomic number, indeed we find that in this area, as in many others, the terms "inorganic" and "organic" are not very helpful, and instead imply heavy differences in composition and behaviour that simply do not exist.[6]

---

[6] While pursuing this line of thought it is worthwhile to note that the original definition of "organic" is a substance isolated from a living source (J.J. Berzelius). Thus, most petroleum based organic chemistry could be argued to be inorganic…

Nets as mathematical constructs have a special chapter. We will try to show the usefulness and the beauty of this approach without making any heavy demands on the reader's (and our own) mathematical abilities.

We have at this point seen many examples of interpenetration, that is when two or more nets are intertwined or entangled. Chapter 11 will deal with this in more detail, both in terms of definitions, so that we can readily speak of this phenomenon, and in terms of an analysis of both the nets themselves (for example length of the links between the nodes, and the porosity of the nets) and of the voids within them.

Thus far, we have concentrated on the descriptive use of nets. Now we will turn to synthesis. In our context, this is not only "supramolecular" that is forming bonds (on purpose) that go beyond the molecule, but also in some sense "molecule neutral" chemistry. Our goals will thus *not* be specific molecules or ions and the excitement of finding out what they can do together, *the synthetic target will be a specific 3D-net*. The problem we have to solve is: *Which* specific sets of molecules can we use to synthesis the net we have chosen? The molecule will be the end to a means, the secondary target only (as long as it possesses the useful property we seek). As the variation of reaction parameters such as temperature, concentration and solvents may result in different products and nets we will also briefly deal with *polymorphism* or perhaps more often *pseudo-polymorphism*. That is, we may encounter different nets having different solvents co-crystallised, thus not "real" polymorphs since the composition does in fact differ between the materials.

Last, there is a brief chapter on computational tools used in crystallography, visualising and for net-nomenclature, the reason for the brevity being the short half-life of this information.

At the end of the book we have compiled what we hope to be a handy set of pictures, space groups, coordinates etc for all the nets encountered in the book.

1.4. Some words, their origin and current usage

Please feel free to skip this section until later and move ahead to Chapter 2!

Misunderstandings are a major source of human conflict. Likewise, new terminology can cause much frustration, especially if it is considered (though often unfairly) as redundant, "new words for old things", or even on the border of Orwellian "newspeak" [34].

Thus, we have to exercise extreme care when coining and promoting new terms and phrases. We have to be aware of its effects not only on the community it is primarily conceived for, but also for its effects on "outsiders", particularly if these words already have a different meaning (although this is not always easy to find out!).

For example, phrases such as "...a new tecton to be used in self assembly of a supramolecular coordination polymer..." is clear to most people in the field, although some would probably have phrased it differently. However, to a polymer chemist it may seem confusing, and one might even take offence over the use of such "meaningless jargon". He or she might have heard about coordination polymerisation (which in polymer chemistry is a polymerisation catalysed by coordination complexes [35]) but will certainly not associate the crystalline and often brittle compounds we refer to as coordination polymers with "polymers".

In this section we will explore the origins (etymology), uses, and possible causes of conflict of certain words currently in much use in the "crystal engineering" community. In doing so we do not proclaim ourselves as a self-appointed task force for linguistic purity, rather we would like to make you aware of the pros and cons of certain words and possibly make some humble suggestions. Because finally, the language we use is a very personal thing, and up to ourselves to judge.

Perhaps after reading this you can make a more informed choice, we certainly feel that we can after having written it!

In this small survey, we have mainly used the Chemical Abstracts database to get some clarity concerning the origin of some words and how they have evolved. Some terms, such as *coordination polymer*, have been around for a while. Others have just started to appear, one example being *metal-organic frameworks*. Note that our statements of "first usage" or similar phrases means that they have appeared in the title, abstract, index terms or equivalent searchable information. These words most likely appeared earlier in the bulk text of articles but this is much more difficult to trace.

1.4.1. Tectons and building blocks

The molecules used in the making of a net are often called *building blocks* but sometimes also *tectons*. Tecton is derived from the Greek "tekton" meaning builder and these words are therefore synonyms. However, there is a slight tendency to associate "building block" with very rigid and hard objects, such as bricks, whereas molecules are often quite flexible and "soft". Therefore it could perhaps be fitting to reserve "building block" for really rigid and strongly bound nets and use tecton as a more general term. The problem with "tecton" is that it may give other associations to outsiders whereas the meaning of "building block" is intuitively clearer.

In its present meaning, "tecton" was first used in the early 1990's, [36-38] and the occasional earlier findings of the word in chemical context is associated with geochemistry and the earth's *tectonic plates*. It is used sparingly, some 10-15 citations a year.

As we are aiming for a wider audience, we will avoid using "tecton" in this book and in general we will refer to the constituents of a net as "building blocks".

1.4.2. Nets of coordination polymers and metal-organic frameworks

These building blocks can be combined into *nets* (or *networks*) where the net, whether 1-, 2- or 3-dimensional, is sometimes called a *coordination polymer* if it is held together by metal-ligand or coordination bonds. For the 3D-case, *metal-organic frameworks* can be used. The acronym "MOF" for metal-organic framework should be avoided since the parent term is not yet well established and there seem to be no great need for an abbreviation in the context where it is used.

One might object to the term "coordination polymer" on the grounds that these compounds do not share the usual "plastic" characteristics of, for example, polyethylene. However, they do have this in common with biopolymers as DNA and RNA. In fact, the first use of the term "coordination polymer" predates present day polymer chemistry (Staudinger in the 1920's [39]) by almost a decade, [40] and it has been in use since the 1950's, [41] see Figure 1.4, although in some of the early occurrences it is used for metal ions coordinated to polymers [42].

The conclusion is that, whether we like it or not, "coordination polymer" is a term we have to live with. In this context we should also point out that searching the chemical literature is made difficult if old terms are continuously exchanged for new fancier words; indeed, the main argument for language conservatism in general is that we should be able to read older texts in our own language without a dictionary.

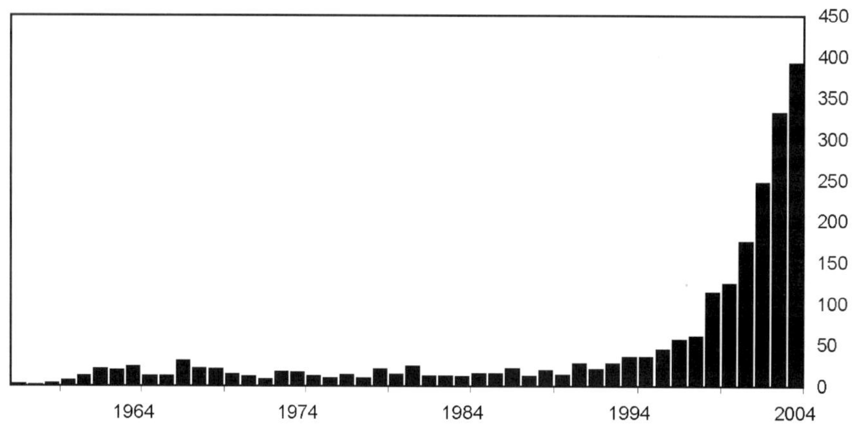

Figure 1.4 The use of "coordination polymer" through the last 45 years. The first occurrence of this term in the Chemical Abstracts is in 1916 [40].

Note also that a hydrogen bonded network of coordination complexes is not a coordination polymer – the polymerisation must be by coordination bonds.

"Metal organic framework" is a recent invention, [43] and, as with "building block", it invokes the picture of something hard and rigid and it has been suggested that it should be reserved for structures that have such characteristics [4]. The combination "metal-organic" means that we are dealing with a classical coordination compound, and not an organometallic species (normally containing a metal-carbon bond). While "coordination compound" and "complex" are useful and well known concepts within (inorganic) chemistry, it is nevertheless a term that may be puzzling to outsiders, and "metal-organic compound" is a more intuitively clear term although there is a small risk of confusion with the term "organic metal".

We will occasionally refer to "coordination polymer", as opposed to hydrogen bound nets, but "framework" and "metal-organic" will rarely appear. However, as scientific descendants of a long line of inorganic chemists, we will use "complex" and "coordination" throughout the text.

1.4.3. Crystal engineering and reticular chemistry

The main objective in *crystal engineering* [2,44-48] is to assemble a target structure in the solid-state, which has a specific ordering of the molecular building blocks, often providing some additional property not present in a random ordering of the same molecules. Although often attributed to G. M. J. Schmidt in 1971, [46,49] the term *crystal engineering* is actually first encountered in an abstract from the proceedings of the American Physical Society meeting in Mexico City, 1955, where Pennsylvania State University Crystallographer Ray Pepinsky wrote:

"Crystallization of organic ions with metal-containing complex ions of suitable sizes, charges and solubilities results in structures with cells and symmetries determined chiefly by packing of complex ions. These cells and symmetries are to a good extent controllable: hence crystals with advantageous properties can be 'engineered'..." [50] In retrospect, this statement seems a bit optimistic, but undoubtedly good progress has been made during these 50 years.

It has also been argued that the practice of crystal engineering has been around much longer than the term itself [51]. Nevertheless, crystal engineering is today a wide field where the deliberate construction of 3D-nets is a subfield. To classify this a new term has been put forward, i.e. *reticular*[7] *chemistry* [4,5,30,54]. The definition reads: "...the practice of logical synthesis

---

[7] If you do not understand the word "reticular" you are not alone, we can assure you! It resembles "retina" but has nothing to do with sight. The Oxford English Dictionary states that reticular is something "resembling a net in appearance or construction" or "net-like", as in the giraffe sub-species "reticulated giraffe" *Giraffidae Giraffa, camelopardalis reticulata*. It also appears as a verb; "to reticulate", to form a net. (The Oxford English Dictionary contains some 600 000 words in comparison with for example French where estimates from the French Academy are 100 000 – 200 000 words) [52,53].

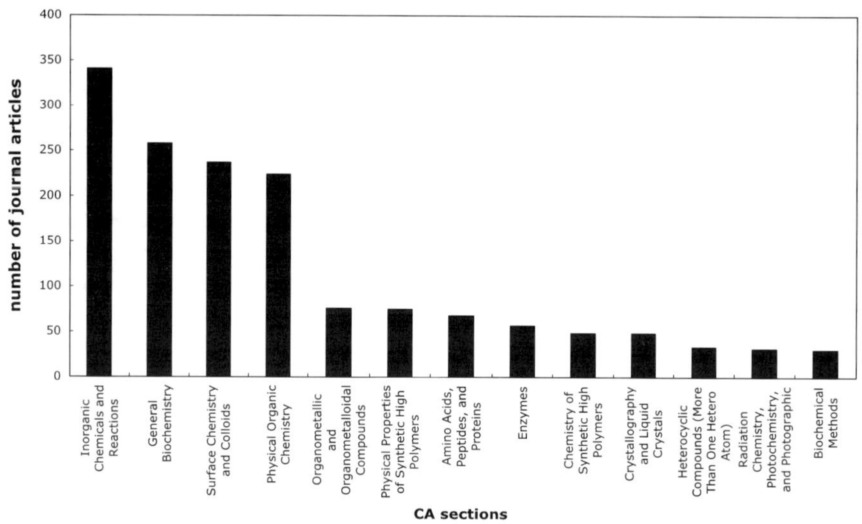

Figure 1.7 Papers in "self-assembly" 1993-2003 by Chemical Abstract index sections from "high profile" (high impact factor) journals; Journal of the American Chemical Society, Chemistry a European Journal, ChemComm., Angewandte Chimie, Journal of Biological Chemistry, Biochemistry, Physical Reviews Letters, Nature, and Science.

1.4.5. Glossary of terms

Binodal	Containing two topologically different nodes
Building block	Neutral or charged molecular unit interacting to form an extended structure (1D, 2D or 3D)
Complex	Metal ion and attached ligands.
Coordination compound	Metal ion and attached ligands.
Coordination polymer	Extended structure containing unbroken chains of -metal-ligand-metal-ligand- units.
Crystal engineering	Solid-state supramolecular chemistry with a strong emphasis of design and structural understanding.
Entangled	Forming structures in which at least two rings are interlocked.
Framework	A rigid 2D or 3D net.

Framework	A rigid 2D or 3D net.
Interpenetrated nets	One or more nets growing in the voids and channels of another net, and that cannot be separated without breaking any links in the net.
Interwoven	Two or more interpenetrated nets growing together displaying some kind of attractive interaction like double-helix formation. May also be used for non-interpenetrated ways on interaction between 1D or 2D polymers.
Metal-Organic Framework (MOF)	A more rigid form of 3D coordination polymer.
Molecular Magnetism	The study of the magnetic properties of molecular based compounds having unpaired electrons.
Molecule-Base Materials	Materials in which a molecule with a specific property give additional properties by its solid state arrangement.
Net	An extended ordered structure of nodes and connections between them.
Network	Net
Node	Vertex, point where links meet in a net or polyhedron.
Open Framework	A 3D net containing accessible channels.
Reticular Chemistry	Chemistry of net-forming molecules and ions.
Secondary Building Unit (SBU)	The molecular unit defining the geometry at a node.
Self-Assembly	More than two molecular units interacting via a predefined pattern of coordination bonds, hydrogen bonds or weaker interactions.
Supramolecular	At least two molecular units interacting via coordination bonds, hydrogen bonds or weaker interactions.
Synthon	The bonding interaction or pattern between two molecules in a supramolecular unit.
Tecton	Building block with some flexibility
Uninodal	Containing only one topological type of node.
Vertex	Node

References

[1] N. J. Burke, A. D. Burrows, M. F. Mahon, S. J. Teat, Crystengcomm 6 (2004) 429.

[2] D. Braga, L. Brammer, N. R. Champness, Crystengcomm 7 (2005) 1.

[3] Encyclopædia Britannica Online, Encyclopædia Britannica Inc., 2005,

[4] O. M. Yaghi, M. O'Keeffe, N. W. Ockwig, H. K. Chae, M. Eddaoudi, J. Kim, Nature 423 (2003) 705.

[5] M. O'Keeffe, O. M. Yaghi, Reticular Chemistry Structure Resource, Tucson, Arizona State University, 2005, http://okeeffe-ws1.la.asu.edu/RCSR/home.htm

[6] G. Rhodes, Stereo Viewing, Department of Chemistry, University of Southern Maine, USA, 1997, http://www.usm.maine.edu/~rhodes/0Help/StereoView.html#con

[7] O. M. Yaghi, C. E. Davis, G. Li, H. Li, J. Am. Chem. Soc. 119 (1997) 2861.

[8] C. S. Abrahams, R. L. Collin, W. N. Lipscomb, Acta Cryst. 4 (1951) 15.

[9] A. F. Wells, Acta Cryst. 7 (1954) 535.

[10] A. F. Wells, Three-dimensional nets and polyhedra, John Wiley & Sons, New York, 1977.

[11] A. F. Wells, Further Studies of Three-Dimensional Nets, Polycrystal book service, Pittsburgh, 1979.

[12] A. F. Wells, Structural Inorganic Chemistry, 5th ed. Clarendon Press, Oxford, 1984.

[13] M. C. Etter, J. Phys. Chem. 95 (1991) 4601.

[14] G. R. Desiraju, Angew. Chem. Int. Ed. 34 (1995) 2311.

[15] B. F. Hoskins, R. Robson, J. Am. Chem. Soc. 112 (1990) 1546.

[16] M. O'Keeffe, B. G. Hyde, Crystal Structures I: Patterns and Symmetry, Mineral Soc. Am., Washington, 1996.

[17] L. Öhrström, K. Larsson, S. Borg, S. T. Norberg, Chem. Eur. J. 7 (2001) 4805.

[18] E. Koch, W. Fischer, Z. Kristallogr 210 (1995) 407

[19] L. Öhrström, K. Larsson, Dalton Trans. (2004) 347.

[20] S. R. Batten, R. Robson, Angew. Chem. Int. Ed. 37 (1998) 1461.

[21] S. R. Batten, Crystengcomm (2001) 1.

[22] A. Erxleben, Coord. Chem. Rev. 246 (2003) 203.

[23] S. L. James, Chem. Soc. Rev. 32 (2003) 276.

[24] S. Kitagawa, K. Uemura, Chem. Soc. Rev. 34 (2005) 109.

[25] W. Lin, H. L. Ngo, in: Chemistry of Nanostructured Materials, P. Yang (Ed.), 2003, pp. 261.

[26] L. J. May, G. K. H. Shimizu, Zeitschrift Fur Kristallographie 220 (2005) 364.

[27] B. Moulton, M. J. Zaworotko, Chem. Rev. 101 (2001) 1629.

[28] N. L. Rosi, M. Eddaoudi, J. Kim, M. O'Keeffe, O. M. Yaghi, Crystengcomm 4 (2002) 401.

[29] J. L. C. Rowsell, O. M. Yaghi, Microporous Mesoporous Mater. 73 (2004) 3.

[30] N. W. Ockwig, O. Delgado-Friedrichs, M. O'Keeffe, O. M. Yaghi, Acc. Chem. Res. 38 (2005) 176.

[31] V. A. Blatov, L. Carlucci, G. Ciani, D. M. Proserpio, Crystengcomm 6 (2004) 377.

[32] L. Carlucci, G. Ciani, D. M. Proserpio, Coord. Chem. Rev. 246 (2003) 247.

[33] R. Robson, J. Chem. Soc., Dalton Trans. (2000) 3735.

[34] G. Orwell, 1984, 1949.

[35] W. Kuran, Principles of Coordination Polymerisation, Wiley, New York, 2001.

[36] M. Simard, D. Su, J. D. Wuest, J. Am. Chem. Soc. 113 (1991) 4696.

[37] X. Wang, M. Simard, J. D. Wuest, J. Am. Chem. Soc. 116 (1994) 12119.

[38] D. Su, X. Wang, M. Simard, J. D. Wuest, Supramol. Chem. 6 (1995) 171.

[39] A. J. Ihde, The Development of Modern Chemistry, Dover Publications, New York, 1984.

[40] Y. Shibata, CAN 11:5339, Journal of the College of Science, Imperial University of Tokyo 37 (1916) 1.

[41] S. Kanda, Y. Saito, Bull. Chem. Soc. Jpn 30 (1957) 192.

[42] F. W. Knobloch, W. H. Rauscher, J. Polym. Chem. 38 (1959) 261.

[43] O. M. Yaghi, G. Li, H. Li, Nature 378 (1995) 703.

[44] D. Braga, G. Desiraju, J. Miller, A. Orpen, S. Price, Crystengcomm (2002) 500.

[45] C. V. K. Sharma, J. Cryst. Growth Des, (2002).

[46] D. Braga, Chem. Commun. (2003) 2751.

[47] L. Brammer, Chem. Soc. Rev. 33 (2004) 476.

[48] D. Braga, F. Grepioni, Coord. Chem. Rev. 183 (1999) 19.

[49] G. M. J. Schmidt, Pure Appl. Chem. 27 (1971) 647.

[50] R. Pepinsky, Phys. Rev. 100 (1955) 971.

[51] P. Erk, H. Hengelsberg, M. F. Haddow, R. v. Gelder, CrystEngComm 6 (2004) 474.

[52] The Oxford English Dictionary, 2nd Oxford University Press, Oxford, 1989.

[53] Académie Française, 2005, http://academie-francaise.fr/

[54] M. Eddaoudi, J. Kim, N. Rosi, D. Vodak, J. Wachter, M. O'Keefe, O. M. Yaghi, Science 295 (2002) 469.

[55] E. J. Corey, C. X.-M., The Logic of Chemical Synthesis, John Wiley & Sons, New York, 1989.

[56] S. L. Keely, Jr., F. C. Tahk, J. Am. Chem. Soc. 90 (1968) 5584.

[57] F. H. Allen, W. D. S. Motherwell, Acta Cryst. B 58 (2002) 407.

[58] A. G. Orpen, Acta Cryst. B 58 (2002) 398.

[59] F. H. Allen, Acta Cryst. B 58 (2002) 380.

[60] K. S. Hagen, J. G. Reynolds, R. H. Holm, J. Am. Chem. Soc. 103 (1981) 4054.

[61] G. M. Whitesides, B. Grzybowski, Science 295 (2002) 2418.

[62] I. Dance, New J. Chem. 27 (2003) 1.

[63] J. M. Lehn, Science 260 (1993) 1762.

Chapter 2

Why bother with nets ?

In the beginning there was close packing, of atoms and ions, of apples and oranges, of stone upon stone, to give cathedrals, mosques and bridges. Then came the engineers. New types of steels, daring framework constructions, bridges spanning unprecedented gaps using intricate three-dimensional nets, spires and office buildings aiming for the sky. A new world was built.[1]

In chemistry we now see the deliberate construction of molecule-based frameworks, 3D-nets and networks. Are we on the dawn of a materials age built upon the nano equivalent of the framework steel bridge? Or are all we see useless but beautiful molecular graphics equivalents of the Eiffel tower (Figure 2.1)?

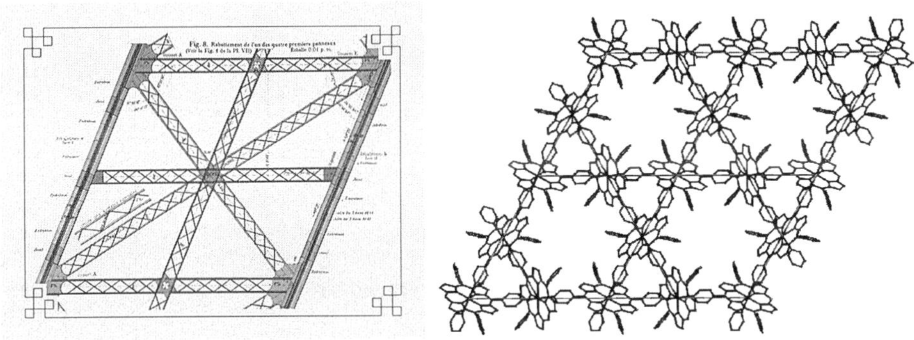

Figure 2.1 Left: one of Gustave Eiffel's blueprints for the framework of the Eiffel Tower in Paris, France [4] Right: A 3D zinc-tetrakis(4-pyridyl)porphyrine framework containing a four-connected diamond (**dia**) net [5].

2.1. Possible impact on society

The possible impact on society is still too early to judge. Certainly, the analysis of molecular (or supra-molecular) crystal structures in terms of three-dimensional nets has provided a very useful way of thinking and looking at many of these compounds [6]. There have also been deliberate and successful

---

[1] Please note that our intent is not to diminish the importance of "close packing", indeed, exciting new developments are taking place in this area too [1-3].

designs of 3D-systems forming the desired predefined crystal structures, [7-11] although one suspects that some of these reports may be tainted by hindsight. Especially, we have to mention the rigid porous networks constructed from small metal ion clusters and organic di-acids as spacers, see Figure 2.2 [12-19].

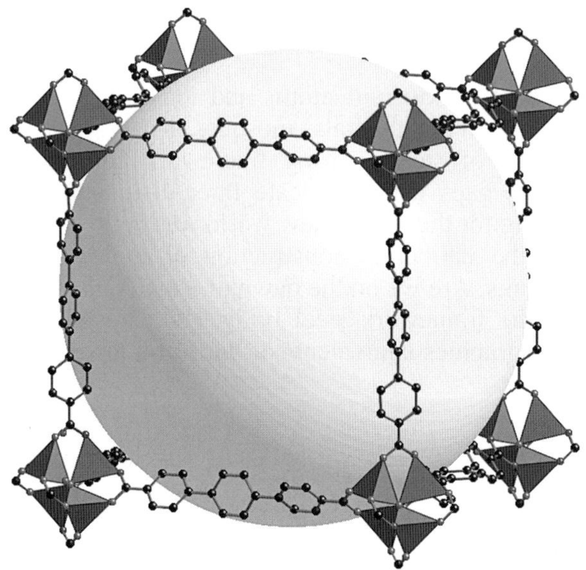

Figure 2.2 [Zn$_4$O(terphenyl-4,4'-dicarboxylate)$_3$] a "metal-organic framework" with very high free volume. In this family of compounds, free volumes up to 91.1% have been recorded [13,20]. This net is built from four ZnO$_4$ tetrahedra (in the picture the Zn(II) ions in the centre of the tatrahedra are not displayed) sharing one oxygen corner and having bridging carboxylate ligands. The result is a net that has four-rings as shortest circuits and that is six-connected, thus a (4,6) net. The short code for this net is **pcu** (primitive cubic packing). See further section 5.2.8 for this net. Figure reproduced with permission from M. O'Keeffe.

Patent literature may say something about the transformation of science into commercial technology, although the analysis of such data is a complex issue [21-23]. In Figure 2.3 we see that in the last few years the numbers of patents in the field seem to be rising, although the absolute numbers are still very small (patent specialists claim that about 95% of all patents are commercialy worthless) and do not suggest any immediate impact on society. As a comparison we note that in the area of fuel cell technology development, a technology that appears to be nearing commercial introduction, some 6000 patents have been issued [23].

Nevertheless, in this chapter we will try to show how "net-forming" is indeed an organising principle followed by nature when there are strong directional intermolecular forces in play and how "net-identification" can help us undeerstand this. We will also deal with how this organising principle can be put to work to make useful compounds for the future science and industry.

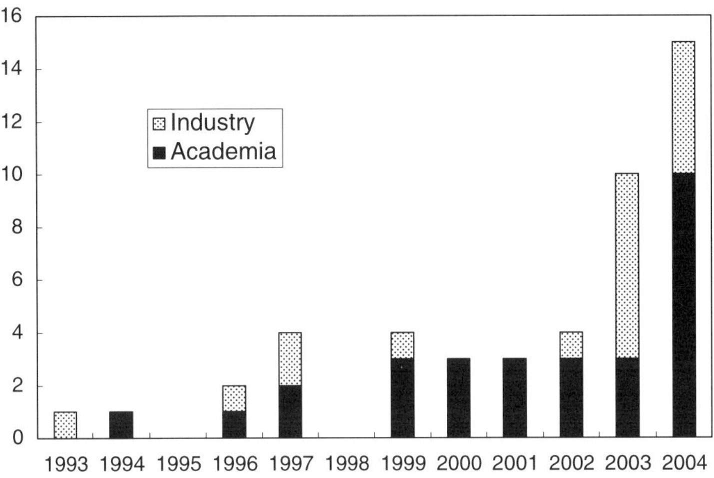

Figure 2.3 The number of patents and patent applications related to molecular 3D-nets the last 12 years. Please note that in general, about 95% of all patents are useless, and these data do not show that a breakthrough for this science in technology is imminent. They merely suggest that the interest in applying this chemistry to real problems is increasing. See also section 2.2.3

2.2. A fruitful way of looking at molecular crystals with strong, directed intermolecular bonding

2.2.1. The analysis of a crystal structure

The first crystal structures determined in the beginning of the last century could readily be interpreted as close packing of hard spheres, and this is also how we encounter the solid state in most contemporary textbooks. This is undoubtedly a good working model, as it reproduces many experimental data and can be used in the design of new solid phases of ionic compounds.

However, it has been pointed out that the success of this method in reproducing experimental data cannot be taken as universal proof of its physical correctness. The lattice enthalpies of lithium metal can, for example, be calculated by modelling this material as small Li^+ cations surrounded by large, close packed Li^- anions, and the agreement with experiment is good, although most people would consider Li(s) a metal, not a salt [24].

When the structures of larger molecular compounds came within reach, focus was on the connectivity and geometries of these individual components.

First, there is the rather pragmatic objective to get a mental picture for oneself of the particular structure, especially if it is important to understand the molecular arrangement in three-dimensions. For example, plotting the (expanded) unit cell contents of the resorcinol structure (Figure 2.6) does not help us a great deal in understanding its three-dimensional packing unless we mentally multiply the unit cells in all three dimensions. If we instead see a 3D-net as the underlying structural theme (Figure 2.4) it becomes much clearer.

Thus, if a relatively small number of 3D-nets can be understood (and it seems that most such structures contain the 10-15 most common nets) we can then use them to rationalise many molecular arrangements in the solid state. This gives us good help in understanding structures, similar to the great aid we have from envisaging closely packed anions with small cations in tetrahedral and octahedral holes, although we know that this is in many ways an incomplete model. Wells himself considered that in this way 3D-nets could be considered more fundamental than space groups, as the latter would be a consequence of the former [28].

On a second level we use the 3D-net to understand why certain molecules form a particular structure. Then we need some criteria for what is and is not a net. It is generally not enough to find a pattern; this pattern has also to be crucial for the formation and stability of the structure. Obviously, in many cases this is not straightforward, since the intermolecular interactions do not carry labels telling us about their strengths. Nets built by coordination bonds seem a bit easier to define, but frequently have counter ions outside the net, complicating the issue. We will pursue this question further in Chapter 3.

Disregarding this set of hurdles, a second-level net analysis should give us a good tool for envisaging both the structure and the forces that are crucial for its formation.

The final level has to do with a more active understanding that can help us in our preparative labours in the laboratory. Analysing nets and the interactions within them should give us information that can be put to use in our pursuit to achieve exact positioning of molecular building blocks to give solid state compounds with predefined properties.

In a way, this is one possible definition of *crystal engineering*, [29,30] and indeed in a 2002 "perspective" on the field C. V. K. Sharma of the Eastman Kodak company writes: "…the greatest achievement(s) of crystal engineering in the past three decades can be summarized in a phrase, i.e., 'viewing crystal structures as networks'." [6].

2.2.3. Seeing the structure in all these structures

All three levels of understanding will greatly profit from an analysis of the vast amount of three-dimensional net structures already in the literature. For the molecular chemistry (that is organic compounds and coordination compounds), these are collected in the Cambridge Structural Database, CSD, and a recent survey showed that within the class of *coordination polymers* (thus excluding for example hydrogen bond nets) there are about 12 000 entries in the database, 10% of those forming 3D-nets [12]. This study also showed the tremendous growth of this field, the doubling time for 3D nets being 4 years compared to the total doubling time of the number of entries in the database that is 9 years [12]. We expect that much will be learnt from these and similar studies. For example, the interpenetration phenomenon (two or more nets interlaced in a structure) was investigated in a related manner by Proserpio et al. [31]

2.3. Synthetic targets for molecular nets: magnetism, chiral channels, gas storage and more

The word engineering is associated with applications and usefulness in general, thus *crystal engineering*, for some people, means that the crystal in question has been prepared for a specific purpose with some specified properties. A less restrictive interpretation that we prefer is that crystal engineering is a chemical preparation with a *specific plan* to form a *certain type of solid state arrangement* of the product, for example a hydrogen bond patter, a 1D coordination polymer or a 3D net built from hydrogen bonds [11,29,30,32]. At the limit, a failed plan giving other products could also be considered as engineering, provided that we can learn something from it.[2] An important point is that a driving force for crystal engineering is that many useful properties are driven by the molecular packing in the solid. If we can control this packing, we can control these properties.

However, we will here deal with the crystal engineering of nets to give specific properties.

2.3.1. Intrinsic properties of 3D-nets

We thus want to ask: In which cases would it be useful for the molecule-based materials designer to consider preparing 3D-nets? Or, in other words, are there any intrinsic properties of 3D-nets that are useful, irrespective of the molecular constituents?

We can think of three such intrinsic properties: (see also Figure 2.6)

- A 3D-net may have strong interactions in three dimensions.
- A 3D-net is by definition not close-packed, and thus has potentially large voids and channels.
- A 3D-net has exact positioning of the molecules and coordination complexes used as building blocks.

---

[2] Engineering has certainly learnt a lot from both small and spectacular failures, such as the Scottish Tay Bridge disaster in 1879 (to continue the historical references to frameworks).

Moreover, certain nets may have additional features related to their symmetry:

- Nets may be used to create multi-component systems with exact ordering.
- Non-centrosymmetric nets may be used for non-linear optical materials.
- Chiral nets may be used for enantiomeric separation and enantioselective synthesis and catalysis.
- Chiral nets may also be used to obtain conglomerates (or spontaneous resolution) from racemic mixtures.

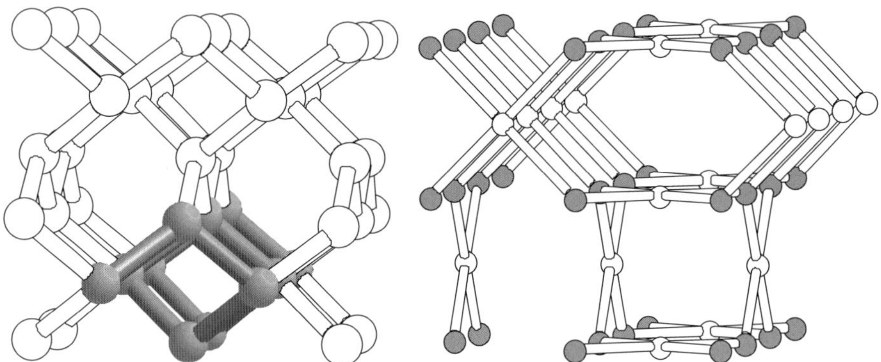

Figure 2.6 Two 3D-nets demonstrating: a) Generic properties: empty channels and "exact" positioning of building-blocks b) The specific properties of certain nets: the chirality of the quartz **qtz** net (left, note that the four-fold helices have the same handedness) and the possibility to assemble two different types of building blocks in to the binodal three and four connected **pto** net (from $Na_xPt_3O_4$ also called (8, 3;4)-a (right).

As for the composition of the nets, this is closely related to the specific application; however, it is appropriate to note that:

- As nets have been demonstrated for a large variety of molecular building blocks, we can envisage including any type of molecular based functionality in a net.
- We can also use any kind of molecular building block, organic, coordination, organometallic, polyfluorinated, DNA-based, or polypeptide based, providing that the correct net-forming features are present.

2.3.2. Applications

Please note that the following sections are not meant to be complete mini-reviews of these subjects. References included are in most cases examples only and the listing is not comprehensive. Some recent reviews are more detailed [33-35].

2.3.2.1. Molecule based magnetic materials

Molecular based magnetic materials are man-made materials with different properties compared to metallic magnets [36-43]. Current synthetic approaches for their preparation include organic chemistry, [44,45] organometallic chemistry, [46,47] coordination chemistry [48-51] or combinations of organic radicals and metal (paramagnetic) ions [52-57]. It would clearly be unreasonable to suggest that these materials will ever replace metallic bulk magnets. However, magnets are ubiquitous in modern society and if we can prepare and design these entirely new types of materials, the prospects of combining magnetism with other properties (low-density, insulation, transparency, non-linear optical effects) may open doors to new applications.

Magnetism is a bulk property dependent on strong interactions in all three dimensions in the solid. Despite that conflicting models have been proposed, knowledge of the factors that favour parallel spin-alignment between the discrete spin carrying units in pairs, chains or larger clusters has advanced considerably in recent years. The strategies to obtain ferromagnetic couplings condense down to: adjacent spins in orthogonal orbitals (in the same spatial region), orthogonal adjacent delocalised or polarised spin densities, and configuration interactions involving different types of excitations from one radical centre to the other [58].

A large number of molecule-based materials with magnetic ordering at low temperature have been prepared, [43] but also a few room-temperature magnets [59]. A common denominator for the low-temperature compounds is that strong ferromagnetic couplings in chains, stacks or sheets are supported only by weak inter-chain interactions. Thus the magnetism is quenched at higher temperatures when thermal motion will break the weak force aligning the spins.

In contrast, for the high temperature materials three-dimensional covalent networks have been found or inferred. As a consequence, much attention is now devoted to the incorporation of organic or inorganic spin carriers into 3D-nets [60-63]. The rationale behind this is that the strong magnetic interactions demands close connections between the spins in all three dimensions. Although there are examples of strong intermolecular magnetic couplings between π-stacked radicals, [64][3] closely positioned spins, and even hydrogen bonds, the prevailing idea today is that covalent (coordinative or other) bonds are needed to transmit this interaction.

The prime example of this is the many magnets based on the Prussian Blue family (Figure 2.7) [63,65-68]. It is also believed that the other example of a room temperature "molecule based magnet", $V(\text{tetracyanoethylene}){\cdot}x(CH_2Cl_2)$, has a structure formed by a three dimensional irregular net (this compound is not crystalline) [69].

---

[3] The net formed by the magnetic J couplings may thus be different and have a different dimensionality from the net built from chemical bonds.

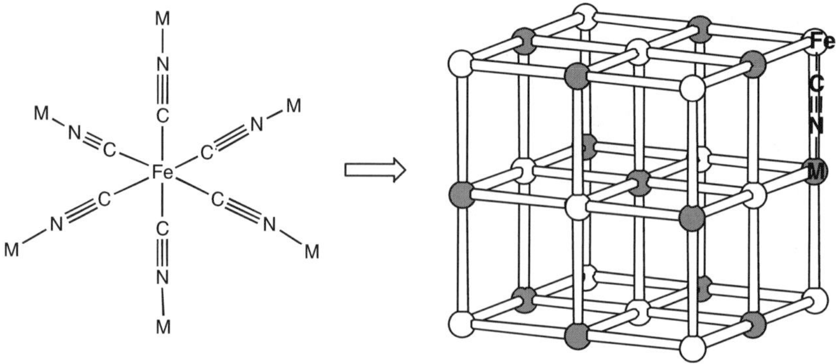

Figure 2.7 Bimetallic Prussian Blue-like α-polonium or **pcu** nets (compare Figure 2.2) based on hexacyanometallate building blocks form an important family of molecule-based magnets.

2.3.2.2. Spin crossover or spin transition materials

Somewhat related to magnetic materials since both are concerned with the spin of unpaired electrons, are the so-called spin crossover or spin transition materials [70]. The phenomenon in itself is not new and can be viewed as having a purely molecular origin. It is observed for may different systems of prevalently the first row transition metals, but Fe(II) systems are the most studied. In these we can have either spin pairing of all six d-electrons in the three slightly stabilised d-orbitals (t_{2g}^{6}) or, if the energy difference to the pair of d-orbitals with higher energy (e_g) is small, a high spin configuration $t_{2g}^{4}e_{g}^{2}$ with four unpaired electrons.

Spin transition occurs when the spin pairing energy is matched to the orbital energy difference, and such compounds may be switched between the two spin states using heat, pressure or light.

The connection to networks is the cooperativity or communication between the metal centres demonstrated by a hysteresis loop for the spin transition. The two spin states will have different metal-ligand bond lengths and if all the coordination complexes are joined in a 3D-net a spin state change for one centre will push or pull at all other centres and thus affecting their spin states too.

Consequently, the construction of spin transition 1D, 2D or 3D coordination polymers has attracted attention in recent years, and may form the base in new devices for memory, display and other magneto-optical applications [71-79].

2.3.2.3. Non-linear optical (NLO) materials

An area where the positioning of the "active" molecules is also of great importance is the preparation of non-linear optical (NLO) materials, that is, materials that will change the wavelength of an incoming light so that the ray that comes out on the other side of the device has, for example, doubled its frequency. Such materials exist today in many commercial applications, but there is a need to improve and extend the properties of these, and thus new NLO materials are a very active area of research.

What one need for a NLO molecule-based material is first of all a molecule that is a strong dipole. However, this is not enough, since dipoles have a tendency to assemble so that their dipole moments cancel, see Figure 2.8, and such an arrangement also cancels the NLO effect.

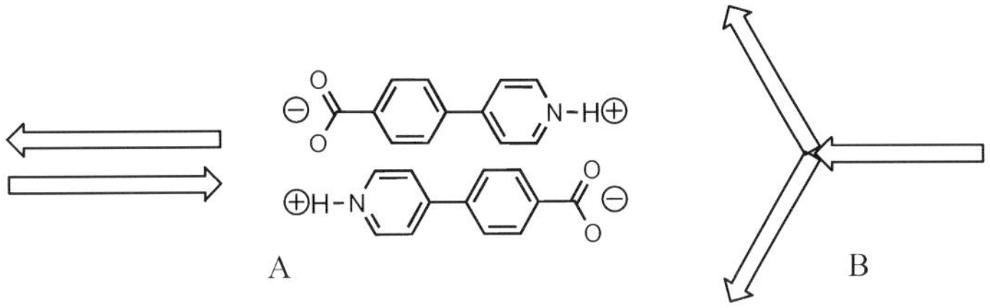

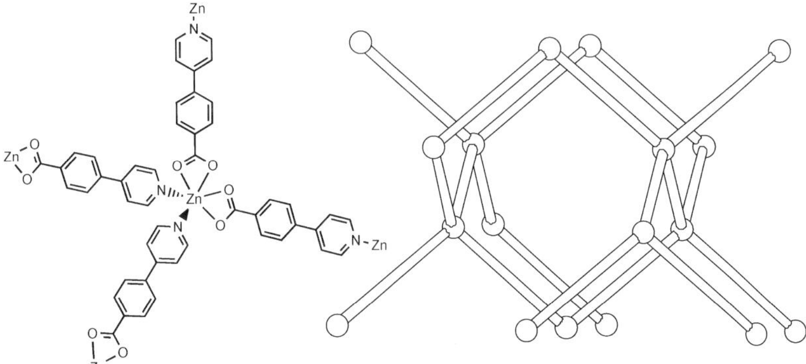

Figure 2.8 The construction of non-linear optical (NLO) materials. In A the dipoles cancel and there is no NLO effect. In B the dipoles do not cancel, and a NLO effect is possible. In crystallographic terms, compounds with NLO effect need to have non-centrosymmetric structures (Note the centres of symmetry in A).

The network approach might therefore be attractive when organic or coordination compounds are used for NLO materials [80]. An example is shown in Figure 2.9 [81].

Figure 2.9 A material with non-linear optical (NLO) effect. This compound (and its Cd(II) analogue) have NLO effects close to lithium niobate used in commercial applications [81]. This structure contains eight interpenetrated diamond (**dia**) nets.

2.3.2.3. Conductivity, superconductivity, semiconductors, luminescence

Without going into details we will just cite a number of research groups that recently have investigated superconductivity, [82] conductivity, [83] semiconductors, [84] and luminescence [85-95] in connection with 3D-nets based on coordination polymers or hydrogen bonding. All these properties are to a larger or lesser extent dependent, not only on the properties of the specific molecules, but on their interactions with neighbouring molecules, and for the formulation of such materials the "structural network approach" is probably suitable.

2.3.2.4. Porous materials

Just as magnets, porous solids are ubiquitous in modern society; from the catalytic supports in car exhaust systems and in the chemical industry, to the zeolites used as water softeners in washing powder. The quest for such new porous materials is currently a very active area, incorporating classical zeolite-type materials, [96] inorganic [97] and organic chemistry [98] and even proteins [99].

Molecular sieves and other porous materials based on inorganic aluminosilicates, aluminophosphates and related frameworks can be prepared with pores of different diameters, [10,100] and are used both in large scale industrial applications as well as in the laboratory [101]. However, these are not the subjects of this book, although they have been (and are) important in the development of theory and nomenclature for 3D-nets [102,103].

We are concerned with molecule-based materials, organic, organometallic or metal-organic. Porous materials formed by such building blocks are attractive since they may afford a greater flexibility and different properties from traditional zeolites [13,104,105]. A brief listing of the possible advantages[4] of such materials include:

- The different surface properties of the channels in such porous materials compared to aluminosilicates and aluminophosphates.
- The possibility to tailor-make the surface for specific applications.
    Hydrogen bonding sites, accepting or donating
    Polyfluorinated surfaces
- The frameworks can be positively charged, neutral, or negatively charged.
- The possibility to tailor-make the size of the channels.
- The possibility to choose the geometrical properties (symmetry) of the channels, especially to prepare chiral channels to get enantiomerically pure porous materials.
- The relative ease with which such nets can be destroyed by acid or base. This opens for a possibility to use them as moulds for polymers [106] that could give (when the mould has been disassembled) robust, covalently bound, porous polymers with a very well defined pore size.

---

[4] Please note that *some* of these properties also apply to the "inorganic" crystalline porous materials but many of the *combinations* are unique for the "organic" or "metal-organic" materials.

- The possibility to use biocompatible materials such as amino acids, nucleotides and proteins. For example, there is currently strong interest in the preparation of new drug-delivery materials, [107] among those 3D cross-linked gels, [108,109] and porous 3D nets built from hydrogen bonded biocompatible molecules could perhaps function in similar ways but with greater control of the void space and thus the encapsulated molecules.

Some of these properties have been used to create materials for specific applications; others have not yet been realized. Examples of such applications are:

- Hydrogen gas storage [110-112]
- Gas adsorption and separation [113-115]
- Catalysts for organic synthesis [116,117]
- Catalysts for inorganic synthesis [118]
- Sorption of organic molecules [119,120]
- Materials for enantiomeric separation and catalysis [121-123]
- Materials for the selective detection of organic and inorganic compounds [74,124].

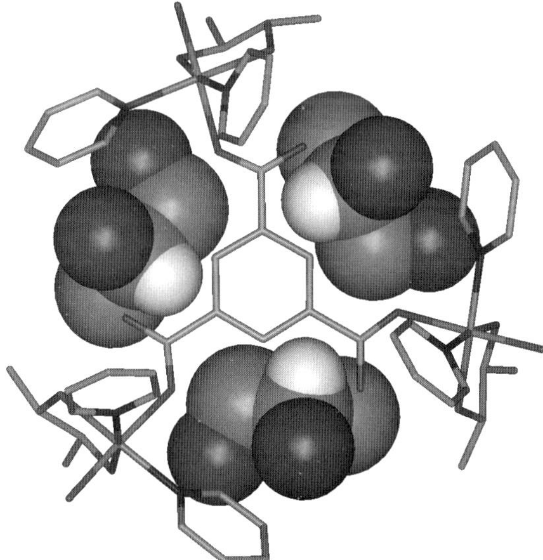

Figure 2.10 1,2-propanediol enantioselectively trapped inside a porous (10,3)-a or **srs** net (only one node and three links are shown) in the compound [Ni₃(1,3,5-benzenetricarboxylate)₂ (pyridine)₆(1,2-propanediol)₃]·11(1,2-propanediol)·8(H₂O) with 51% of solvent accessible volume [121]. Picture reproduced with permission from M. J. Rosseinsky.

2.4. Patents related to molecular 3D-nets

In this section we list the patents that form the basis of Figure 2.2. Due to the rather different way that patents are written and keywords are used, this list is probably not exhaustive. There is also our subjective judgement as to which patents fit the description "related to molecular 3D-nets". Nevertheless, we hope that this will give some ideas of future applications. Note, however, that "95% of all patents are worthless".

This list will conclude chapter 2. In chapter 3 we will move on to some definitions of nets.

1. *O. M. Yaghi, A. J. Matzger and J. L. C. Rowsell,* Implementation of a strategy for achieving extraordinary levels of surface area and porosity in crystals, (The Regents of the University of Michigan, USA). **2004**, 2004-841983, 2004225134
2. *S. Takamizawa,* Preparation of stable microporous materials comprised of self-assembled metal complexes containing regularly-distributed pores useful as catalysts, (Yokohama TLO Company, Ltd., Japan). **2004**, 2003-141474, 2004043454
3. *K. Seki and S. Kitagawa,* Preparation of three-dimensional metal complexes as adsorbents and separators for adsorption and separation apparatus, (Osaka Gas Co., Ltd., Japan). **2004**, 2002-329499, 2004161675
4. *U. Muller, O. Metelkina, H. Junicke, T. Butz and O. M. Yaghi,* Manufacture of hydrogen peroxide from the elements using a metal-organic framework catalyst, (BASF Akiengesellschaft, Germany; Michigan State University). **2004**, 2002-280013, 2004081611
5. *T. J. Marks and P. Zhu,* Vapor deposited electro-optic films self-assembled through hydrogen bonding, (Northwestern University, USA). **2004**, 2004-US6249 2004079434
6. *W. Lin, A. Hu and H. L. Ngo,* Chiral porous metal phosphonates for heterogeneous asymmetric catalysis, (The University of North Carolina at Chapel Hill, USA). **2004**, 2004-US8797 2004084834
7. *J. Li, L. Pan and X. Huang,* Porous polymeric coordination compounds, (Rutgers State University, USA). **2004**, 2003-US37269, 2004045759
8. *S. Kitagawa and T. Kobayashi,* Method for alignment of gaseous molecules and materials for their holding, (Kyoto University, Japan). **2004**, 2002-367343 2004196594
9. *S. Kitagawa, T. Uemura, K. Uemura, M. Arakawa and S. Tanabe,* Selective adsorbents for isoprene gas, their manufacture, and their use in isoprene gas separation, (Nard Institute Ltd., Japan). **2004**, 2003-122867, 2004323455
10. *M. Fujita, M. Yoshizawa and M. Nagao,* Preparation of multimetal complexes with two- or three-dimensional structures utilizing coordination-restricting ligands, (University of Tokyo, Japan). **2004**, 2003-39947, 2004250345
11. *T. Endo, F. Hamada, Y. Aoyama and H. Suzuki,* Process for preparation of anthracenecarboxylic acid derivatives as organic zeolite for gas absorption, (Nissan Chemical Industries, Ltd., Japan). **2004**, 2003-51536, 2004256487
12. *T. Endo, F. Hamada, Y. Aoyama and H. Suzuki,* Preparation of hydrogen-bonding network-type crystals of phosphoryl anthracenes and their use for inclusion of volatile organic substances and metals, (Nissan Chemical Industries, Ltd., Japan). **2004**, 2003-69229 2004277316

13. *C.-M. Che and S.-C. Yu*, Electroluminescent metallo-supramolecules with terpyridine-based groups, (Peop. Rep. China). **2004**, 2002-290120 2004086744

14. *J. L. Atwood and L. R. Macgillivray*, Substantially spherical supramolecular assemblies based on Platonic and Archimedian solids, their preparation from calixarenes and other multifunctional compounds, and their uses, (USA). **2004**, 2003-408605, 2004014963

15. *J. L. Atwood, L. J. Barbour and A. Jerga*, Calixarene-based guest-host assemblies for guest storage and transfer, (USA). **2004**, 2002-286179, 2004087666

16. *K. Seki*, Three-dimensional porous metal complex, gas adsorbent, and separation material, Osaka Gas Co., Ltd., Japan, **2003**, JP 2002-149556 20020523.,

17. *T. M. Nenoff and M. D. Nyman*, Niobate-based octahedral molecular sieves, (Sandia Corporation, USA). **2003**, 2001-876903, 6596254

18. *U. Muller, M. Stober, R. Ruppel, E. Baum, E. Bohres, M. Sigl, L. Lobree, O. M. Yaghi and M. Eddaoudi*, Process for the alkoxylation of organic compounds in the presence of novel framework materials, (Germany). **2003**, 2001-39733, 2003078311

19. *U. Mueller, G. Luinstra and O. M. Yaghi*, Producing polyalkylene carbonates and catalysts, (BASF Aktiengesellschaft, Germany; University of Michigan). **2003**, 2002-279940, 6617467

20. *U. Mueller, L. Lobree, M. Hesse, O. M. Yaghi and M. Eddaoudi*, Process for epoxidation of organic compounds with oxygen or oxygen-delivering compounds using catalysts containing metal-organic framework (MOF) materials, (BASF Aktiengesellschaft, Germany; The Regents of the University of Michigan). **2003**, 2002-157494, 6624318

21. *U. Mueller, L. Lobree, M. Hesse, O. Yaghi and M. Eddaoudi*, Shaped bodies containing metal-organic frameworks, (BASF Aktiengesellschaft, Germany). **2003**, 2002-157182, 2003222023

22. *J. S. Miller and K. I. Pokhodnya*, Low temperature chemical vapor deposition of thin film magnets, University of Utah Research Foundation (Salt Lake City, UT), **2003**, 089480,

23. *M. Maekawa, M. Arakawa and S. Tanabe*, Ethylene gas adsorbents as food preservatives., Nard Institute Ltd., Japan, **2003**, JP 2003-122868 20030425.CAN 141:410168,

24. *S. Kitagawa, H. Yamamoto and J. Tatsumi*, Porous coordinative unsaturated metal complex catalyst., Nippon Shokubai Co., Ltd., Japan, **2003**, EP 2003-10912 20030515. Priority: JP 2002-140495 20020515; JP 2003-25182 20030131.

25. *G. BASF AG*, Device for storage, absorption, and discharge of gases using new porous metal organic scaffolding materials for use in fuel cells., BASF AG, Germany, **2003**, DE 2002-20210139 20020701, US 2002-61147 20020201,

26. *M. J. Zaworotko and B. Moulton*, Nanoscale faceted polyhedra, (USA). **2002**, 2002-83781 2002120165

27. *O. M. Yaghi, M. Eddaoudi, H. Li, J. Kim and N. Rosi*, Preparation of isoreticular metal-organic frameworks and systematic design of pore size and functionality, with application for gas storage, (The Regents of the University of Michigan, USA). **2002**, 2002-US13763, 2002088148

28. *U. Mueller, M. Hesse, L. Lobree, M. Hoelzle, J.-D. Arndt and P. Rudolf*, Manufacture of coordinated metal salts of dicarboxylic aromatic acids as porous materials., BASF Aktiengesellschaft, Germany, **2002**, WO 2002-EP2523 20020307. Priority: DE 2001-10111230 20010308.,

29. *K.-J. Lin*, Organometallic zeolite based on metal porphine complexes, (Academia Sinica, Taiwan). **2002**, 99-342304 6355793

30. *K. Ikeda, S. Ogoshi and K. Hashimoto*, Nonlinear optical ferromagnetic materials, their manufacture, and magnetooptical devices using them, (Kanagawa Academy of Science and

[36] O. Kahn, Molecular magnetism, VCH, Weinheim, 1993.

[37] J. S. Miller, A. J. Epstein, Angew. Chem. Int. Ed. 33 (1994) 385.

[38] J. S. Miller, Adv. Mater. 6 (1994) 322

[39] J. S. Miller, A. J. Epstein, MRS Bull. 25 (2000) 21.

[40] J. S. Miller, A. J. Epstein, Coord. Chem. Rev. 206 (2000) 651.

[41] J. S. Miller, M. Drillon (Eds.), Magnetism: Molecules to materials I-III, Wiley-VCH, Weinheim,2001.

[42] J. S. Miller, Inorg. Chem. 39 (2000) 4392.

[43] S. J. Blundell, F. L. Pratt, J. Phys.: Condens. Matter 16 (2004) R771.

[44] A. Rajca, Chem. Eur. J. 8 (2002) 2835.

[45] A. Rajca, J. Wongsriratanakul, S. Rajca, Science 294 (2001) 1503.

[46] F. Zuo, S. Zane, P. Zhou, A. J. Epstein, R. S. McLean, J. S. Miller, J Appl Phys 73 (1993) 5476.

[47] T. Takenobu, D. H. Chi, S. Margadonna, K. Prassides, Y. Kubozono, A. N. Fitch, K.-i. Kato, Y. Iwasa, J. Am. Chem. Soc. 125 (2003) 1897.

[48] Z. M. Wang, B. Zhang, H. Fujiwara, H. Kobayashi, M. Kurmoo, Chem. Commun. (2004) 416.

[49] D. Maspoch, D. Ruiz-Molina, J. Veciana, J. Mat. Chem. 14 (2004) 2713.

[50] H. J. Chen, Z. W. Mao, S. Gao, X. M. Chen, Chem. Commun. (2001) 2320.

[51] C. M. Liu, S. Gao, D. Q. Zhang, Y. H. Huang, R. G. Xiong, Z. L. Liu, F. C. Jiang, D. B. Zhu, Angew. Chem. Int. Ed. 43 (2004) 990.

[52] K. Fegy, D. Luneau, T. Ohm, C. Paulsen, P. Rey, Angew. Chem. Int. Ed. 37 (1998) 1270.

[53] T. M. Barclay, R. G. Hicks, M. T. Lemaire, L. K. Thompson, Inorg. Chem. 42 (2003) 2261.

[54] G. Ballester, E. Coronado, C. Gimenez-Saiz, F. M. Romero, Angew. Chem. Int. Ed. 40 (2001) 792.

[55] M. Minguet, D. Luneau, E. Lhotel, V. Villar, C. Paulsen, D. B. Amabilino, J. Veciana, Angew. Chem. Int. Ed. 41 (2002) 586.

[56] H. Zhao, M. J. Bazile, J. R. Galan-Mascaros, K. R. Dunbar, Angew. Chem. Int. Ed. 42 (2003) 1015.

[57] E. B. Vickers, T. D. Selby, M. S. Thorum, M. L. Taliaferro, J. S. Miller, Inorg. Chem. 43 (2004) 6414.

[58] L. Öhrström, Compt. Rend. Chimie (2005) in press

[59] K. I. Pokhodnya, A. J. Epstein, J. S. Miller, Advanced Materials 12 (2000) 410.

[60] V. Tangoulis, C. P. Raptopoulou, V. Psycharis, A. Terzis, K. Skorda, S. P. Perlepes, O. Cador, O. Kahn, E. G. Bakalbassis, Inorg. Chem. 39 (2000) 2522.

[61] F. Mathevet, D. Luneau, J. Am. Chem. Soc. 123 (2001) 7465.

[62] S. Konar, P. S. Mukherjee, E. Zangrando, F. Lloret, N. R. Chaudhuri, Angew. Chem. Int. Ed. 41 (2002) 1561.

[63] T. E. Vos, Y. Liao, W. W. Shum, J. H. Her, P. W. Stephens, W. M. Reiff, J. S. Miller, J. Am. Chem. Soc. 126 (2004) 11630.

[64] R. G. Hicks, M. T. Lemaire, L. Öhrström, J. F. Richardson, L. K. Thompson, Z. Q. Xu, J. Am. Chem. Soc. 123 (2001) 7154.

[65] T. Mallah, S. Thiébaut, M. Verdaguer, P. Veillet, Science 262 (1993) 1554.

[66] V. Gadet, T. Mallah, I. Castro, M. Verdaguer, P. Veillet, J. Am. Chem. Soc. 114 (1992) 9213.

[67] W. E. Buschmann, J. S. Miller, Inorg. Chem. 39 (2000) 2411.

[68] J. S. Miller, MRS Bull. 25 (2000) 60.

[69] J. M. Manriques, G. T. Yee, R. S. McLean, A. E. Epstein, J. E. Miller, Science 252 (1991) 1451.

[70] P. Gütlich, P. J. van Koningsbruggen, F. Renz, in: Optical Spectra and Chemical Bonding in Transition Metal Complexes, T. Schönherr (Ed.) Structure and Bonding vol 107, Springer, 2004, pp. 27.

[71] J. A. Real, A. B. Gaspar, M. C. Muñoz, Dalton Trans. (2005) 2062

[72] O. Kahn, C. J. Martinez, Science 279 (1998) 44.

[73] V. Niel, J. M. Martinez-Agudo, M. C. Munoz, A. B. Gaspar, J. A. Real, Inorg. Chem. 40 (2001) 3838.

[74] G. J. Halder, C. J. Kepert, B. Moubaraki, K. S. Murray, J. D. Cashion, Science 298 (2002) 1762.

[75] A. Galet, V. Niel, M. C. Munoz, J. A. Real, J. Am. Chem. Soc. 125 (2003) 14224.

[76] V. Niel, A. Galet, A. B. Gaspar, M. C. Munoz, J. A. Real, Chem. Commun. (2003) 1248.

[77] A. Galet, M. C. Munoz, V. Martinez, J. A. Real, Chem. Commun. (2004) 2268.

[78] Y. Garcia, V. Niel, M. C. Munoz, J. A. Real, in: Spin Crossover in Transition Metal Compounds I, P. Gütlich, H.A. Goodwin (Eds.) Topics in Current Chemistry 233, 2004, pp. 229.

[79] K. S. Murray, C. J. Kepert, in: Spin Crossover in Transition Metal Compounds I, P. Gütlich, H.A. Goodwin (Eds.) Topics in Current Chemistry 233, 2004, pp. 195.

[80] O. R. Evans, W. B. Lin, Acc. Chem. Res. 35 (2002) 511.

[81] W. B. Lin, L. Ma, O. R. Evans, Chem. Commun. (2000) 2263.

[82] J. A. Schlueter, U. Geiser, J. L. Manson, Journal de Physique IV 114 (2004) 475.

[83] Y. Morita, T. Murata, K. Fukui, S. Yamada, K. Sato, D. Shiomi, T. Takui, H. Kitagawa, H. Yamochi, G. Saito, K. Nakasuji, J. Org. Chem. 70 (2005) 2739.

[84] J. J. Zhang, T. L. Sheng, S. M. Hu, S. Q. Xia, G. Leibeling, F. Meyer, Z. Y. Fu, L. Chen, R. B. Fu, X. T. Wu, Chem. Eur. J. 10 (2004) 3963.

[85] H. F. Zhu, Z. H. Zhang, W. Y. Sun, T. Okamura, N. Ueyama, Cryst. Growth Des. 5 (2005) 177.

[86] J. P. Zhang, Y. Y. Lin, X. C. Huang, X. M. Chen, J. Am. Chem. Soc. 127 (2005) 5495.

[87] E. Y. Lee, S. Y. Jang, M. P. Suh, J. Am. Chem. Soc. 127 (2005) 6374.

[88] S. Hu, M. L. Tong, Dalton Trans. (2005) 1165.

[89] Z. He, E. Q. Gao, Z. M. Wang, C. H. Yan, M. Kurmoo, Inorg. Chem. 44 (2005) 862.

[90] L. Han, R. H. Wang, D. Q. Yuan, B. L. Wu, B. Y. Lou, M. C. Hong, J. Mol. Struct. 737 (2005) 55.

[91] P. Y. Feng, X. H. Bu, N. F. Zheng, Acc. Chem. Res. 38 (2005) 293.

[92] D. T. de Lill, N. S. Gunning, C. L. Cahill, Inorg. Chem. 44 (2005) 258.

[93] C.-D. Wu, H. L. Ngo, W. Lin, Chem. Commun. (2004) 1588.

[94] W. Dong, Y. Q. Sun, B. Yu, H. B. Zhou, H. B. Song, Z. Q. Liu, Q. M. Wang, D. Z. Liao, Z. H. Jiang, S. P. Yan, P. Cheng, New J. Chem. 28 (2004) 1347.

[95] W. Chen, J. Y. Wang, C. Chen, Q. Yue, H. M. Yuan, J. S. Chen, S. N. Wang, Inorg. Chem. 42 (2003) 944.

[96] D. E. Akporiaye, Angew. Chem. Int. Ed. 37 (1998) 2456.

[97] C. L. Bowes, G. A. Ozin, Adv. Mater. 8 (1996) 13.

[98] P. J. Langley, J. Hulliger, Chem. Soc. Rev. 28 (1999) 279.

[99] L. Z. Vilenchik, J. P. Griffith, N. St Clair, M. A. Navia, A. L. Margolin, J. Am. Chem. Soc. 120 (1998) 4290.

[100] M. E. Davis, Chem. Eur. J. 3 (1997) 1745.

[101] W. Büchner, R. Schliebs, G. Winter, K. H. Büchel, Industrial Inorganic Chemistry;, VCH, Weinheim, 1989.

[102] J. V. Smith, Chem. Rev. 88 (1988) 149.

[103] O. Delgado-Friedrichs, A. W. M. Dress, D. H. Huson, J. Klinowski, A. L. Mackay, Nature 400 (1999) 644.

[104] D. Bradshaw, J. B. Claridge, E. J. Cussen, T. J. Prior, M. J. Rosseinsky, Acc. Chem. Res. 38 (2005) 273.

[105] M. J. Rosseinsky, Microporous Mesoporous Mater. 73 (2004) 15.

[106] Q. Wei, S. L. James, Chem. Commun. (2005) 1555.

[107] G. Orive, R. M. Hernandez, A. R. Gascon, A. Dominguez-Gil, J. L. Pedraz, Curr. Opin. Biotechnol. 14 (2003) 659.

[108] N. Jibry, R. K. Heenan, S. Murdan, Pharm. Res. 21 (2004) 1852.

[109] T. Coviello, F. Alhaique, C. Parisi, P. Matricardi, G. Bocchinfuso, M. Grassi, J. Controlled Release 102 (2005) 643.

[110] N. L. Rosi, J. Eckert, M. Eddaoudi, D. T. Vodak, J. Kim, M. O'Keeffe, O. M. Yaghi, Science 300 (2003) 1127.

[111] B. Kesanli, Y. Cui, M. R. Smith, E. W. Bittner, B. C. Brockrath, W. Lin, Angew. Chem. Int. Ed. 44 (2004) 72.

[112] X. B. Zhao, B. Xiao, A. J. Fletcher, K. M. Thomas, D. Bradshaw, M. J. Rosseinsky, Science 306 (2004) 1012.

[113] T. Dueren, L. Sarkisov, O. M. Yaghi, R. Q. Snurr, Langmuir 20 (2004) 2683.

[114] M. Otake, Catal. Surv. Jpn. 2 (1998) 209.

[115] A. J. Fletcher, E. J. Cussen, D. Bradshaw, M. J. Rosseinsky, K. M. Thomas, J. Am. Chem. Soc. 126 (2004) 9750.

[116] J. S. Seo, D. Whang, H. Lee, S. I. Jun, J. Oh, Y. J. Jeon, K. Kim, Nature 404 (2000) 982.

[117] K. Endo, T. Koike, T. Sawaki, O. Hayashida, H. Masuda, Y. Aoyama, J. Am. Chem. Soc. 119 (1997) 4117.

[118] U. Muller, O. Metelkina, H. Junicke, T. Butz, O. M. Yaghi, Application: US US, (BASF Akiengesellschaft, Germany; Michigan State University). 2004, 2002-280013 2004081611

[119] E. J. Cussen, J. B. Claridge, M. J. Rosseinsky, C. J. Kepert, J. Am. Chem. Soc. 124 (2002) 9574.

[120] O. M. Yaghi, G. Li, H. Li, Nature 378 (1995) 703.

[121] C. J. Kepert, T. J. Prior, M. J. Rosseinsky, J. Am. Chem. Soc. 122 (2000) 5158.

[122] B. Kesanli, W. Lin, Coord. Chem. Rev. 246 (2003) 305.

[123] D. Bradshaw, T. J. Prior, E. J. Cussen, J. B. Claridge, M. J. Rosseinsky, J. Am. Chem. Soc. 126 (2004) 6106.

[124] G. A. Ozin, Supramol. Chem. 6 (1995) 125.

Chapter 3

What is a net?

Depending on your previous experience, this question may seem a bit odd: You solve your crystal structure, you see molecules acting as connection points, and you see other molecules acting as rods connecting these nodes. If the resulting geometric pattern extends infinitely in the x-, y- and z- directions, then you have a 3D-net. What more is there to say?

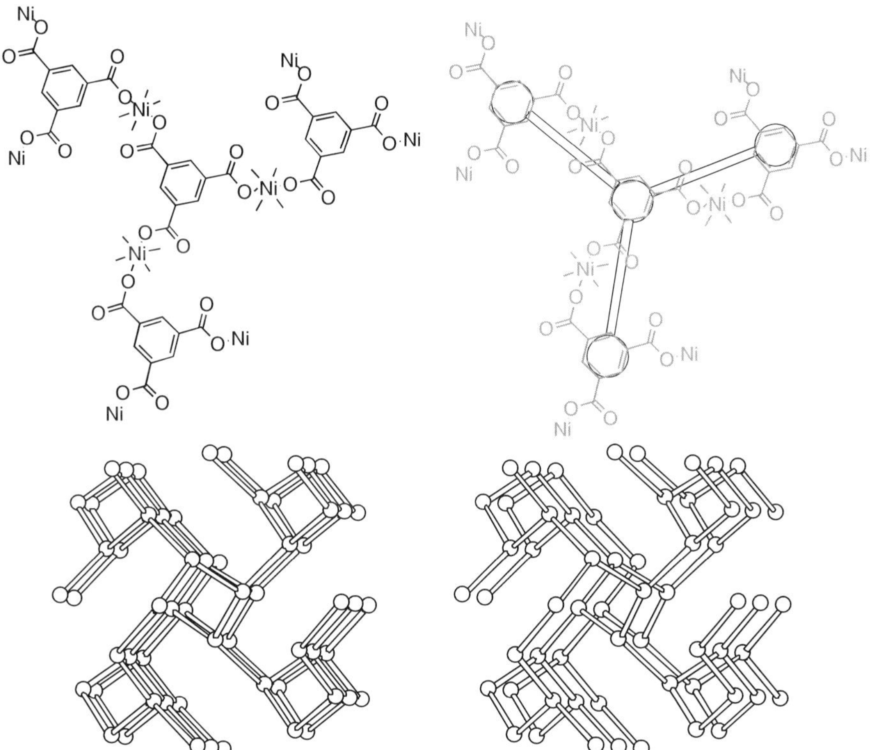

Figure 3.1 The molecular constituents of a net from the structure of $Ni_3(btc)_2(3\text{-pic})_6(1,2\text{-pd})_3$, $[(1,2\text{-pd})_9(H_2O)_{11}]$ [1] (above left), an overlay of the net structure (above right), and a stereo-drawing of the final (10,3)-a or **srs** net (below).

Of course there are many such examples, especially in the class of coordination polymers, see Figure 3.1. However, there are also many not quite so clear-cut cases, and there are no laws that will tell you when you have the right to draw a bond or a dotted line. In fact, especially the latter type may cause heated debates at conferences and seminars, because the meaning we attach to this dotted line may be very different.[1]

In the words of Gautam Desiraju: "First we did not draw any lines, then we got close packing. Then we started to draw lines and got molecules. We drew more lines and got networks. If we continue to draw more and more lines, then we get back to close packing again!" [2]

In this Chapter, we will not tell you when you can or cannot draw bonds or dotted lines. Rather, we will try to discuss when it is *meaningful* to describe a structure in terms of a net and let you come to your own conclusions.

3.1. Definitions

There are purely mathematical definitions of a net[2], but they need not concern us for the moment. For the chemist it is important that analysing nets and the interactions within them should help us in our pursuit to achieve exact positioning of molecular building blocks to give solid-state compounds with predefined properties. Thus, the net should clearly be the organising principle of the structure. Figure 3.2 illustrates some relative energy relations used in net formation.

Finding out if this is the case is often easier said than done. However, we suggest some general guidelines:

- Follow the path of covalent bonds.
- Follow the path of coordinative bonds.
- Be careful with bond definitions suggested by different software. For example, nitrogen bases coordinated to transition metals giving octahedral coordination are clear cut cases, while highly asymmetric coordination figures obtained by Ag(I) and oxy-anions need to be carefully examined for their plausibility.
- Follow the path of strong hydrogen bonds.
- Follow the path of weaker hydrogen bonds.
- Take well known, weaker but directional, interactions into account like "π-π stacking", π-σ interactions and multiple phenyl embraces.

---

[1] It has been suggested that every figure containing dotted lines should be accompanied by the appropriate definition used when drawing this line.

[2] The mathematical term seems to be *net* and this is one reason we mostly use this term instead of the more or less synonymous "network".

- Do not take for granted that a short cation-anion interaction implies a strong directionality along such a "bond" axis. The columbic forces are long distance and affect several shells around a specific ion.

In designating[3] your net you should convince yourself that you are not ignoring any stronger interactions than those linking the nodes of the net. In short, we propose the following definition:

A path along the net should follow the strongest *directional* intermolecular interactions (or bonds) and be significantly stronger than all other directional short-range intermolecular interactions

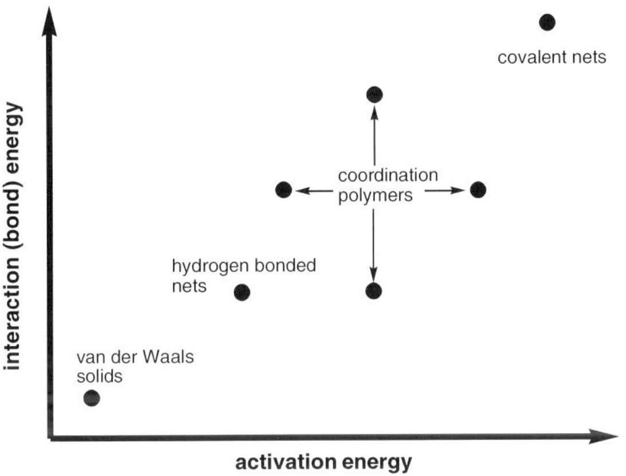

Figure 3.2 A schematic illustration of the energy range of bonding interactions for different types of nets (*y*-axis) and the activation energies for network link formation (*x*-axis). The scale is not linear.

Although it need not concern us in detail until the next chapter, we need to include a definition of when two nets are the same:

Two nets are the same if one can be transformed into the other (by squeezing, bending, sheering etc) without breaking any of the linkages. (Figure 3.3.)

This will in due course lead us to the topology discussion in Chapter 5, since a definition of topology is: "(topology) …studies those properties an object retains under deformation—specifically, bending, stretching and squeezing, but not breaking or tearing" [3].

---

[3] Designing your net is a different story that will be treated in later chapters.

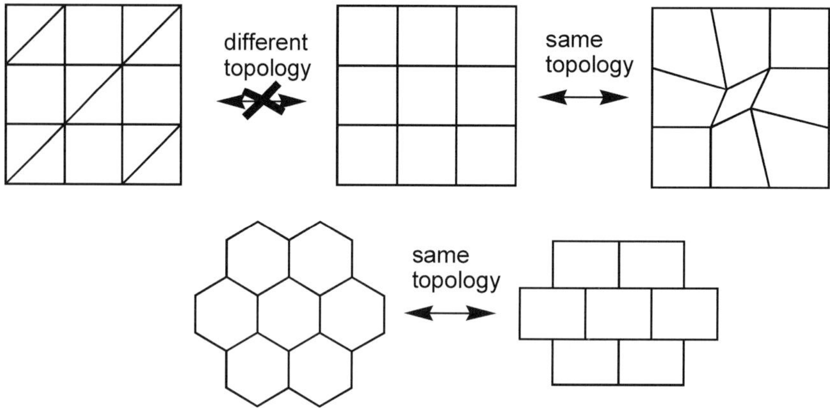

Figure 3.3 An illustration of idenitical and different topologies. If no bonds are broken or formed when a nets is transformed into another net, then these nets have the same topology.

3.2. Short survey of network interactions

While covalent and coordinative bonds and intermolecular interactions are thoroughly covered in all sorts of textbooks and more advanced text, we will nevertheless make a short survey of them here. The reason is their paramount importance for our conception of nets, and the more general role they play in crystal engineering, [4,5] see Figure 3.2.

3.2.1. Covalent bonds

We will not waste much time on covalent bonds, they are well known to the chemically educated reader. However, it may be prudent to point out two things; first, there are still a considerable number of elements in the periodic table that have never been connected to each other by a covalent bond, or where the examples are very few. In such cases there are no accepted definitions of "normal" bond lengths and angles, and controversy may result when proposing such bonds.[4]

Secondly, the formation of covalent bonds is usually irreversible (but see below) and proceeds through transition states and meta-stable intermediates with very different geometry from the starting material and product. For example, in nucleophilic substitution reactions on sp^3 carbons we begin and end with a tetrahedral geometry while in-between we may have either triangular planar or penta-coordinated geometries. This may have dire consequences for

---

[4] An example is the reports of Ga-Ga and Ga-Fe triple bonds in 1997 [6,7] that upset a few people, including Texas A&M professor F.A. Cotton who told *Chemical Engineering News*, "That's no more a triple bond than I'm the Dalai Lama!" [8]. For a conclusion see [9].

the closure of rings in nets if the bond-forming intermediates and the geometry restrictions imposed by the net are incompatible.

A covalently bonded net is shown in Figure 3.4.

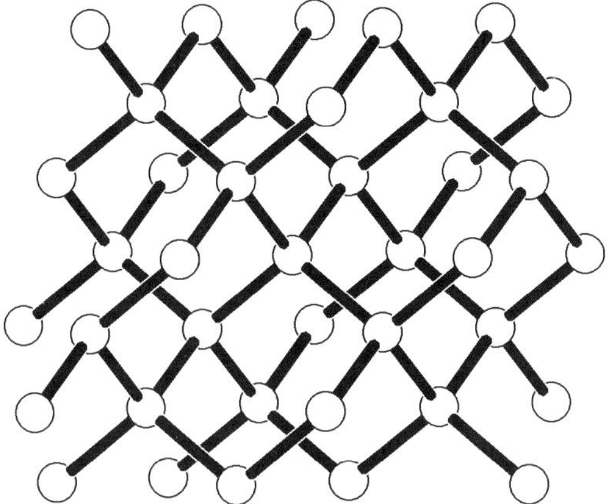

Figure 3.4 The diamond phase of carbon(s), a **dia** net constructed with covalent bonds.

3.2.2. Coordinative bonds

There is a well-known definition of supramolecular chemistry as "the chemistry of the non-covalent bond", and a lot of supramolecular chemistry has been done using coordination chemistry and the coordinative bond. However, it has been pointed out, [10] and is clear from most textbook treatment of ligand field theory, [11] that this is contradictory; the coordinative bond is indeed a covalent bond.

However, there are good reasons for treating this type of bond separately. One is that this bond is often reversible, that is there are low energy barriers for bond breaking and bond formation. This is however not always the case, and for some metal ions in certain oxidation states bond forming and breaking is extremely slow, for example the $[Ir(H_2O)_6]^{3+}$ ion has the slowest self exchange kinetics of any homoleptic[5] mononuclear metal centre with a half-life of coordinated waters of more than 3 years [12].

As a rule, water molecules bound to metal ions exchange with bulk water quite rapidly, for $[Fe(H_2O)_6]^{3+}$ the mean lifetime (τ) is 10^{-2} seconds. A comparison of τ values for $[M(H_2O)_6]^{n+}$ will thus give general ideas about the ease of ligand exchange for a certain metal ion M^{n+} [13]. Note, though, that this should not be confused with low bond energy or weak bonds, that is, the thermodynamic stability of the bond, see Figure 3.5.

---

[5] Having only one kind of ligand.

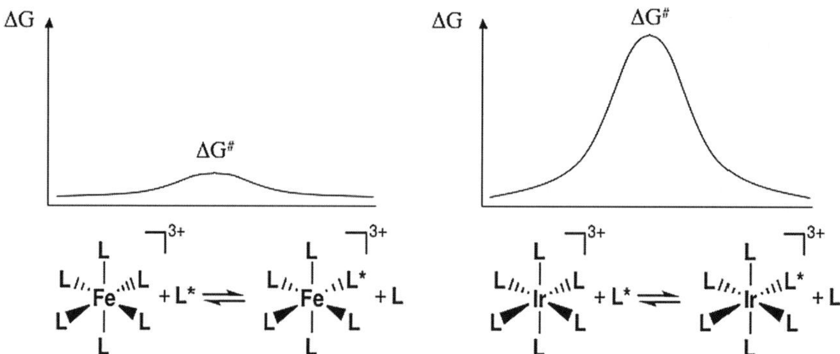

Figure 3.5 Coordination compounds are normally characterised by low kinetic barriers to bond breaking and bond formation (left) but especially transition metal ions with d-electron configurations d^3, d^6 and d^8 display slow or very slow ligand exchange kinetics (inert complexes, right). This should not be confused with bond energies, which in either case may be high or low and depend on the relative ΔG (or ΔH) values of the complexes and the free ligand.

Not surprisingly, since we are dealing with a large portion of the periodic system, from the light alkaline metal ions to the rare earth metal ions there are no general rules for bond energies, bond lengths and coordination geometries and if you are not familiar with a specific system you will do well to consult the abundant specialist literature [13,14]. The Cambridge Structural Database (CSD) [15,16] and specific surveys based on searches therein are also invaluable sources [17,18].

You may even then find yourself in situations where it will be difficult to judge if a certain interaction between a metal ion and a Lewis base is a bond, and if it can be considered as forming a net structure or not.

For example, the anions of strong oxy-acids like HNO_3 are extremely weak bases and are normally considered as non-coordinating or weak coordinating ligands. In compounds where these anions are charge balancing Ag(I) complexes, known to have somewhat irregular coordination geometries, there will probably be, even if there is no electron pair sharing by the anion and the Ag(I) ion, Ag…O distances that are shorter than the combined van der Waals radii [19]. Whether this is a bond or not, and if it is bridging or not, will be up to your judgment.

Another situation where software-defined bonds may cause trouble is Jahn-Teller distorted Cu(II) that can often initially be assigned as square-planar.

A possible guide may be to look at the presumptive net you have constructed. If this net is a high symmetry common 3D-net (these two factors are related) [20] then it seems likely that a net description of your structure is appropriate. If, on the other hand, your net is highly irregular and the only example of such a net hitherto found in the literature, then be very careful.

Often, these types of nets are referred to as coordination polymers, and an example is shown in Figure 3.6 [21]. In this net Cu^{2+} ions coordinate neutral 4,4′-bipyridine ligands and we thus have a cationic net. With coordination polymers we may also build neutral or anionic nets.

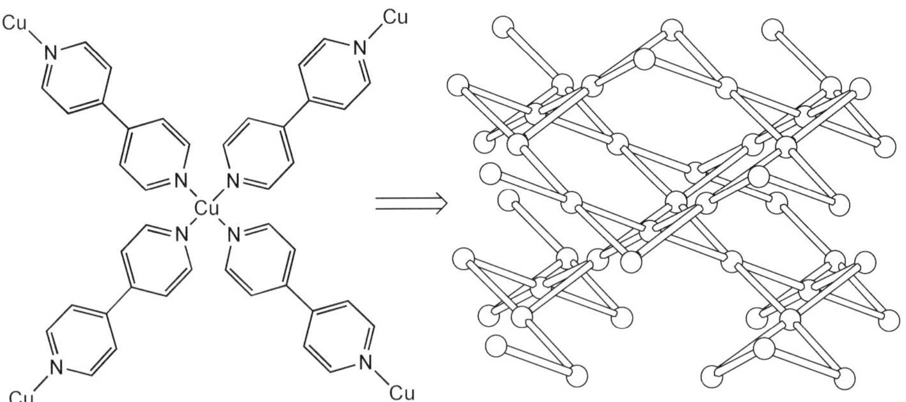

Figure 3.6 The 4-connected 3D net **lvt** formed by the coordination polymer [Cu(4,4′-bipy)$_2$ (CF$_3$SO$_3$)$_2$]·2CH$_2$Cl$_2$·H$_2$O [21] Nodes are placed at the Cu atoms.

3.2.3. *"Normal" hydrogen bonds*

Hydrogen bonding is a highly directional interaction between a hydrogen, which is bonded to an atom X with high electronegativity, and an acceptor, A, likewise with high electronegativity, and is schematically written as X—H···A, see Figure 3.7 in which some of the geometrical parameters are also defined.

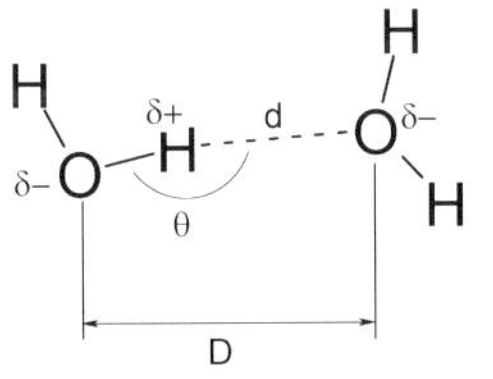

Figure 3.7 A classical hydrogen bond. Strong hydrogen bonds have short d values but also θ values close to 180°.

Since the definition is geometric rather than based on a specific type of interaction, the complexity of this type of bonding is high, ranging from covalent to weakly electrostatic and dipolar to dispersive, although the classical bond is mainly electrostatic. In Table 3.1, data are listed for some hydrogen bonds, as well as a classification system [22]. It is important to remember that these values are approximate, since there are no clear borders between strong, moderate, and weak interactions.

Table 3.2 Classification of hydrogen bonds (uncharged species). Adapted from J. C. Jeffery, An Introduction to Hydrogen Bonding, Oxford University Press, Oxford, **1997** [22].

	Strong	*Moderate*	*Weak*
Interaction type	strongly covalent	mostly electrostatic	electrostatic/disperse
Bond energy [kJ mol^{-1}]	60-170	15-60	<15
Bond length, D X···A [Å]	2.2-2.5	2.5-3.2	>3.2
Bond length, d H···A [Å]	1.2-1.5	1.5-2.2	>2.2
Bond angle, θ[°]	170-180	>130	>90

However, in standard textbooks X and A are often required to be either, F, N or O, [23] and we will treat these under the heading "normal hydrogen bonds". A hydrogen bonded 3D-net of this type is shown in Figure 3.8 [24].

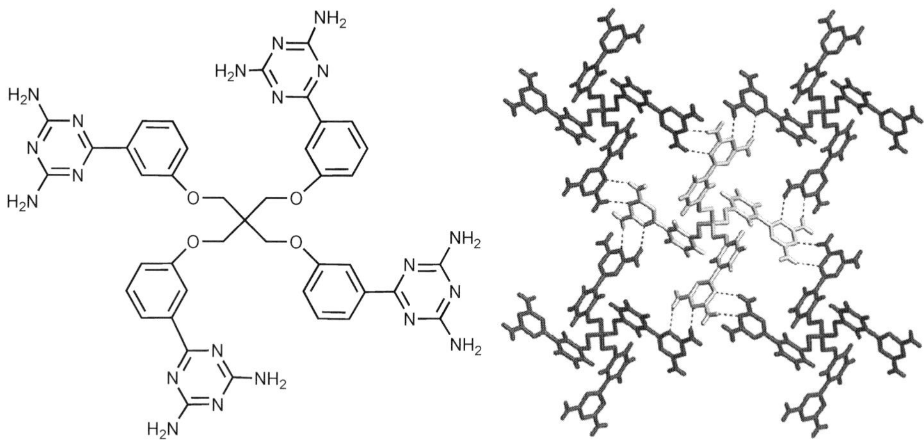

Figure 3.8 A 3D hydrogen bonded net with neutral components (a pentaerythrityl derivative). After Laliberté et al. [24] Reprinted with permission from [24]. Copyright 2004 American Chemical Society.

Furthermore, when comparing hydrogen bonds it is important to differentiate between charged and uncharged species, e.g., between X—H^{+}···A^{-} and X—H···A. This is because the charges change the electronic environment (polarisation) around the bond, so as to increase the repulsion or attraction, sometimes called "charge assisted hydrogen bonding" [4] and an example is shown in Figure 3.9. We should take care and differentiate this effect from the overall charge-charge (coulombic) interactions introduced compared to a neutral system. In general, ionic systems have increased stability (higher melting points for example) but it is important to realise that this is a global effect in the crystal; the coulombic attractions and repulsions are long range (distance dependence 1/r) and cannot be condensed down to involve only the shortest cation-anion interaction [25-27].

Hydrogen bond patterns are often classified with *graph set analysis*, a notation describing the type of interaction and number of atoms involved, a subject we will briefly treat in Chapter 4 [28,29]. It is useful in describing the synthon used in linking the net together.

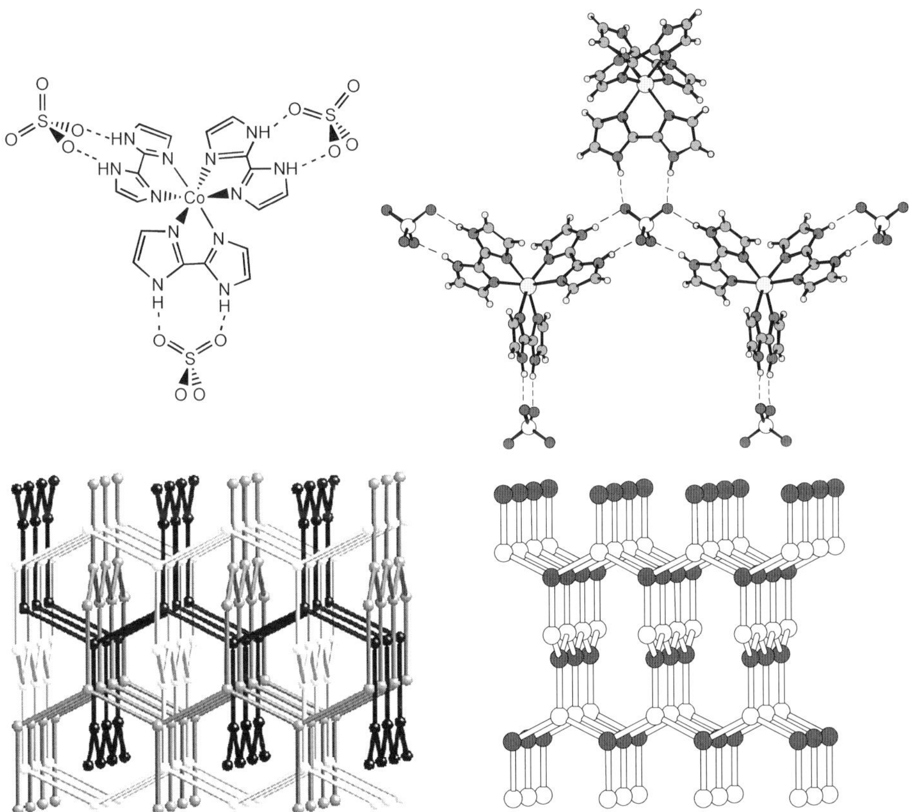

Figure 3.9 Portion of the charge assisted hydrogen bonded 3D net formed by Δ-[Cr(2,2'-biimidazole)$_2$(2,2'-biimidazolato)](SO$_4$)·H$_2$O [30] An isomorphous structure is found for [[Fe(2,2'-biimidazole)$_3$]CO$_3$·MeOH [31]. In these structures the net is triply interpenetrated.

3.2.4. "Weak" hydrogen bonds

The hydrogen bond acceptor and donor atoms are, however, not limited to oxygen, nitrogen and fluorine [32,33]. All we need for the acceptor is a free electron pair, and for the hydrogen to be bonded to an atom with higher electronegativity than itself. I.e., C-H donors are possible, and are increasingly recognized as playing an important part in solid state organic chemistry. An acceptor emerging as an important structural factor in supramolecular coordination chemistry is the coordinated chlorine ion, for example in MCl_4^{2-}. At the limit, we may even consider π-systems as acceptors.

Two factors are important when we discuss the weaker interactions. The first is that they become important in the absence of stronger interactions, and secondly that they can make up in number what they lack in energy. Thus, even in the presence of a strong hydrogen bond, the weaker forces may determine the overall structure if they are more numerous.

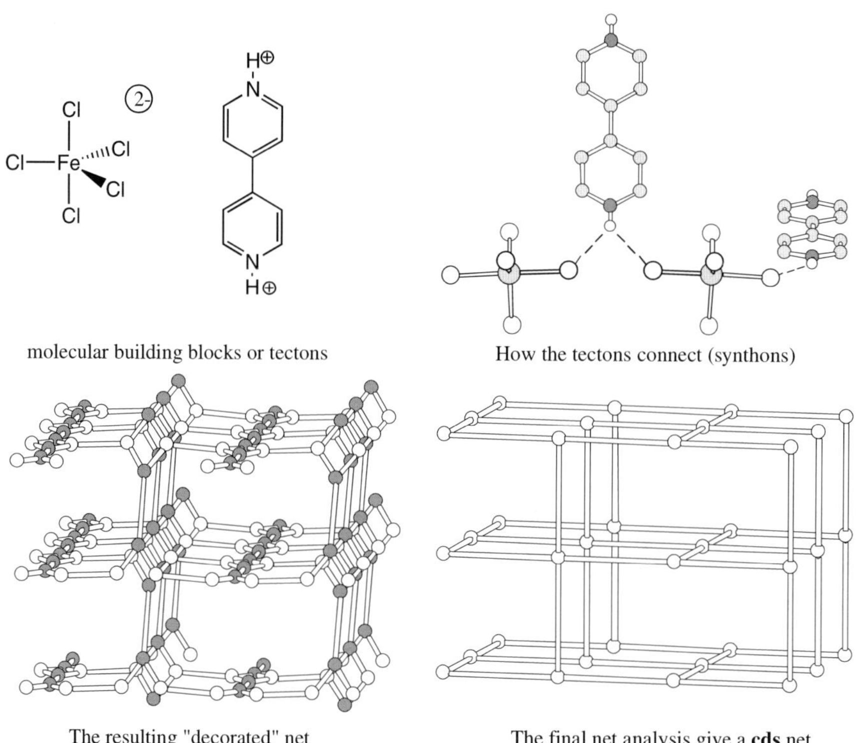

molecular building blocks or tectons   How the tectons connect (synthons)

The resulting "decorated" net   The final net analysis give a **cds** net

Figure 3.10 An example of a net built from "weak" hydrogen bonds, [H₂bipy][FeCl₅] [34]. This picture also illustrates the words tecton as a synonym for building block, synthon as a word for the connections, and decorated net for a net containing extra rings that can be reduced to a more basic net.

An example of a net built from such weaker, or non-classical, hydrogen bonds is shown in Figure 3.10 [34]. This Figure also illustrates the words *tecton* as a synonym for building block, *synthon* as a word for the connections, and *decorated net* for a net containing extra rings that can be reduced to a more basic net.

3.2.5. π–π and π–σ interactions

The π–π and π–σ interactions, which are also called phenyl factors in some systems, occurs between aromatic π-systems. The interaction is the result of a dispersion interaction and a columbic $C^{\delta-}/H^{\delta+}$ polarization. In energy terms, these interactions can be as strong as weak hydrogen bonds. Two categories are usually distinguished, C-H···π and π···π, which are shown in Figure 3.11. The "π···π"-interaction often involves an offset of the aromatic rings. This is probably because it allows a hydrogen in a C-H group to be located close to the centre of the π-system, thereby increasing the electrostatic interaction. The offset is even more pronounced in heteroaromatic compounds. Investigations of this offset, using the CSD, give typical values for nitrogen-containing aromatics of 1.3 Å for the offset and a centroid-centroid[6] distance of 3.7 Å [35].

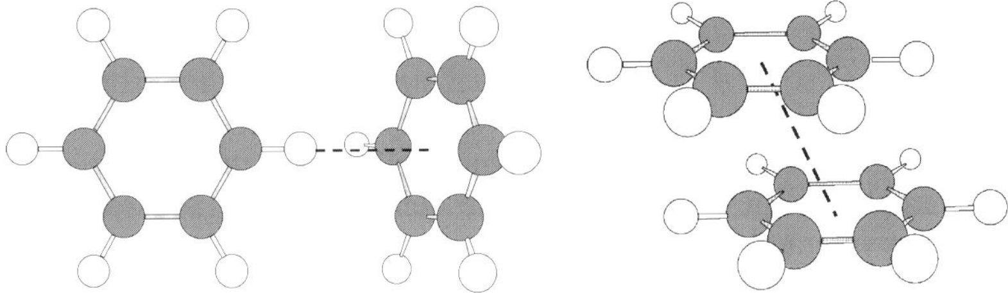

Figure 3.11 Two categories of "π"-interactions, the edge-to-face C-H···π and the face-to-face π···π.

It is less evident to apply the net formalism to "π-stacked" systems since it has to involve the use of multiple centroids on the aromatic rings as nodes, but it may be worth a try. For example, a compound sharing many characteristics of the 3D nets built from hydrogen or coordination bonds is [Cu(2,5-bis(4-pyridyl)-1,3,4-oxadiazole)$_2$(CH$_3$CN)]ClO$_4$·CH$_3$CN·1.5H$_2$O. It has large channels, it contains solvent molecules and the network is apparently stable to solvent loss [36]. It is formed by a 1D coordination polymer and two sets of close "π···π"-interactions giving a four-connected net, see Figure 3.12.

---

[6] The term 'centroid' refers to a point that is located in the centre of a ring closure.

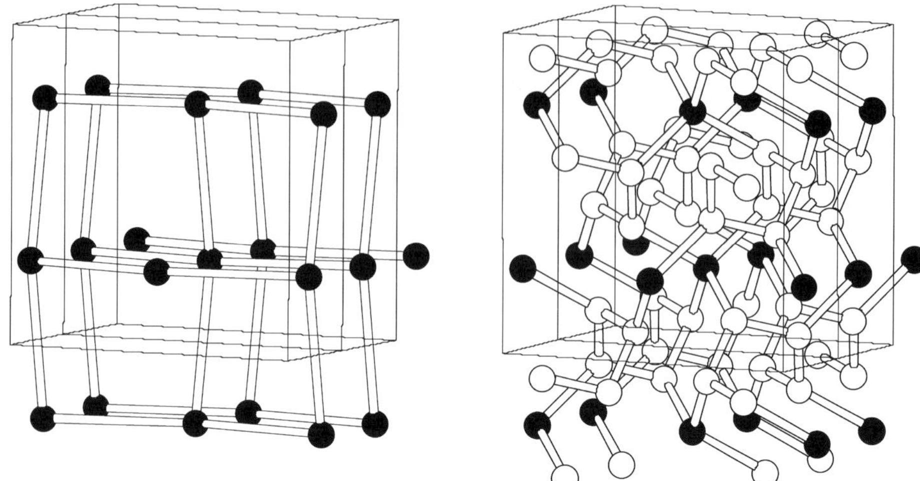

Figure 3.15 Three dimensional representation of the two nets formed by the nodes in Figure 3.13. To the left we have a diamond net **dia**, slightly squashed in one dimension, representing the connectivity of the molecular building blocks. To the right we get a binodal three- and four-connected net called **jph** (see chapter 8) that corresponds better to the arrangement of atoms in the structure (see also Figure 3.14).

While the net formed by the node designation to the left in Figure 3.14 is a true representation of the overall intermolecular connectivity of the molecules in the structure, it can nevertheless be regarded as only a mathematical construction. We prefer that the links in the net follow as closely as possible the actual intermolecular and covalent bonds that hold the net together. Then the net will also be a simplified physical picture of the structure, and will for example give some ideas about the porosity of the compound, see Figure 3.15. If, in this process, we have to assign several nodes to one particular molecular building block, so be it.

This makes the node assignment somewhat subjective, but we find this to be better than rules set in stone that sometimes give misleading results, see also a recent discussion of this by Carlucci et al. [42]

The requirement that the links somehow follow the molecules of the structure does not necessarily mean that the nodes themselves are centred on the building blocks or parts of them. In Figure 3.16 we see the net formed by the co-crystallisation of Co(III)(2,2′-biimidazole)$_3$ complexes with phthalates. One biimidazole and two phthalates are joined together, and the natural choice in this case is to assign one of the three nodes to the centre of this cluster of hydrogen bonds.

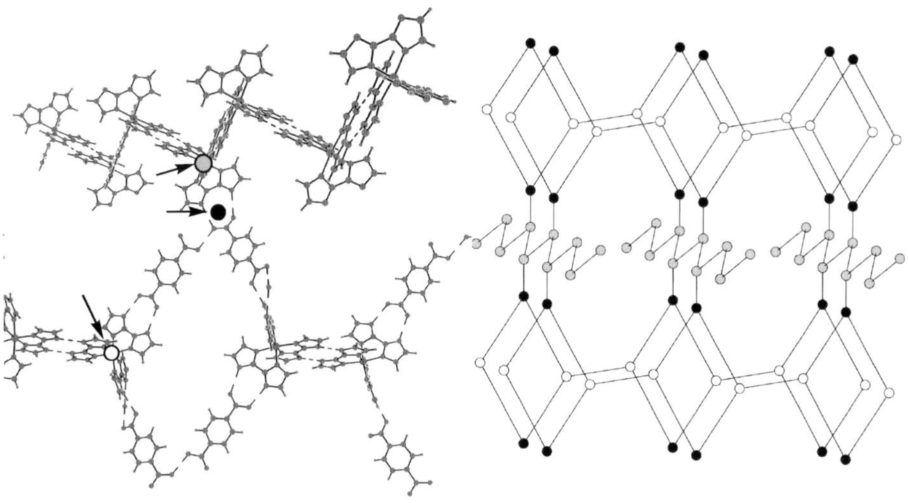

Figure 3.16 Left: The node assignment in the nets of [Co(Hbiim)$_2$(H$_2$biim)]$_2$(*para*-OOCC$_6$H$_4$COOH)$_2$·H$_2$O with arrows denoting cobalt centred nodes (white dot) and the terephthalate-terephthalate-biimidazole cluster defining the second type of three connected nodes (black dot) bridging these chains with the zigzag chains containing the third type of node (grey dot). Right: A simplification showing the resulting 3D-net with the same colour coding of the nodes [27]. Reproduced by permission of The Royal Society of Chemistry. H$_2$biim = 2,2-biimidazole. This net is referred to as "Wells net 10" [43].

3.4. Interpenetration

We conclude this chapter with a brief discussion of interpenetration (see Figure 3.8) since this phenomenon is a frequent companion to 3D nets. In a later chapter we will discuss this topic more thoroughly.

3.4.1. "Nature abhors a vacuum"

"Nature abhors a vacuum" is a citation attributed to Aristotle [3] and in the context of this book the significance is that the nets built up from molecular starting materials often have large empty spaces in them, indeed this is one of the reasons for our interest in them. However, these voids have to be filled with something, at least initially during the formation of the crystalline state, and more often than not the "filling" is one or more identical nets [42,44-47].
Interpenetration does not seem to be dependent on the type of chemical bond that holds the net together, that is interpenetration appears to be as frequent for coordination polymers as for hydrogen bonded nets. It raises interesting questions about the nucleation and driving force for 3D-net formation: Does one net template the formation of the other? Or is the structure formed by the

packing of the individual building blocks of the different nets and the net formation simply a consequence of adjusting the packing to maximal interaction energy?

Obviously there are many more things to say and discuss about interpenetration, and so later in the book an entire chapter is devoted to this theme. Here we will simply conclude by showing a picture of the interpenetrated nets of [Cu(4,4′-bipyridine)$_2$(CF$_3$SO$_3$)$_2$]·2CH$_2$Cl$_2$·H$_2$O [21] (a single net is shown in Figure 3.6), see Figure 3.17. Note that the structure in Figure 3.10 is also doubly interpenetrated, the [Ag(4,4′-bipyridine)]NO$_3$ [40,41] compound of Figure 3.13 is triply interpenetrated, and to get the complete structure of the compound in Figure 3.16 we need five nets (all these interpenetrated nets were avoided in the figures for simplicity).

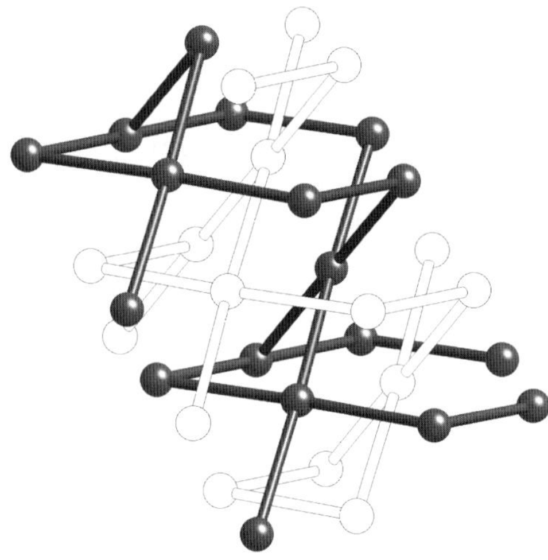

Figure 3.17 The two interpenetrated 4-connected **lvt** nets formed by the coordination polymer [Cu(4,4′-bipy)$_2$(CF$_3$SO$_3$)$_2$]·2CH$_2$Cl$_2$·H$_2$O [21]

References

[1] D. Bradshaw, T. J. Prior, E. J. Cussen, J. B. Claridge, M. J. Rosseinsky, J. Am. Chem. Soc. 126 (2004) 6106.
[2] G. R. Desiraju at the RSC Crystengcom Discussion Meeting, Nottingham, 2004.
[3] Encyclopædia Britannica Online, Encyclopædia Britannica Inc., 2005,
[4] D. Braga, F. Grepioni, Acc. Chem. Res. 33 (2000) 601.
[5] I. Dance, Crystengcomm 5 (2003) 208.
[6] J. R. Su, X. W. Li, R. C. Crittendon, G. H. Robinson, J. Am. Chem. Soc. 119 (1997) 5471.
[7] J. R. Su, X. W. Li, R. C. Crittendon, C. F. Campana, G. H. Robinson, Organometallics 16 (1997) 4511.
[8] R. Dagani, Chemical & Engineering News 76 (1998) 31.
[9] G. H. Robinson, Chem. Commun. (2000) 2175.
[10] I. Dance, New J. Chem. 27 (2003) 1.
[11] D. F. Shriver, P. W. Atkins, Inorganic Chemistry, 3rd ed. Oxford University Press, Oxford, 1999.
[12] A. Cusanelli, U. Frey, D. T. Richens, A. E. Merbach, J. Am. Chem. Soc. 118 (1996) 5265.
[13] F. A. Cotton, G. Wilkinson, Advanced Inorganic Chemistry, 4th ed. Wiley, New York, 1989.
[14] N. N. Greenwood, A. Earnshaw, Chemistry of the Elements, 2nd ed. Pergamon Press, Oxford, 1997.
[15] F. H. Allen, Acta Cryst. B 58 (2002) 380.
[16] F. H. Allen, O. Kennard, Chem. Design Auto. News 8 (1993) 31.
[17] F. H. Allen, W. D. S. Motherwell, Acta Cryst. B 58 (2002) 407.
[18] A. G. Orpen, Acta Cryst. B 58 (2002) 398.
[19] M. Winter, WebElementsTM, the periodic table on the WWW, The University of Sheffield and WebElements Ltd, UK, 2003, http://www.webelements.com/
[20] N. W. Ockwig, O. Delgado-Friedrichs, M. O'Keeffe, O. M. Yaghi, Acc. Chem. Res. 38 (2005) 176.
[21] L. Carlucci, N. Cozzi, G. Ciani, M. Moret, D. M. Proserpio, S. Rizzato, Chem. Commun. (2002) 1354.
[22] J. C. Jeffery, An Introduction to Hydrogen Bonding, Oxford University Press, Oxford, 1997.
[23] P. W. Atkins, L. Jones, Chemical Principles, the Quest for Insight, 3rd ed. W.H. Freeman, 2004.
[24] D. Laliberte, T. Maris, J. D. Wuest, J. Org. Chem. 69 (2004) 1776.
[25] D. Braga, F. Grepioni, E. Tagliavini, J. J. Novoa, F. Mota, New J. Chem. 22 (1998) 755.
[26] J. J. Novoa, I. Nobeli, F. Grepioni, D. Braga, New J. Chem. 24 (2000) 5.
[27] K. Larsson, L. Öhrström, Crystengcomm 6 (2004) 354.
[28] M. C. Etter, J. C. Macdonald, J. Bernstein, Acta Cryst. B 46 (1990) 256.
[29] J. Bernstein, R. E. Davis, L. Shimoni, N. L. Chang, Angew. Chem. Int. Ed. 34 (1995) 1555.
[30] K. Larsson, L. Öhrström, Crystengcomm 5 (2003) 222.
[31] M. A. M. Lorenet, F. Dahan, Y. Sanakis, V. Petroules, A. Bousseksou, J. P. Tuchagues, Inorg. Chem. 34 (1996) 5346.
[32] G. R. Desiraju, T. Steiner, The Weak Hydrogen Bond in Structural Chemistry and Biology, Oxford University Press, Oxford, 2001.
[33] T. Steiner, Angew. Chem. Int. Ed. 41 (2002) 48.
[34] B. Dolling, A. L. Gillon, A. G. Orpen, S. Jonathan, W. Xi-Meng, Chem. Commun. (2001) 567–568.
[35] C. Janiak, J. Chem. Soc., Dalton Trans. (2000) 3885.

[36] M. Du, C. K. Lam, X. H. Bu, T. C. W. Mak, Inorg. Chem. Com. 7 (2004) 315.

[37] I. Dance, M. Scudder, J. Chem. Soc., Chem. Commun. (1995) 1039.

[38] M. Scudder, I. Dance, Crystengcomm 3 (2001) 46.

[39] I. Dance, New J. Chem. 27 (2003) 22.

[40] O. M. Yaghi, H. Li, J. Am. Chem. Soc. 118 (1996) 295.

[41] F. Robinson, M. J. Zaworotko, J. Chem. Soc., Chem. Commun. (1995) 2413.

[42] L. Carlucci, G. Ciani, D. M. Proserpio, Coord. Chem. Rev. 246 (2003) 247.

[43] A. F. Wells, Acta Cryst. 7 (1954) 535.

[44] S. R. Batten, Crystengcomm (2001) 1.

[45] V. A. Blatov, L. Carlucci, G. Ciani, D. M. Proserpio, Crystengcomm 6 (2004) 377.

[46] S. R. Batten, R. Robson, Angew. Chem. Int. Ed. 37 (1998) 1461.

[47] S. R. Batten, Curr. Opin. Solid State Mat. Sci. 5 (2001) 107.

Chapter 4

Naming the nets and finding them

In order to talk about things, we have to give them names, otherwise we will be lost in endless battles of definitions. Nomenclature is just "name giving", the Latin roots of this word can be loosely translated as "calling of names", and is thus not only the tedious creation of systematic names according to rules set by the IUPAC committees.

However, we should not scorn the efforts of such committees even though we will continue to call CH_3COOH "acetic acid" and not "ethanoic acid". It is indeed very important, except perhaps in public relations and advertising, that the names we use are unambiguous, so that it is clear what we are talking about. Trivial names are also part of the nomenclature but they need to refer back to a systematic naming system so that doubts about their meaning can always be settled.

4.1. A state of some concern, but with a solution?

Unfortunately there are no IUPAC or IUCr recommendations,[1] or even a consensus among the scientists in the field, about the nomenclature of 3D-nets, and several naming systems are currently in use. As noted by O'Keeffe et al. "some have many names and symbols…other structures have no names at all" and they give the example of the **srs** net that is also know as: "(10,3)-a", "Laves net", "Y*", "3/10/c1", "SrSi$_2$" and "labyrinth graph of the gyroid surface" [1,2]. It can also be described by various sets of numbers, the most complete being $10_5 \cdot 10_5 \cdot 10_5$, referred to as the *vertex symbol*[2]. Unfortunately, there is no present nomenclature that creates a set of numbers for each net that can be proven to be unique.

This Babylonian naming situation is of course a state of some concern, but there may be an acceptable solution at hand by the use of the shorthand notation, lower case three letter codes, proposed by O'Keeffe et al. [1,3,4] and, when appropriate, the vertex symbol [5] also advocated by O'Keeffe.

---

[1] IUPAC is the International Union of Pure and Applied Chemistry, and IUCr is the International Union of Crystallography.

[2] 10 refers to the size of the smallest rings and five to the number of rings. Vertex is synonymous to node.

In this chapter, we will provide an introduction to the different types of nomenclature used by chemists when designating a net. As we have described in Chapter 3, a net consists of connected points forming rings of different sizes and these are used to create a set of identifying numbers, see Figure 4.1.

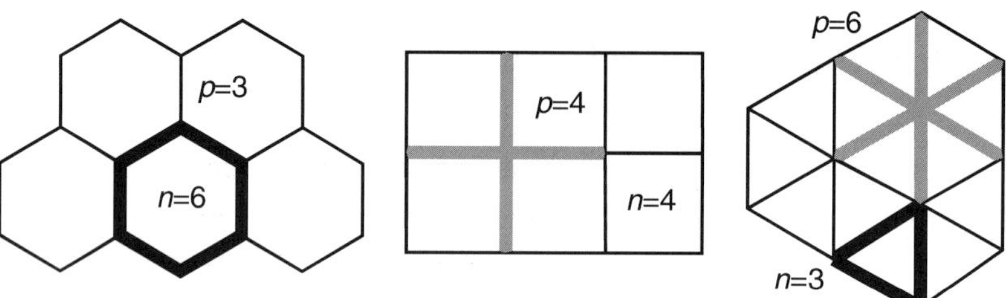

Figure 4.1 Some 2D-nets illustrating the use of *p* and *n*; *n* designates the shortest circuit found in the net and *p* is the connectivity at each node. From left to right these are assigned as: (6,3), (4,4) and (3,6) nets.

4.2. Nomenclature

Some basic information is in order before we describe the different types of notation in net nomenclature. A net, as shown in Figure 4.2, is an assembly of links, meeting at intersections called nodes or vertices. A path is formed by connecting a continuous sequence of nodes, and a path with *n* number of nodes starting and ending at the same node is called an *n*-circuit (we can also say that this circuit forms an *n*-gon).

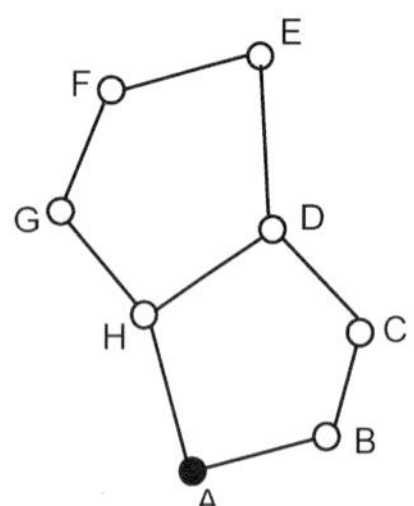

Figure 4.2. The difference between a circuit (ABCDEFGH) and a fundamental ring (ABCDH). The latter cannot be divided into smaller rings.

The methods for describing a net are all based on determining the size of the shortest circuits going through a node, the so called *fundamental* [6] rings. A fundamental ring is a circuit that cannot be divided into two smaller ones, see Figure 4.2. In the simplest case, all nodes are equivalent and only the rings going through one node need to be counted. This is known as a *uninodal* net, (see Figure 4.3). However, many nets have more than one "type" of node and in this case the fundamental rings going through each of the different nodes have to be counted.

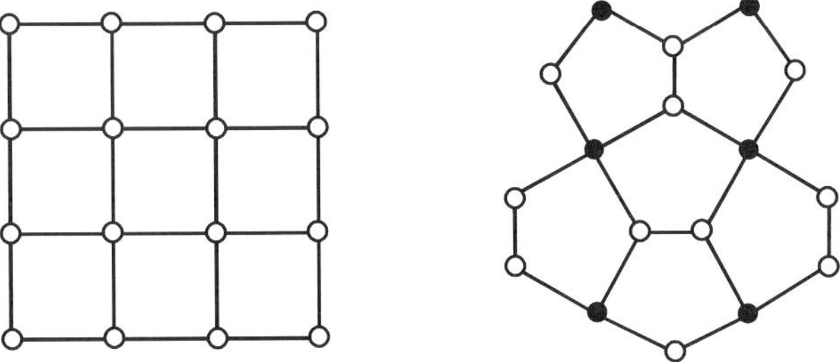

Figure 4.3 Fragments of a uninodal (left) and multinodal (right) net. The value of p is 4 for net 1 and 3 and 4 for net 2.

As an example we will use the (10,3)-a or **srs** net in Figure 4.4.

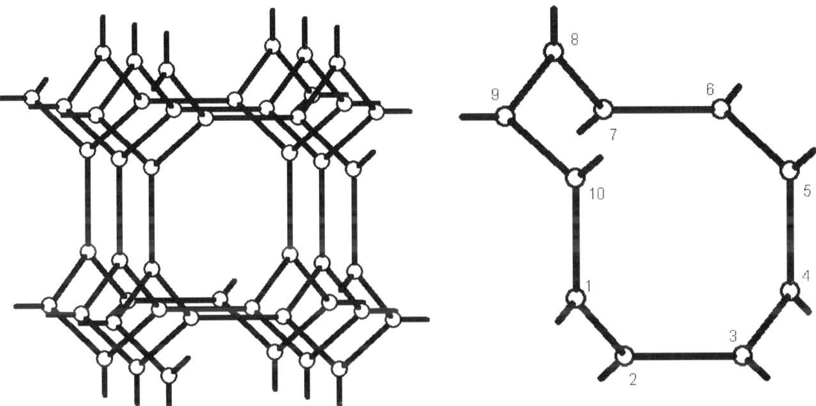

Figure 4.4 Left: The **srs** or (10,3)-a net also known as "Laves net", "Y*", "3/10/c1", "SrSi$_2$" and "labyrinth graph of the gyroid surface" [1]. Right: A 10-ring contained in this net.

Nets often have trivial names based on an inorganic structure and in this particular case the net in is also called an SrSi$_2$-net. When using these kinds of names it is important to remember that we sometimes mean the net formed by only one particular type of atom in such a structure, in the case of **srs** the net is formed by the Si-atoms [7]. Note that this is very different from the "type structure" nomenclature in inorganic chemistry, when, for example, having a "rock-salt structure" means that the complete arrangement of atoms in the structure is very similar to the NaCl-structure. We would call a net based on the NaCl structure for a **pcu** net or α-polonium net (**pcu** stands for primitive cubic packing).

4.2.1. Nomenclature according to Wells

In the simplest form, for a net containing only one type of ring and node, the notation used by Wells [7,8] is (n,p) where the number p is the number of links at the node and n is the smallest ring. These nets are called *uniform* nets. It is worth mentioning that this type of notation can be used for both 2- and 3-D nets. Thus, graphite is a 2D (6,3) net and diamond a 3D (6,4) net. Our example net, **srs**, is a (10,3)-net with ten nodes in the fundamental ring and three links at each node, as shown in Figure 4.4

This notation can be extended to include nets with more than one type of node and ring. In the general form this would be:

$$\begin{pmatrix} m & & p \\ n & , & q \\ \dots & & \dots \end{pmatrix}$$

where all combinations can be included. Thus, a net with two types of rings and one node reduces the notation to:

$$\begin{pmatrix} m & & \\ & , & p \\ n & & \end{pmatrix}$$

and a net with one ring and two nodes to

$$\begin{pmatrix} & & p \\ n & , & \\ & & q \end{pmatrix}$$

Although these special cases are sometimes encountered, another notation can used when describing more complex nets. To differentiate between nets with the same values of n and p, Wells added a letter after the notation. Starting from –a, the nets were ordered from high to low symmetry. Returning to our example in Figure 4.4, this is a (10,3)-a net since it is the most symmetric of the three-connected nets with 10-gons as smallest rings.

4.2.2. Schläfli symbol

To describe more complex nets, a different nomenclature can be used. In this case, also according to Wells, a set of numbers is assigned to each type of node in a net. These are in the form $A^{x_1}.B^{x_2}.C^{x_3}...$ where the letters A, B, C... are the size of the smallest circuit originating from that node. The circuits should be ordered as $A < B < C$. The superscripts x_1, x_2... are the number of circuits (of type A, B, C...) at the node. Although the connectivity of the node, p, is not included, it can be deduced from the sum of the superscripts:

$$\sum x_i = \frac{p(p-1)}{2} \qquad \textit{which gives} \qquad (4.1)$$

$$p = \frac{1}{2} + \sqrt{\frac{1}{4} + 2\sum x_i} \qquad (4.2)$$

The total number of circuits to count is determined by all the possible link pairs that can be formed at the node. For example, a 3-connected node with the links a, b and c has 3 pairs: ab, bc, and ac, and in the general case the number of linked pairs is equal to $p(p-1)/2$ as in equation 4.1 and the connectivity p is given by equation 4.2.

This type of notation is sometimes called *Wells point symbol, Schläfli*[3] *symbol* or *short symbol,* and applying this to our example we will get a 10^3 net as shown in Figure 4.5.

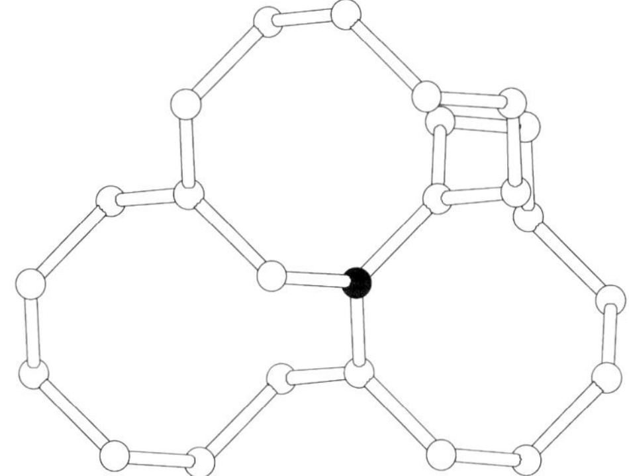

Figure 4.5 The **srs** net showing one node in black and three 10 membered rings for every link pair. This net thus has the Schläfli or short symbol 10^3.

Although the Schläfli symbol gives somewhat more information, both types of notation result in there being more than one net with the same set of numbers. The added letter in Wells notation does of course take care of this, but "new" nets are hard to designate since there is no central committee handing out extra letters.

4.2.3. Rings and circuits

A complication that may arise is that the shortest path from link *a* to link *b* has a shortcut through link *c* which leads back to the node. Obviously, this path is not a fundamental ring since it can be divided into two smaller rings. To

---

[3] After Ludwig Schläfli (1814-1895) professor of mathematics at the university of Bern.

avoid confusion, Smith (applying this to zeolites) introduced a distinction between smallest *rings* and *circuits*, see Figure 4.6 [2].

- The smallest rings are always fundamental *rings*. A shortest path going from link *a* to link *b* that has a shortcut via link *c* is not counted as a ring
- All shortest paths connecting any link pairs are called *circuits*.
- All the smallest *rings* are therefore also the shortest *circuits*, but not all shortest *circuits* coincide with the smallest *rings*.

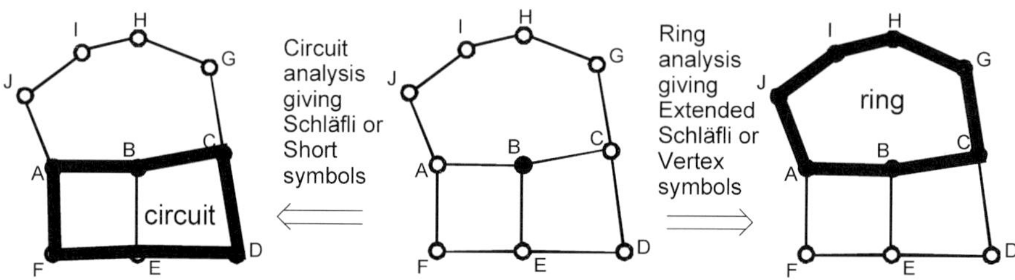

Figure 4.6 Difference between a ring and a circuit. For the central node B the shortest ring for the link pair BA-BC is AFEDC (left). However, as this path has a shortcut back to B via the BE link this *circuit* is not the shortest *ring*. The shortest ring is instead BCGHIJA (right). The circuit analysis used for the short symbol gives $4^2.6$ and the ring analysis gives two 4-rings and one 7-ring.

This means that for any link pair there is always a shortest circuit, but there may be link pairs that have no shortest rings if all circuits contain shortcuts back via another link, see Figure 4.7.

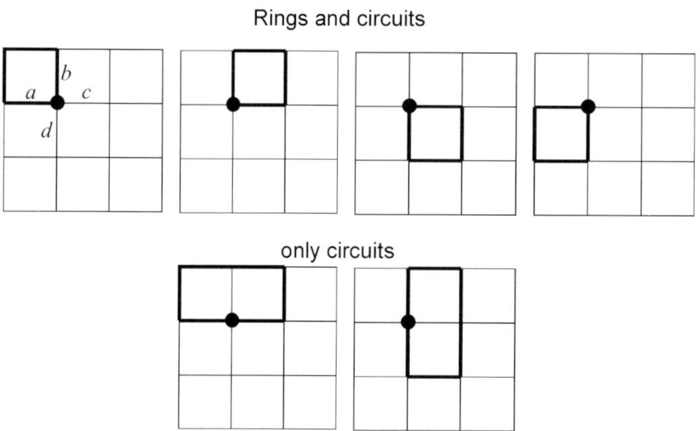

Figure 4.7 For the four-connected square grid net we get Schläfli or short symbol $4^4.6^2$ using circuits, but we count only four 4-rings since the last four pairs of links (*ac* and *bd*) all contain shortcuts and there are no alternative longer paths without shortcuts giving vertex symbol 4·4·4·4·*·* (see section 4.2.4).

4.2.4. Extended Schläfli or vertex symbols

The extended Schläfli notation has been put forward by O'Keeffe [5,9] and is, as the name indicates, an expansion of the Schläfli notation. This notation is also called *vertex symbol* or *long notation*. Now we will not only note the size of the smallest rings of a link pair, but we will also count how many of these there are.

Thus, all possible ways to unite a link pair using smallest *rings* are counted and then added as a subscript index to the ring size as $M_{y_1} \cdot N_{y_2} \cdot O_{y_3}...$ The connectivity p is now deduced from the number of link pairs, each of these having a M_{y_1} symbol assigned to it. Thus summing the *number of link pairs* is equal to $p(p-1)/2$ (equation 4.1). For example: for a three-connected net there will be three link pairs ($M_{y_1} \cdot N_{y_2} \cdot O_{y_3}$) and for a four-connected net there will be six ($M_{y_1} \cdot N_{y_2} \cdot O_{y_3} \cdot P_{y_4} \cdot Q_{y_5} \cdot R_{y_6}$). The subscript is used to avoid confusion with the Schläfli notation.

The order of the link pairs are the same as for the short (Schläfli) notation when $p=3$ and $p>4$. However, for nets where the nodes have $p=4$ a different order is used. The link pairs are paired opposite to each other and ordered with the lowest numbers first. In these pairs, each link is used once. Designating the links as a, b, c and d, the order will be (ab,cd)(ac,bd)(ad,bc).

For link pairs without smallest rings, that is all connections contain a short cut back to the node, we write "*" or "∞") and thus for the square-grid net in Figure 4.7 the vertex symbol will be 4·4·4·4·*·*.

Unfortunately, this notation can also result in more than one net with the same set of numbers. To further distinguish different nets we therefore look at the *coordination sequence* and we count the number of bonded nodes up to the tenth coordination layer, the *C10-value*, or *TD10 value* if it is an average over all types of nodes, see Figure 4.7.

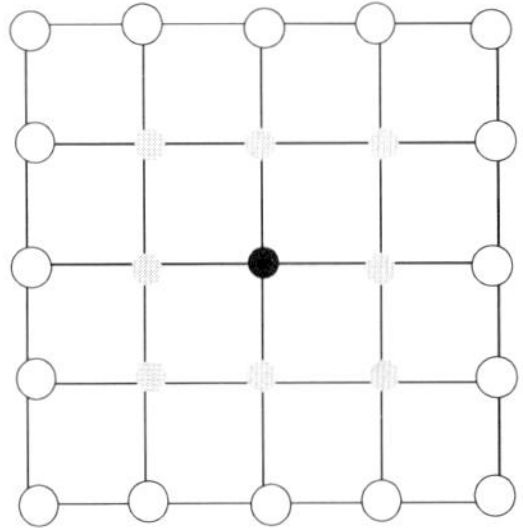

Figure 4.7 First (grey) and second (white) coordination shell for the square-grid net, the C2 value is 1+8+14=23.

Although there seems to be a theoretical possibility of having nets with the same set of vertex symbol and C10 value, for practical purposes this it is an adequate nomenclature, albeit somewhat cumbersome to communicate, especially for multinodal nets with higher connectivity.[4]

Using the vertex symbol notation the **srs** net is a $10_5 \cdot 10_5 \cdot 10_5$ net, where we count all five 10-membered circuits for each link pair. Two of the five circuits at a node can be found in Figure 4.8. The C10 value for this net is 529.

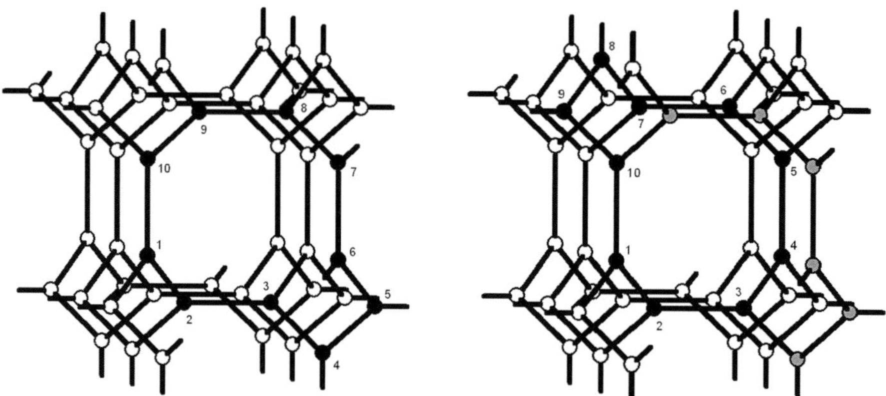

Figure 4.8 Two of the five 10 membered rings in the $10_5 \cdot 10_5 \cdot 10_5$ **srs** net. The first is shown on the left while the second is on the right. The grey nodes in the left picture are the nodes from only the first circuit.

This enumeration may seem forbidding to work out "by hand", especially if we consider nets with more than one type of node, but fortunately, there is now both free and commercial software that will do the ring counting more or less automatically (depending on your type of net-connection) [11-14]. A brief introduction will be given in Chapter 13.

O'Keeffe and co-workers have recently published a free-access web-based searchable database with both hypothetical and synthesised 3D-nets [3]. All nets are listed with their long (vertex) notation, as well as all the types of nodes and links in each net and a number of other parameters. This is a valuable help for the identification of nets once the ring analysis giving the vertex symbols is finished.

4.2.5. Zeolite-like abbreviations

As was mentioned in section 1.2, O'Keeffe, Yaghi and co-workers have also adopted three letter codes for the naming of each net, corresponding to a similar

---

[4] Note that the C10 values may differ with one unit depending on if the first node is counted or not. We are not aware of any observed nets having different topology and the same C10 value but recently it has been shown that there exists two topological distinctive tetrahedral based nets with the same vertex symbols and identical coordination sequences up to the 16th shell [10].

system used for a long time in zeolite research. To avoid confusion with zeolites, lower case letters are used. The letters are not set after rules like the numbering scheme we have described so far. Instead, the maintainers of the database have assigned codes from trivial names of different types, for example, the SrSi$_2$-net is named **srs** and the quartz net **qtz**. Other rationalisations for certain codes also exist, such as **pcu** (from **p**rimitive **cu**bic packing) for the α-polonium net. The important thing is that these codes form a set of well-defined trivial names, easy to use and easy to retrieve in a database search.

A suffix is added to the three letter code for nets which can be derived from a "basic" net. The following paragraphs will list and exemplify the suffixes using the "base"-net **dia** (diamond) shown in Figure 4.9.

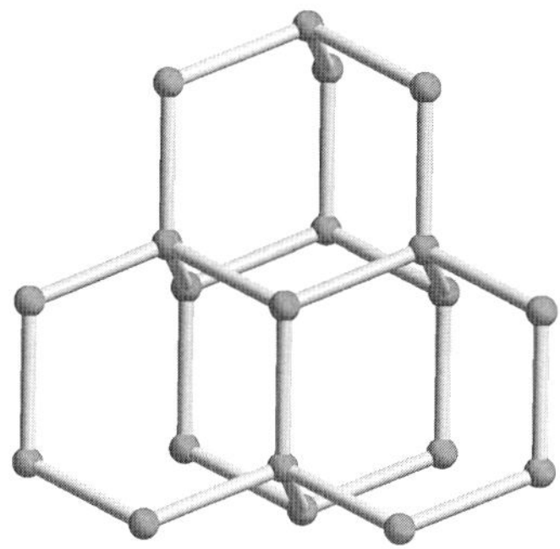

Figure 4.9 The diamond net, **dia**.

-a   **A**ugmented net. In this net, each node has been replaced by a polygon or polyhedron with the same connectivity as the node. Thus in the **dia-a** net, found in Figure 4.10, every node has been replaced by a. tetrahedron.

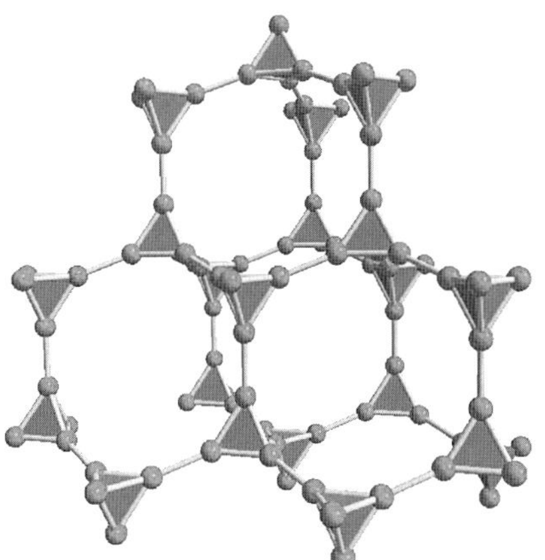

Figure 4.10 The **dia-a** net. The nodes in the **dia** net have been replaced by a tetrahedron, shown in grey.

we should be able to use the "stand alone" three letter code for the most well known nets.

- The vertex (long) symbol should be given in the experimental part together with crystallography data, except for obvious examples of the most common nets when it can be excluded.

4.2.7. Fischer symbol

The Fischer symbol was created for the description of different packings of identical spheres and is consequently only applicable to uninodal nets. It has the format: *p/n/f#*.

The letters *p* and *n* have the same meaning as explained earlier in this chapter. It is the connectivity of the node and rings size. The letter *f* designates the highest possible symmetry for the specific net and the following subclasses are used: cubic (c), hexagonal/trigonal (h), tetragonal (t), orthorhombic (o), monoclinic (m) and triclinic (a, for anorthic). An serial number # is added to distinguish between nets with the same set of *p*, *n* and *f* numbers [15].

4.2.8. Graph set theory

Although only applicable for hydrogen bonded nets, this system is convenient when describing complex arrays of hydrogen bonded entities. It is also a useful tool for identifying the interactions that hold a network together (synthons) and a section in Chapter 14 will describe the RPLUTO software package. In this section we will briefly cover the theory, a more in depth explanation can be found in the articles by Etter and Bernstein [16,17].

The notation in this system uses a designator and one or two indices. The designators are chains (**C**), rings (**R**), intramolecular hydrogen-bonded patterns (**S**) and other finite patterns (**D**). This is followed by a subscript (**d**) and a superscript (**a**) designating the number of hydrogen bonding donor atoms and accepting atoms. In addition, the degree of the pattern (**n**) is added in parenthesis. This is the total number of atoms in the pattern. The sub- or superscript is omitted if there is only one donor or acceptor. Some examples can be found in Figure 4.14.

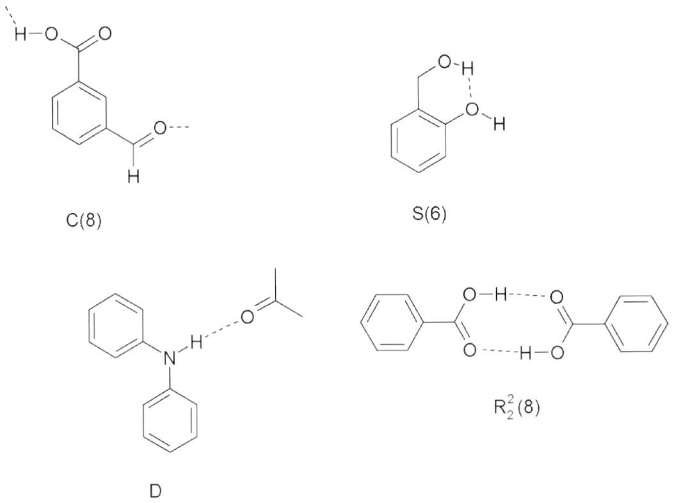

Figure 4.14 Examples of the four different designators, with indices, in Graph Set Theory.

As can be seen, this notation focuses on the bridge between two building blocks. Although the entire network or a hydrogen bonded net can be described with Graph Set Theory (but not the topology), this becomes quite complex and cannot be recommended. However, for an automated search of the CSD for hydrogen bonded nets it would be very useful to have these automatically classified by the Graph Set Theory of the bonding interactions (supramolecular synthons).

4.3. Examples

In this section we will use two nets to show how to count the rings using the Schläfli and the vertex symbol (extended Schläfli) notation described earlier.

4.3.1. Diamond net

The diamond net (**dia**) is uninodal, four connected with tetrahedral nodes and contains six-membered rings. Figure 4.15 shows one node and the four links ABCD.

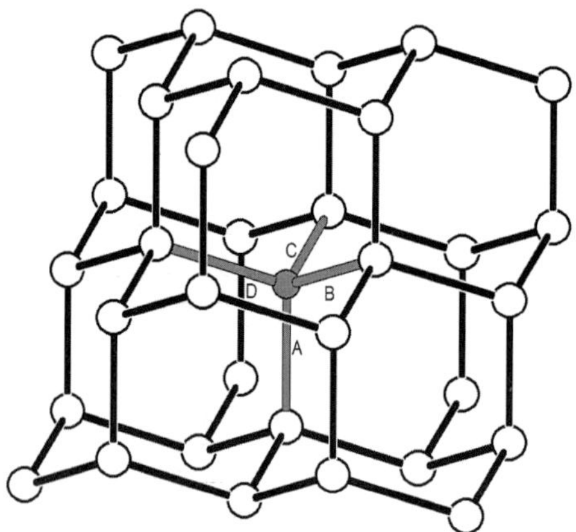

Figure 4.15 The **dia** net with one node and its links ABCD highlighted.

Starting with the Schläfli notation where only one circuit is counted for each link pair we will find a total of six circuit (one each for AB, AC, AD, BC, BD and CD). The resulting symbol is 6^6 and the circuits are shown in Figure 4.16.

If we want to use the vertex symbol, the link pairs have to be arranged with opposing pairs: AB,CD,AC,BD,AD,BC. In this particular case, we will see that there is only one type of ring and the sequence we have chosen here is arbitrary. The resulting vertex symbol for the **dia** net is $6_2 \cdot 6_2 \cdot 6_2 \cdot 6_2 \cdot 6_2 \cdot 6_2$ and the rings are shown in Figure 4.17.

In a net with rings of more than one size, the sequence would be arranged in ascending ring size. For example, let us say that we have a four connected net with ring sizes AB=10, AC=4, AD=10, BC=6, BD=6 and CD=10. The sequence would be: AC,BD,BC,AD,AB,CD.

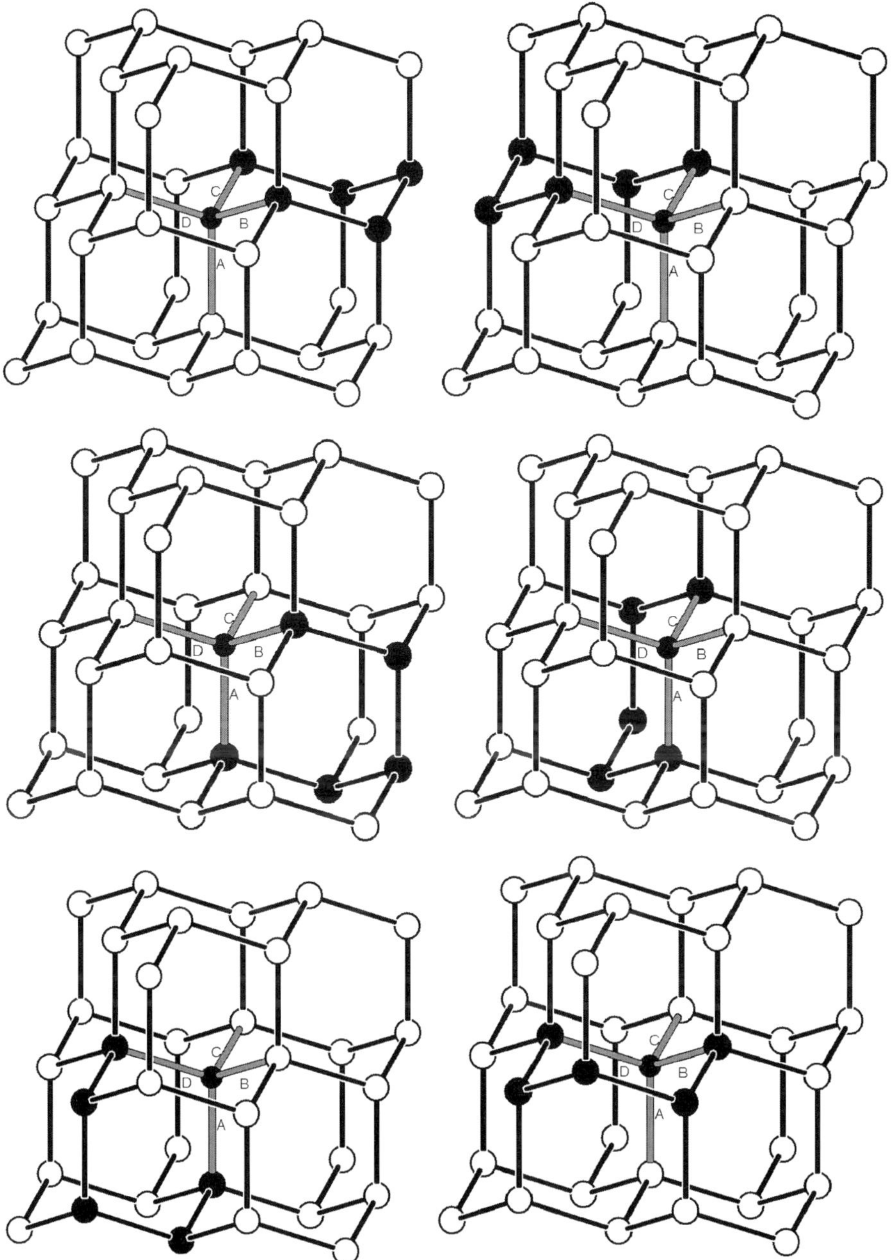

Figure 4.16 The six-rings in the **dia** net using the Schafli notation (6^6)

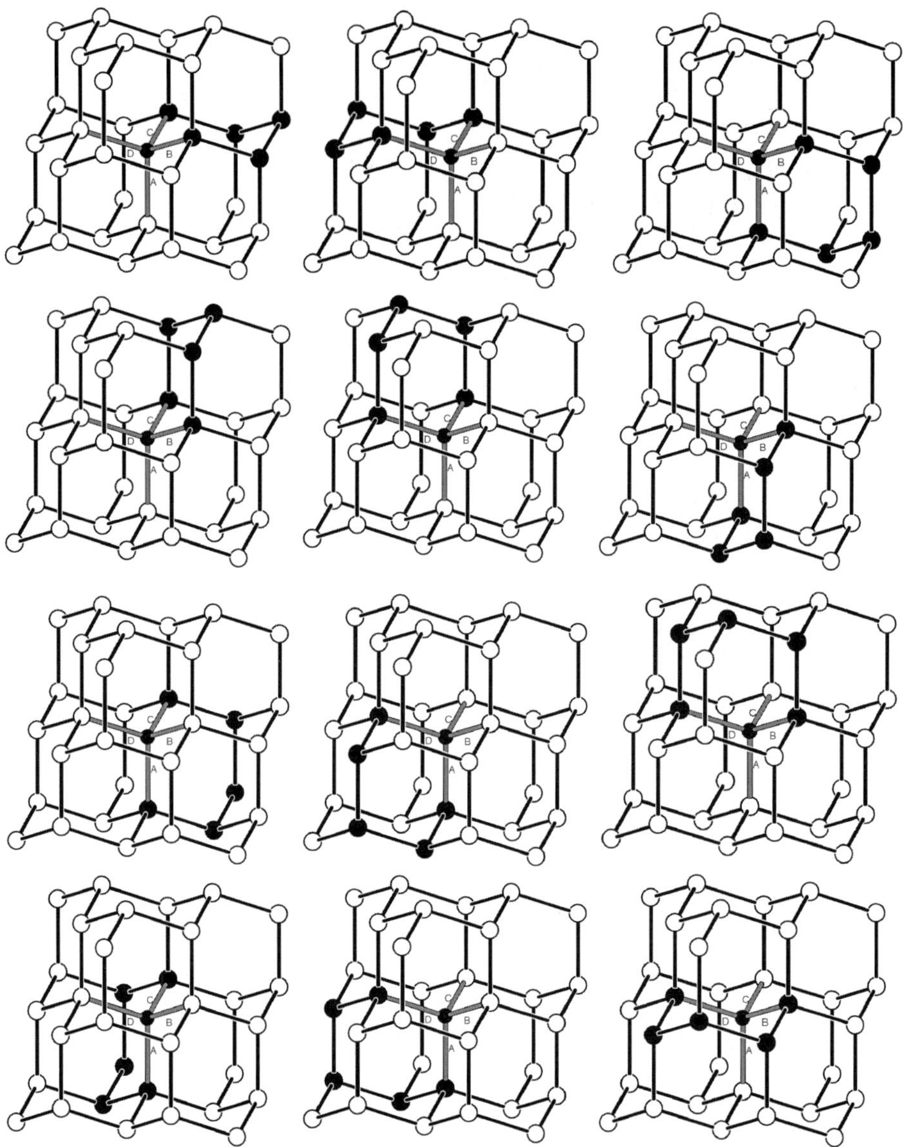

Figure 4.17 All shortest rings in the **dia** net. The resulting vertex symbol is $6_2 \cdot 6_2 \cdot 6_2 \cdot 6_2 \cdot 6_2 \cdot 6_2$.

4.3.2. CdSO₄ net

The CdSO₄ net is abbreviated **cds** and is also a four connected net. However, in this case the nodes are square planar and are arranged alternating in the horizontal and vertical plane as shown in Figure 4.18.

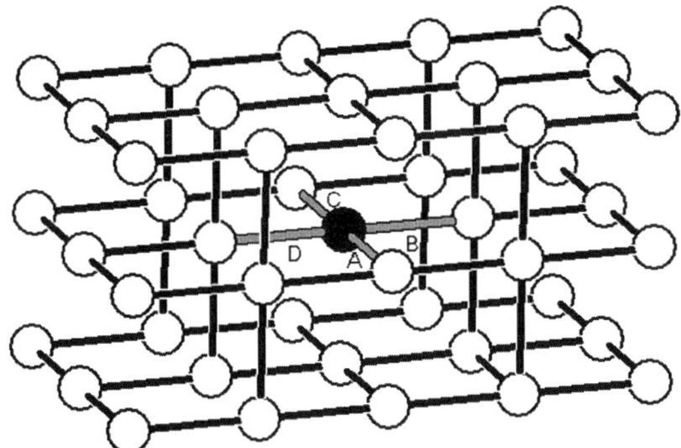

Figure 4.18 The **cds** net.

Since this also is a uninodal net, we only get one set of numbers. The Schläfli symbol is $6^5.8$ and the circuits are shown in Figure 4.19.

However, the result is somewhat different when using the vertex symbols. The 8-membered circuit for the AC pair is a part of two smaller 6-rings in AD and AC. Therefore it is not a fundamental ring and thus not counted. This is denoted by a "∞" or a "*". Pair BD has one extra 6-circuit (shown in Figure 4.20) and the complete symbol is $6·6·6·6·6_2·*$ (or ∞).

4.4. Searching the literature

4.4.1. Using words

An important reason for a consistent nomenclature is that we should be able to search the chemical literature and retrieve the information we need, otherwise we will keep reinventing the wheel.

At the writing of this book, the suggestions by O'Keeffe, Yaghi and co-workers concerning the naming of nets are very recent and therefore rarely encountered in the literature.

On the other hand, the Extended Schäfli or vertex symbol notations have been around for a while, but are not in universal use, probably because the assignments are cumbersome to do by hand and prone to errors.

Thus retrieving publications concerning structures containing 10^3-**srs** nets will not only be incomplete because of the cases where the net was not recognised, the search terms "(10,3)-a", "srs" and "SrSi2" will also yield a lot of irrelevant references.

                                                 L. Öhrström & K. Larsson

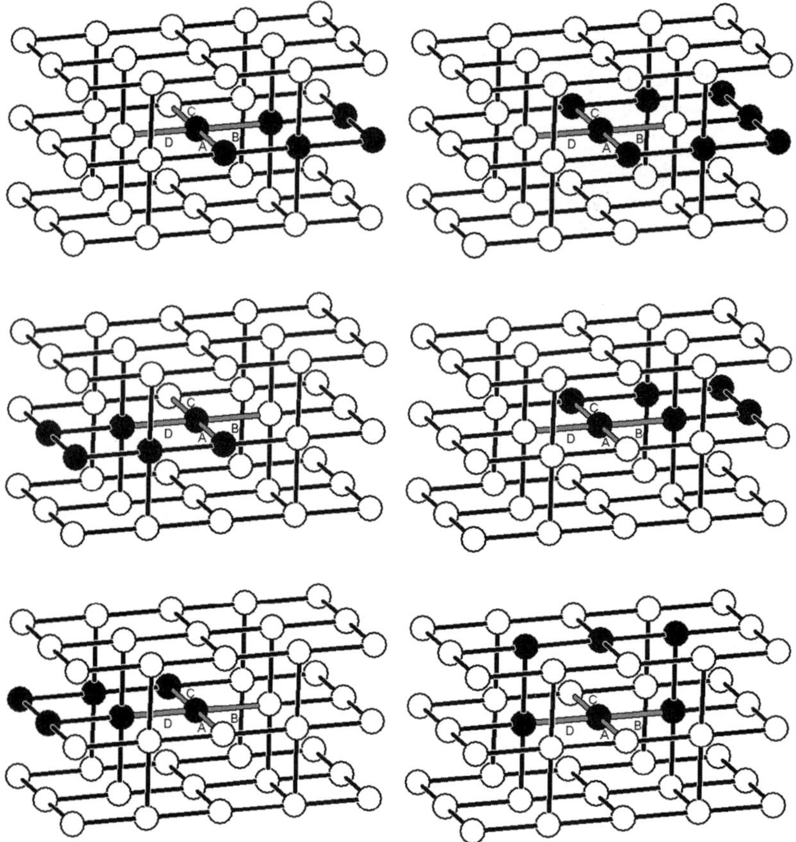

Figure 4.19 The rings in the **cds** net using the Schläfli notation. The Schläfli symbol is $6^5 8$.

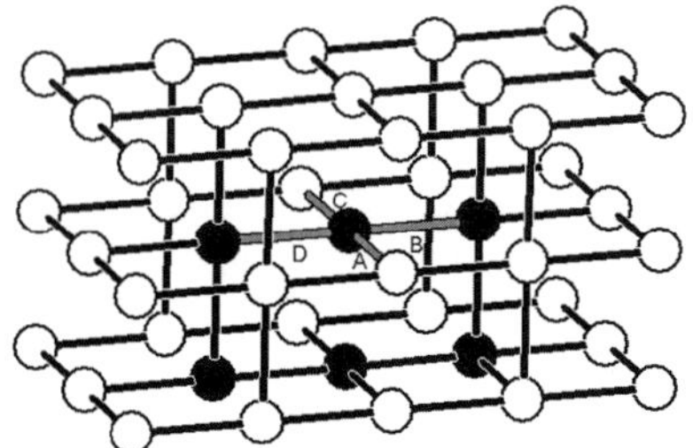

Figure 4.20 The extra 6-ring in **cds** counted when using the vertex symbol notation giving $6 \cdot 6 \cdot 6 \cdot 6 \cdot 6_2 \cdot ^*$.

There are similar problems with interpenetration when searching the literature, as pointed out by Batten [18]. "Interpenetration" will refer to many other phenomena and in addition we will also need to search for terms such as "interwoven", "intercalated", "entangled" and "catenated" although neither of these uniquely identifies interpenetration. In addition, it is possible that the interpenetration has escaped the structural analysis of the authors, in which case the information will be accessible only from compilations such as those by Batten [19] and Proserpio [20] or via searches in the Cambridge Structural Database and subsequent analysis "by hand".

4.4.2. Searching the Cambridge Structural Database

Although the Cambridge Structural Database does not contain (at present) any information of the type discussed in this chapter (Schäfli notation, Vertex symbols, three letter codes) it is appropriate to mention it here as we are briefly dealing with information retrieval [21-25].

As not only interpenetration, but the network itself, may have gone undetected by the scientists responsible for the original chemical and crystallographic work,[5] it is important that we find ways to obtain information, and searching the CSD using the "ConQuest" software is a good option (for Unix the corresponding program is "Quest").

Coordination polymers are easily detected by specifying the metal - donor atom bond as "polymeric" in ConQuest. The problem is then to weed out the 1D and 2D structures from the 3D-nets.

Also hydrogen-bonded nets can be found by specifying the O-H bond as polymeric, but it is also quite easy (and necessary since the polymeric O-H bond will recognise but a fraction of all such interactions) to specify a certain hydrogen bond pattern and search for structures containing it. An example is shown in Figure 4.21.

In both cases the crystallographic data[6] can easily be extracted and analysed by many different commercial available or free software, but perhaps easiest as a first option is Mercury (that works directly with ConQuest), [12] obtainable free of charge for academic users from the Cambridge Crystallography Data Centre (CCDC), the providers of the CSD. From the identification of a candidate structure in the CSD, the determination of its dimensionality using Mercury is often a matter of a few minutes (per compound! A systematic study will take many hours). See further Chapter 13.

---

[5] This should not be seen as an implicit criticism. We should remember that the task of detecting nets was very difficult not so long ago, especially if the focus of the research was something else.

[6] Crystallographic data comprises, among other things, unit cell parameters and atom coordinates and they are contained in a crystallographic information file, CIF [26]. It is thus essential that your software can read and manipulate these files and not only files based on Cartesian coordinates or Z-matrices.

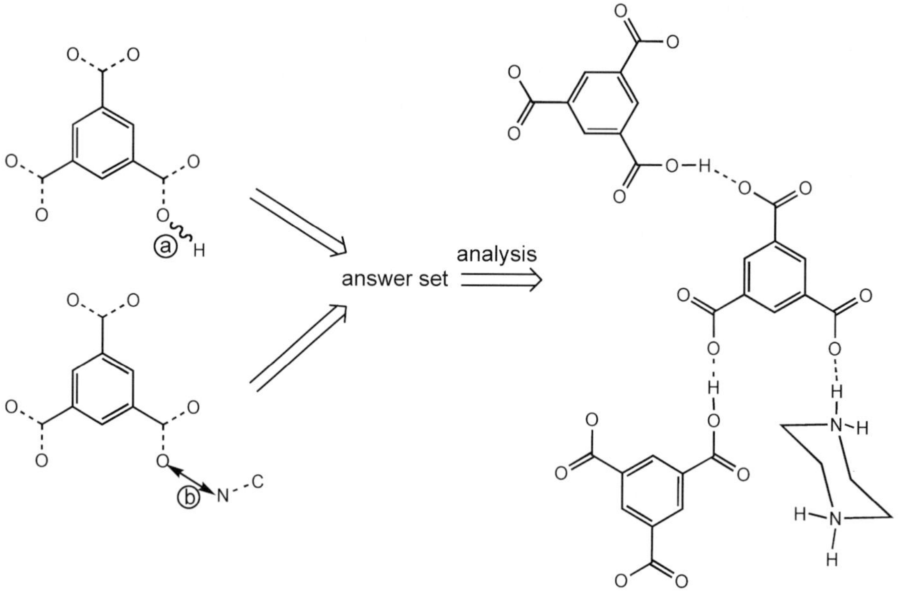

Figure 4.22 Two ways of retrieving hydrogen bonded 3D-nets from the Cambridge Structural Database. Unspecified bonds between C and O enables any kind of carboxylate unit to be found. In addition, in **a** a O-H bond has been specified as "polymeric" and in **b** a non-bonded interaction has been specified between N and O. In both cases the data set has to be further analysed to yield the 3D hydrogen bonded structure [27] shown as the result.

Having said all this, we would like to alert the reader to the fact that the CCDC, is not a simply a body devoted to the permanent storage of crystallographic data but also very much an active research organisation, both in the analysis of structural data and in the development of new research and search tools for the CSD. Thus, for example, it is likely that we in the future will see search tools that can automatically recognise such features as nets, dimensionality and interpenetration. CCDC also have some more recent products called "knowledge bases", IsoStar and Mogul, that may be quite useful as well.

4.5. Recommendations

We would like to recommend the following:

- Analyse the nets in terms of short (Schläfli) symbols and vertex (long) symbols. There is now both free and commercially available software that will do this more or less automatically (depending on your type of net-connection) [11-14]. A brief introduction will be given in Chapter 13. Do not do this by "hand" only!

- Use the three-letter codes from the RCSR database, [28] but be aware that the current problem with these three-letter codes is that they are not well known, and few people will be able to understand them if you refer to them in a research article. Indeed, to most people it will not even be clear that they refer to nets. We therefore propose that until these codes are commonly accepted they be explicitly referred to by a phrase like "in this article we will use the three letter codes from the RCSR database [28]" (including the reference!). If possible use a code without suffix.
- In the section headings in the following chapters we have adopted a system of adding the short (Schläfli) symbol before the three-letter code (as in 10^3-**srs**) and we recommend this for general usage (section 4.2.6). The reason is that even if you are familiar with the code system, the three letters in question will not tell you anything if you do not already known that particular net. Adding the short symbol will give you a rough idea about what kind of net this can be.[7]
- The simultaneous use of any other names that are likely to be understood by many people. These include both Wells names and $ThSi_2$ type names. Please consult the contemporary literature!

4.6. A few words about Alexander F. Wells

As we often refer to the work of Alexander F. Wells in this book, we thought it appropriate to make a small digression from our main subject and write a few words about him. Neither of us had the opportunity to meet him, he belonged to a different generation of scientists, but we have, as many others, been amazed at his ability to analyse 3D-structures, and 3D-nets in particular, and this at a time when the tools were mainly paper, pencil and model building.

He gave his own thoughts on "3D-thinking" in the introduction to "The Third Dimension in Chemistry", [29] writing among other things about our "addiction to flat surfaces": "It seems likely that a young child has a better appreciation of the third dimension before he can read and write than afterwards". Interestingly, this echoes one of his contemporaries obsessed with seeing things as they are and not as we have learnt to see them: Pablo Picasso. Picassso claimed that it took him fifteen years to learn to paint like a man and the rest of his life to learn to paint as a child [30].[8]

The following sketch is based on a recent biographical article by Wells' colleague at the university of Connecticut, John Tanaka [31].

---

[7] A complication here is that the short symbol cannot be derived from the vertex (long) symbol due to the different definitions using circuits and rings and currently the short symbol is not listed in the RCSR database.

[8] Sometimes cited as "As a child I could draw like Leonardo, as an adult I want to paint as a child."

Alexander F. Wells was born in 1912 in London, obtained a scholarship to go to Oxford and after his BA moved on to Cambridge where he gained his PhD in 1937 under the supervision of J. D. Bernal. He spent a few years doing research in Cambridge, but the Second World War soon started and he moved to Birmingham to do defence related work. Remarkably enough, his classic book "Structural Inorganic Chemistry" [7] was written in his spare time during the war and the first edition (four more were to follow) was published in 1945. Wells continued his work in industry and was a Senior Research Associate at Imperial Chemical Industries (ICI) 1944-1968 during which time he developed much of his thoughts on 3D-nets. The first paper in the series "The geometrical basis of crystal chemistry" published in 1954 [32]. Apparently, much of this work was also pursued during evenings and weekends with the help and support from his family. His son Alec recalls how he helped his father building models; "We used an assortment of materials, including punched paper strips, cut-down bicycle spokes, flexible plastic commercial items, and parts of his model kits. Mother did not like blobs of solder on the table..." [31] Wells himself elaborated on model building for teaching, and materials therefore, in a later book: "Models in Structural Inorganic Chemistry" [33].

He crossed over the Atlantic and once again into academic research, first as a NSF Visiting Scientist and later as a professor of chemistry (1968-1980) at the University of Connecticut in Storrs, Connecticut. He moved back to England (where his four children lived) in 1986 and published his last research paper "Relations between dense sphere packings" with B. L. Chamberland, in 1987 [34]. He died in 1994 at the age of 82. His scientific legacy includes 70 research papers, most of them with himself as a single author, five books, [7,8,29,33,35] and some 200 3D-models of inorganic structures and networks still in use at the University of Connecticut.

References

[1] O. Delgado-Friedrichs, M. O'Keeffe, O. M. Yaghi, Acta Cryst. A 59 (2003) 22.

[2] J. V. Smith, Am. Miner. 63 (1978) 960.

[3] C. Bonneau, O. Delgado-Friedrichs, M. O'Keeffe, O. M. Yaghi, Acta Cryst. A 60 (2004) 517.

[4] O. M. Yaghi, M. O'Keeffe, N. W. Ockwig, H. K. Chae, M. Eddaoudi, J. Kim, Nature 423 (2003) 705.

[5] M. O'Keeffe, Z. Kristall. 196 (1991) 21.

[6] L. Stixrude, M. S. T. Bukowinski, Am. Miner. 75 (1990) 1159.

[7] A. F. Wells, Structural Inorganic Chemistry, 5th ed. Clarendon Press, Oxford, 1984.

[8] A. F. Wells, Three-dimensional nets and polyhedra, John Wiley & Sons, New York, 1977.

[9] M. O'Keeffe, M. Eddaoudi, H. L. Li, T. Reineke, O. M. Yaghi, J. Solid State Chem. 152 (2000) 3.

[10] M. M. J. Treacy, I. Rivin, E. Balkovsky, K. H. Randall, M. D. Foster, Microporous Mesoporous Mater. 74 (2004) 121.

[11] O. V. Dolomanov, A. J. Blake, N. R. Champness, M. Schröder, J. Appl. Cryst. 36 (2003).

[12] O. V. Dolomanov, OLEX, http://www.ccp14.ac.uk/ccp/web-mirrors/lcells/index.htm

[13] V. A. Blatov, A. P. Shevchenko, V. N. Serezhkin, J. Appl. Cryst. 33 (2000) 1193.

[14] V. A. Blatov, 2004, http://www.topos.ssu.samara.ru/

[15] E. Koch, W. Fischer, Z. Kristallogr 210 (1995) 407

[16] M. C. Etter, J. C. Macdonald, J. Bernstein, Acta Cryst. B 46 (1990) 256.

[17] J. Bernstein, R. E. Davis, L. Shimoni, N. L. Chang, Angew. Chem. Int. Ed. 34 (1995) 1555.

[18] S. R. Batten, Crystengcomm (2001) 1.

[19] S. R. Batten, Monash University, Australia, 2005, http://web.chem.monash.edu.au/Department/Staff/Batten/Intptn.htm

[20] V. A. Blatov, L. Carlucci, G. Ciani, D. M. Proserpio, Crystengcomm 6 (2004) 377.

[21] F. H. Allen, O. Kennard, Chem. Design Auto. News 8 (1993) 31.

[22] F. H. Allen, Acta Cryst. B 58 (2002) 380.

[23] F. H. Allen, W. D. S. Motherwell, Acta Cryst. B 58 (2002) 407.

[24] A. G. Orpen, Acta Cryst. B 58 (2002) 398.

[25] F. H. Allen, R. Taylor, Chem. Soc. Rev. 33 (2004) 463.

[26] CIF-file information can be obtained from:, International Union of Crystallography, 2005, http://iucr.ac.uk

[27] D. Krishnamurthy, R. Murugavel, Indian J. Chem. 42 (2003) 2267.

[28] M. O'Keeffe, O. M. Yaghi, Reticular Chemistry Structure Resource, Tucson, Arizona State University, 2005, http://okeeffe-ws1.la.asu.edu/RCSR/home.htm

[29] A. F. Wells, The Third Dimension in Chemistry, reprinted 1962, 1968, 3:ed 1970, Clarendon Press, Oxford, 1956.

[30] A Child Prodigy, 1998, http://humanitiesweb.org/

[31] J. Tanaka, J. Chem. Hist. submitted for publication (2005).

[32] A. F. Wells, Acta Cryst. 7 (1954) 535.

[33] A. F. Wells, Models in Structural Inorganic Chemistry, Clarendon Press,, Oxford, 1970.

[34] A. F. Wells, B. L. Chamberland, J. Solid State Chem. 66 (1987) 26.

[35] A. F. Wells, Further Studies of Three-Dimensional Nets, Polycrystal book service, Pittsburgh, 1979.

Chapter 5

The most common 3D-nets

In this chapter we will give examples of the most common 3D-nets prepared from molecular building blocks (tectons), and in the following chapters other types of nets will be presented.

It is possible to arrange such a collection in different ways, for example based on connectivity or the more recent concepts of net *genus* or *transitivity*. While these latter concepts have been introduced rather recently [1-3] and will probably gain in importance, they are not indispensable for the way we present the material in this book, and a detailed discussion of them will be postponed till later. In this and the following chapters we will use genus as an indicator of the symmetry of a net, with lower values being the most symmetric the lowest possible for a 3D-net being three.[1]

The connectivity approach, on the other hand, gives equal weight to all nets, which is clearly unreasonable since some are very common while others have never been observed and are purely mathematical constructions.

We will instead start with the most frequently observed nets, based on two surveys of the Cambridge Structural Database, [3,5] dubbed the "default nets", [6] and then move on to a connectivity based approach. However, we will in general refrain from extensive discussion of hypothetical nets that have not been observed. Not because these are uninteresting, on the contrary they could be important synthetic targets,[2] but because in principle there is an infinite number of these nets. We refer the reader to the growing literature specialising in this area [7-9] and to the RCSR database [4].

For all the nets presented in this Chapter, atomic coordinates, unit cells and space groups can be found in Appendix A and stereo drawings can be found in Appendix B.

However, before we move on to the nets, it seems sensible to give a short introduction to the geometric demands of the nodes in a 3D-net in order to get

---

[1] The general idea of *genus* is to measure the complexity or symmetry of a net irrespectively of its connectivity. Originally, genus is a topological variable that basically gives the number of holes in a three-dimensional body of any shape. The extension of this concept to 3D-nets will be explained in Chapter 10. All values of genus have been taken from the RCSR database [4].

[2] In a related area, new nets based on tetrahedral coordination are interesting for the synthesis of new zeolites materials [7].

some idea of the shape and connectivity of molecules and ions usable for net construction.

5.1. Requirements for a 3D-net

A 3D-net needs building blocks with propagation vectors extending into all three dimensions of space. For each type of connectivity there is a minimal number of nodes in the *repeating unit*, that is the unit that is repeated in the x, y, and z directions, called the Z_t number. This is the minimal unit that needs to be moved around (no rotations or other symmetry operations allowed) in order to create the net, see Figure 5.1.

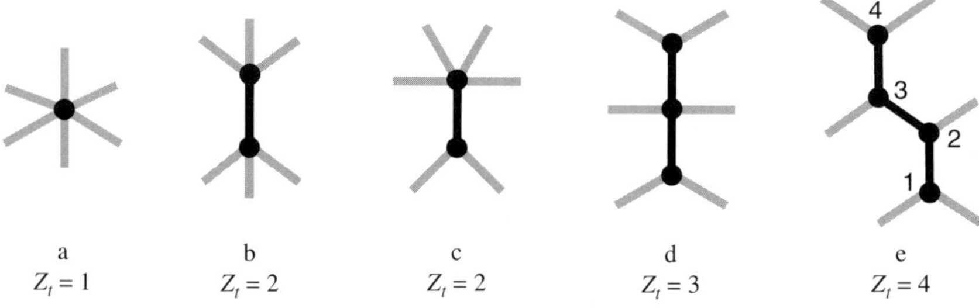

a   b   c   d   e

$Z_t = 1$ $Z_t = 2$ $Z_t = 2$ $Z_t = 3$ $Z_t = 4$

Figure 5.1 The minimal repeating units (Z_t is the number of nodes in this unit) of nets with connection numbers up to six. Note that for each type of net this is the minimal value of Z_t and Z_t increases with decreasing symmetry. The most common nets all have minimal Z_t values.

The nodes may still be symmetry related so that the nets based on three-connected nodes and $Z_t = 4$ may still be uninodal, see Figure 5.2.

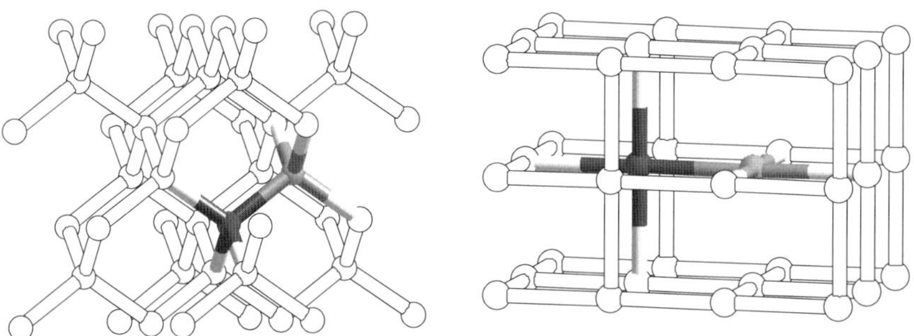

Figure 5.2 The **dia** (left) and **cds** (cadmium sulphate, right) nets have both $Z_t = 2$ since this is the minimal unit that needs to be moved around (no rotations or other symmetry operations allowed) in order to create the net. Both nets have only one type of node, however, since the two centres in the Z_t unit are symmetry related by rotation. Note that it may not be obvious by visual inspection only that these two nets are different, since they both have hexagons as smallest rings. However, detailed counting of the rings reveals that they have different vertex symbols (see Table 5.1), and thus it is not possible to turn one into the other without breaking any links (chemical bonds).

It is also important to realise that the symmetry of the net very likely is different from the actual crystallographic symmetry of the compound in question. For example, we can build diamond nets from two different types of building blocks, so that in the crystal it is impossible to have a symmetry relation between the nodes, but the 3D-net will still be uninodal since this is a property of the **dia** net.

5.2. The most common 3D-nets

In Table 5.1 we list the most frequently encountered 3D nets for the five coordination geometries; trigonal planar, tetrahedral, square planar, trigonal bipyramidal and octahedral, together with vertex symbols (see the preceding chapter) and genus (see Chapter 10). The main reason for the high frequency of these nets is that they can be formed by combining the most symmetrical forms of the basic coordination geometries. In the following text, each net will be discussed in some detail using examples from the literature.

We will encounter nets very close to the "ideal" nets, and some that are severely distorted, so it is important to keep in mind that we are allowed to do this, on condition that we do not break any of the links in the net.

Table 5.1 The most common 3D-nets for coordination polymers [3]).

Net	Wells	Name	p[a]	Geomet.[b]	Vertex symbols	Short symbol[c]	Genus[e]
srs	(10,3)-a	$SrSi_2$	3	Trig. Planar	$10_5 \cdot 10_5 \cdot 10_5$	10^3	3
ths	(10,3)-b	$ThSi_2$	3	Trig. Planar	$10_2 \cdot 10_4 \cdot 10_4$	10^3	3
dia	6^6-(a)	diamond	4	Tetr.	$6_2 \cdot 6_2 \cdot 6_2 \cdot 6_2 \cdot 6_2 \cdot 6_2$	6^6	3
nbo	$8^2 6^4$-4	NbO	4	Sq.Pl.	$6_2 \cdot 6_2 \cdot 6_2 \cdot 6_2 \cdot 8_2 \cdot 8_2$	$6^4.8^2$	4
cds	-	$CdSO_4$	4	Sq.Pl.	$6 \cdot 6 \cdot 6 \cdot 6 \cdot 6_2 \cdot *$	$6^5.8$	3
pts	$4^2 8^4$	PtS	4	Sq.Pl.+Tetr	$4 \cdot 4 \cdot 8_2 \cdot 8_2 \cdot 8_8 \cdot 8_8$ $4 \cdot 4 \cdot 8_7 \cdot 8_7 \cdot 8_7 \cdot 8_7$	$4^2.8^4$ $4^2.8^4$	5
bnn	-	BN	5	Trig.bipyr	$4 \cdot 4 \cdot 4 \cdot 4 \cdot 4 \cdot 4 \cdot 6 \cdot 6 \cdot 6 \cdot *$	$4^6.6^4$	4
pcu	-	α-Po	6	Octahedr.	$4 \cdot 4 \cdot 4 \cdot 4 \cdot 4 \cdot 4 \cdot 4 \cdot 4 \cdot 4 \cdot 4 \cdot 4 \cdot 4 \cdot * \cdot * \cdot *$	$4^{12}.6^3$	3

[a] Connectivity [b] Coordination geometries of the nodes; trigonal planar, tetrahedral, square planar, trigonal bipyramidal or octahedral, [c] Vertex symbol [d] See section 10.2.1

For the examples of compounds forming different types of nets that follow in this and later chapters we will almost never comment on who made the initial net assignment since this is completely irrelevant for the presentation of the material. We should just mention that there are several possible sources: the original article, the CSD searches by Proserpio and co-workers [5] and by O'Keeffe and co-workers [3] and the authors of this book.

5.2.1. The 10^3-srs or (10,3)-a net, also known as the $SrSi_2$ net

The main feature of the **srs** or (10,3)-a net is four-fold helices extending along all axes. The helices are of the same handedness and therefore the whole net is chiral and this has lead several authors to suggest the possibility to use this net in enantioselective synthesis, catalysis and separation.

The net has cubic symmetry in its highest space group, basically meaning that viewed from the x, y or z direction this net will look the same. Two views of this net are shown in Figure 5.3. As we shall see in Chapter 6 some other nets give very similar views if seen from certain directions, but are easily discerned from the **srs** net by rotating the view. Also note that the four-fold helix is a common theme in nets containing three-connected nodes.

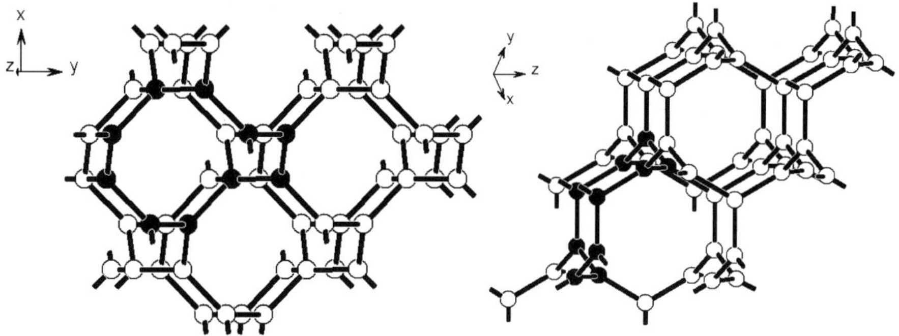

Figure 5.3 Two views of the ideal **srs** or (10,3)-a net. A 10-gon (10-ring) is emphasized in black. Note that the cubic symmetry makes these views equal when x, y and z are interchanged.

The **srs** net can be found in [Zn$_2$(1,3,5-benzenetricarboxylate)(NO$_3$)] ·(H$_2$O)(C$_2$H$_5$OH)$_5$ [10] and a number of related compounds based on ligands with similar trigonal symmetry and other metal ions [11-15].

Trigonal coordination around the metal ion can also lead to **srs** nets, such as in [Ag$_2$(bis(phenylthio)methane)$_3$](ClO$_4$)$_2$, shown in Figure 5.4 [16]. Two things are noteworthy about this latter structure; it is formed by a single net, thus has no interpenetration, and it is formed by achiral molecular building blocks.[3] We will return to the **srs** net in the interpenetration chapter as well as when we discuss 3D nets as specific synthetic targets in Chapter 12.

---

[3] It is not clear from the original report if a single enantiomer was obtained in each preparation or if the compound crystallises as a racemic mixture of enantiomeric crystals. The former case may seem unlikely, but is in fact often found in practice. The optical purity of a batch of crystals can conveniently be analysed by solid state CD-spectroscopy. See further Chapter 12.

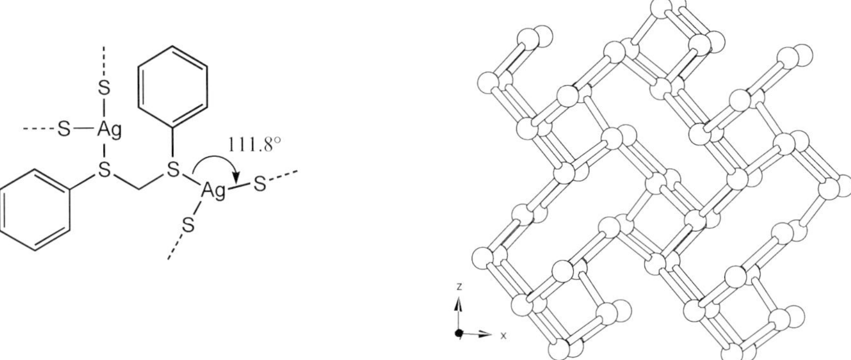

Figure 5.4 The **srs** net in $[Ag_2(bis(phenylthio)methane)_3]_n(ClO_4)_{2n}$ and its molecular constituents [16]. Note that this net is a single enantiomer and that it is formed by achiral molecular building blocks.

A special class of compounds that readily form anionic **srs** nets are tris-oxalate complexes, connected by another metal ion to give $[M(oxalate)_3]_3[M']_2$](cation), [17,18] partly due to the templating effect (see Chapter 12) of $[M^{II/III}(phen)_3]^{2/3+}$ or $[M^{II/III}(bipy)_3]^{2/3+}$ counter ions, but also due to the perfect match between the torsion angles in an ideal **srs** net and the torsion angles obtained when this net is built up from bis-chelating octahedral complexes *of the same chirality (Δ or Λ)*[4] [19].

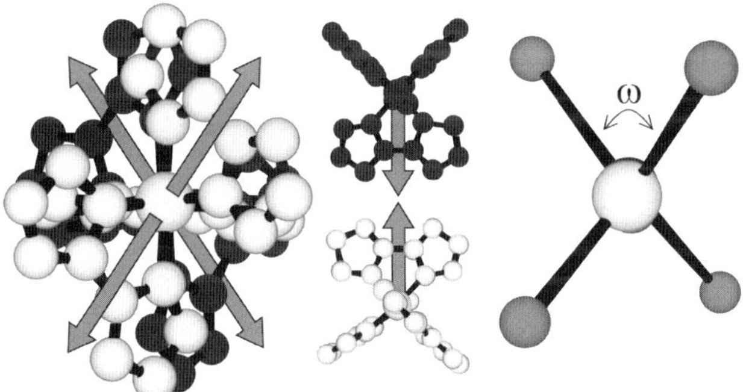

Figure 5.5 Turning of the propagation vectors of two Δ-complexes giving the perfect torsion angle ω for the **srs** ((10,3)-a) net [19]. Reproduced by permission of The Royal Society of Chemistry.

It is perhaps also fitting to mention that the simplest molecular building block giving this net is hydrogen peroxide [20].

---

[4] If these complexes are combined with alternating chirality ($\Delta,\Lambda,\Delta,\Lambda...$) a 2D hexagonal graphite type net is obtained.

5.2.2. The 10^3-*ths* or (10,3)-b net, also called the ThSi₂-net

The second frequently encountered net based on trigonal nodes with angles 120° and all links of equal length is the **ths** net, called (10,3)-b by Wells and ThSi₂ by others. It has a characteristic zigzag strip connecting alternately up and down to the next strip. The layers of strips run perpendicular to each other, see Figure 5.6.

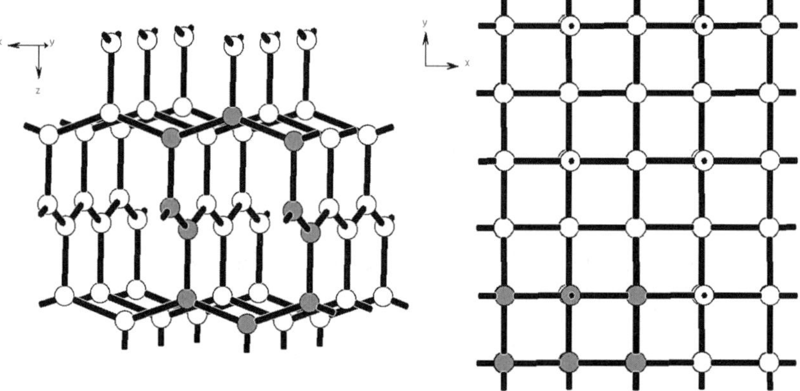

Figure 5.6 Two views of the **ths** (for **ThSi₂**) or (10,3)-b net. A 10-gon (10-ring) is emphasised in black.

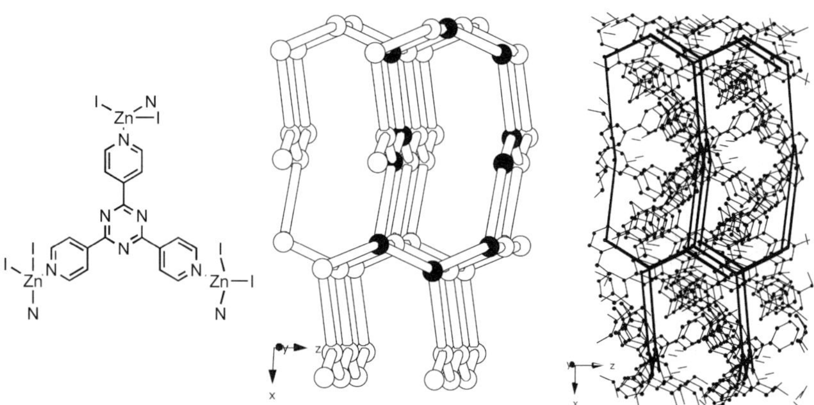

Figure 5.7 ZnI₂ and 2,4,6-pyridyl-triazine (left) give a single **ths** or (10,3)-b net [21] when CBr₄ and MeOH are used as solvents. In the middle the net formed with the centre of the triazine ring as a node is shown. Due to the N-Zn-N angle in the tetrahedral complex, the shortest link connecting these centroids will deviate from the pyridyl-Zn-pyridyl real chemical link, as is shown in the rightmost picture where the net is superimposed on the ligands and Zinc ions.

The possibility of the torsion angle to change between neighbouring nodes giving the net quite different appearances are illustrated by the remarkable example of the folding of the nets based on ZnI_2 and 2,4,6-pyridyl-triazine upon solvent removal [21]. An analogous non-interpenetrated structure from the same study is shown in Figure 5.7. Also the angle between nodes may vary substantially and T-shaped nodes giving **ths** nets are known, see Figure 3.13 (these are sometimes called *rod nets*).

The **ths** net can also be constructed using tris-chelated metal complexes, but in contrast to the enantiopure **srs** nets with Δ_n isomers, the **ths** net will have the sequence $\Delta\Delta\Lambda\Lambda\Delta\Delta$ between the zigzag chains and $\Delta\Lambda$ in the chains and is thus racemic [19,22-24].

5.2.3. The 6^6-*dia* or diamond net

The four-connected **dia** net found in diamond can be made out of perfect tetrahedral sharing vertices as seen in Figure 5.8. We can also see this net as built from adamantine (a polycyclic hydrocarbon) type polyhedra, and the net is thus also called the adamantine net.

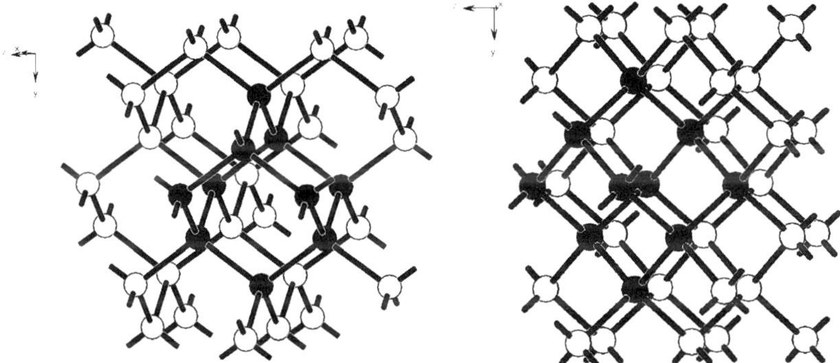

Figure 5.8 Two views of the ideal **dia** or diamond net. To the left the adamantane polyhedron is emphasised in dark grey.

The **dia** net is the most common net among coordination polymers, [3] all categories included, but this may reflect a certain bias in the choice of building blocks by the "constructors". Reviews concentrating on this type of net only have been published, [25] and a selection of molecular building blocks that has yielded diamond nets by either coordination bonds or hydrogen bonding are shown in Figure 5.9.

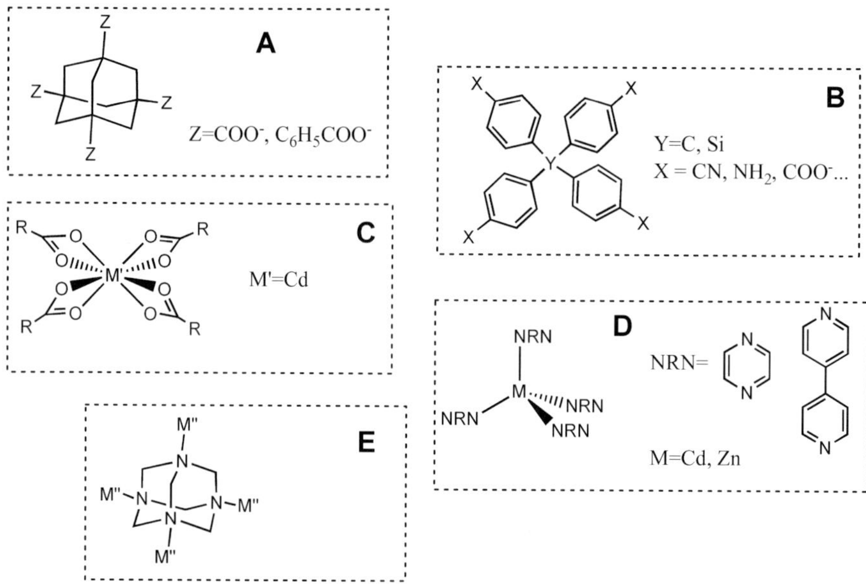

Figure 5.9 Some molecular building blocks that have given diamond nets. Note the adamantane motif in A and E.

As an example, the net in [Cd(3,3'-azodibenzoate)$_2$](H$_2$NMe$_2$)(NH$_4$) is shown in Figure 5.10 [26]. This structure contains six of these nets interpenetrated, and was constructed with non-linear optical properties (NLO) in mind. This requires a non-centrosymmetric space group, that is, there should be no inversion centre in the crystal. In the **dia** net is is impossible to have inversion centres at the nodes, and structures based on this net are therefore often acentric.

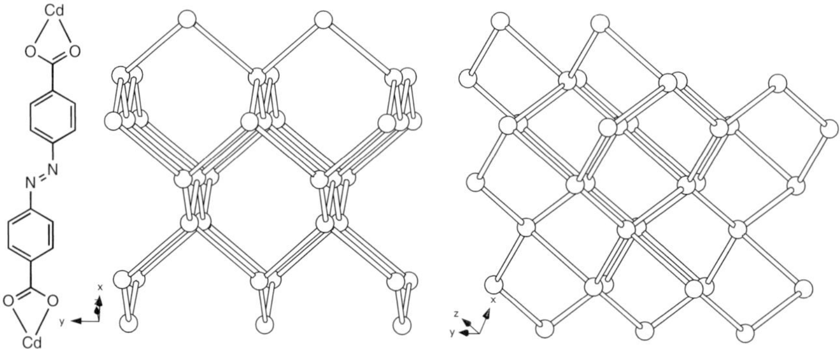

Figure 5.10 The **dia** net in [Cd(3,3'-azodibenzoate)$_2$](H$_2$NMe$_2$)(NH$_4$) is based on motif C in Figure 5.9 [26]. It looks quite symmetric in the left view, but the right view clearly shows the distortion. As in Figure 5.7 the links between the nodes are not as straight as it appears in the simplified net but curved as indicated by the molecular structure (far left).

5.2.4. The $6^4.8^2$-**nbo** or NbO net with square planar coordination

The **nbo** or niobium oxide net is built from perpendicular oriented square planar units, see Figure 5.11. This is the second most common four-connected net, and as the square planar geometry is mostly restricted to the d^8-metal ions Co(I), Rh(I), Ir(I), Ni(II), Pd(II) and Pt(II) this may seem a bit surprising. However, the connectivity at the node is not necessarily the same as the coordination number of the metal atom (if this is a metal ion centred net), and thus this net is frequently encountered for five-coordinated Cu(II) in distorted square-planar pyramidal geometries, [27-30] or in octahedral complexes with two non-bridging axial ligands as in our example shown in Figure 5.12.

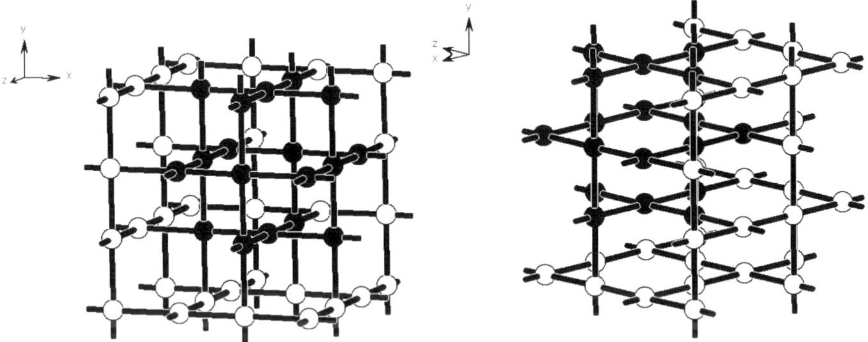

Figure 5.11 Two views of the **nbo** net. The black nodes are enclosing one of the voids in the net.

The selected example, [Fe(NCS)$_2$(3,3,´5,5-tetramethyl-4,4-bipyrazolyl)$_2$], consists of two interpenetrating **nbo** nets [31].

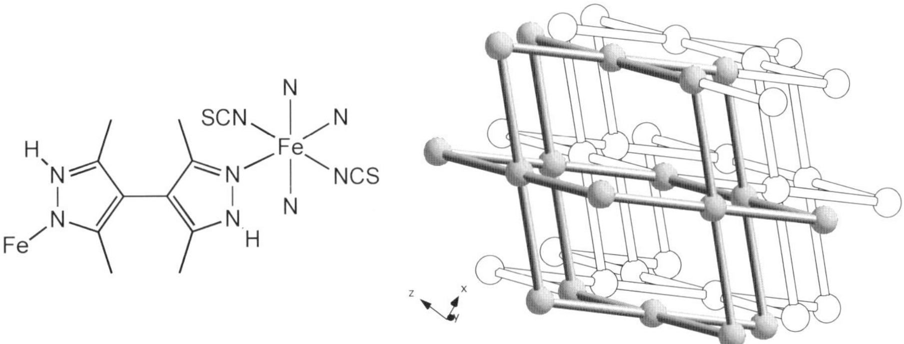

Figure 5.12 [Fe(NCS)$_2$(3,3',5,5'-tetramethyl-4,4'-bipyrazolyl)$_2$] consists of two interpenetrating **nbo** nets [31]. A similar cage as in Figure 5.11 is highlighted.

Usually this type of 3D-net is in competition with the 2D-square-grid (4,4)-net (Figure 5.13A), and it is a general observation that, given a choice, the lower dimensionality usually wins. In the case of $[Fe(NCS)_2(3,3',5,5'\text{-tetramethyl-}4,4'\text{-bipyrazolyl})_2]$ the authors argue that steric effects of the methyl groups both control the relative orientation of the coordination plane and the ligand plane, and the orientation of the two pyrazolyl rings Figure 5.13B.

This is in fact a general problem for three or four connected planar nodes; to be sure to get a 3D net, the torsion angles between the propagation vectors (nodes) need to be controlled. The same effect of the twist in the $3,3',5,5'$-tetramethyl-$4,4'$-bipyrazolyl can be seen in the different hydrogen bonded (10,3)-nets obtained with this building block by Boldog et al. [32]

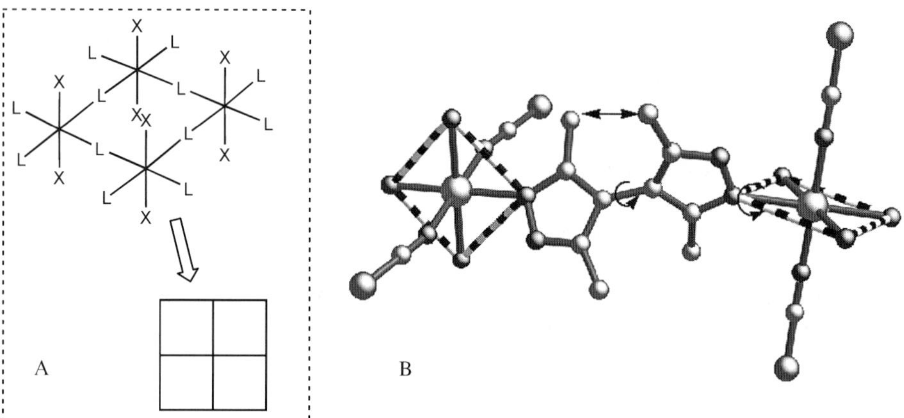

Figure 5.13 A. Square planar coordination with bridging ligands often give the 2D-square grid (4,4)-net. B. To be sure to get a 3D net the torsion angles between the nodes (*i.e.* between the planes defined by the striped "bonds" in the picture) need to be controlled. This is achieved by the bulkiness of the methyl groups giving a twist to the bipyrazolyl ligand, and by the pyrazolyl groups itself severely restricting the rotation around the metal-ligand bond.

5.2.5. The $6^5.8$-**cds** or cadmium sulphate net with square planar coordination

Another possibility of orienting the squares to get a 3D-structure can be found in cadmium sulphate, thus the **cds** abbreviation. Note the differences as compared to the **nbo** net, smallest rings are hexagons, not octagons, and only half of the neighbouring nodes are related by a 90° turn, see Figure 5.14.

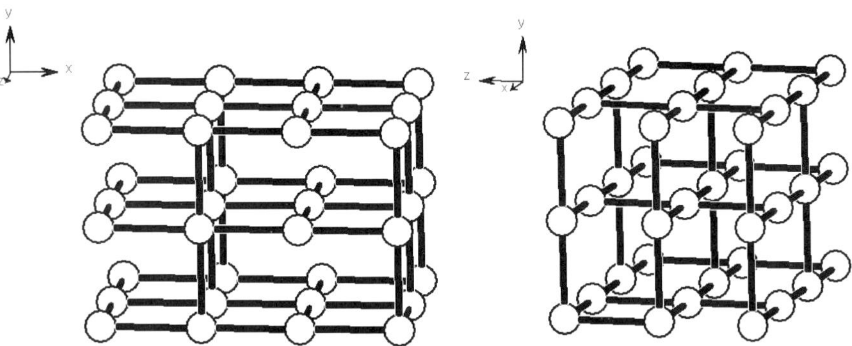

Figure 5.14 Two views of the **cds** or Cadmium sulphate net.

Examples are found among the coordination polymers, predominantly copper-containing polymers.

An interesting pair of compounds are [Fe(pyrimidine)(H_2O)(M(CN)$_2$)$_2$]·H_2O, M = Ag, Au, designed with spin transition properties in mind, see Figure 5.15 [33]. This structure is triply interpenetrated and the compound exhibits spin crossover with hysteresis and dramatic colour change from dark red to pale yellow. The hysteresis (also called "memory effect" since the colour and spin state are dependent on the history of the compound) is almost certainly an effect of all the spin centres being interconnected, thus a change in spin state for one iron centre, and the simultaneous reduction or increase (about 10%) of the bond Fe-N bond lengths, will affect the entire 3D-net.

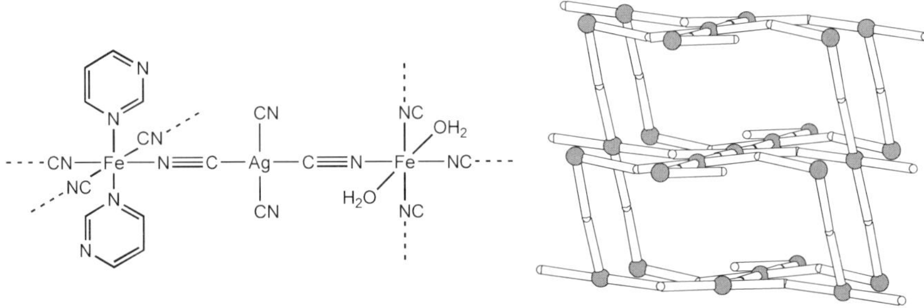

Figure 5.15 [Fe(pyrimidine)(H_2O)(M(CN)$_2$)$_2$]·H_2O, M=Ag, Au is an example of the **cds** (CdSO$_4$) net. The Ag compound is show with iron-centred nodes and Ag ions are shown as nicks on the links. This structure is triply interpenetrated and the compound exhibits spin crossover with hysteresis and dramatic colour change as well as a solid-state ligand substitution reaction with loss of water, making the pyrimidine ligand bridging irons from two different nets [33].

5.2.6. The platinum sulphide, $(4^2.8^4)(4^2.8^4)$-**pts** net, a square and a tetrahedron

In the last example the **cds** net was built from two different types of nodes having the same basic geometry. We now turn to two nets that are constructed from nodes with different geometry, the square and the tetrahedron. First of these is the platinum sulphide or **pts** net, see Figure 5.16.

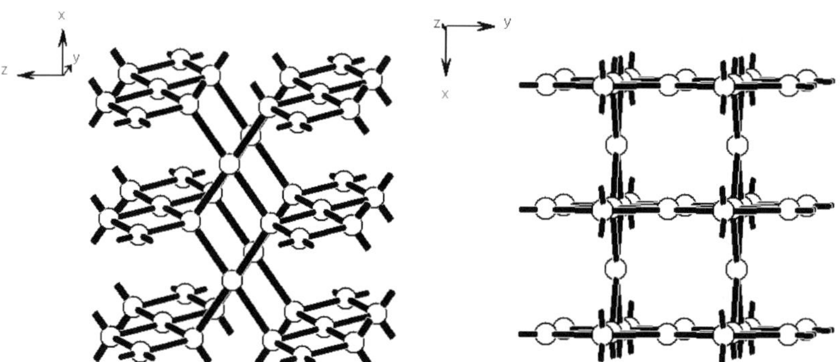

Figure 5.16 The most symmetric form of the platinum sulphide or **pts** net. Grey nodes are tetrahedral, white nodes are square planar.

As expected, this net is formed by square planar and tetrahedral forms such as 1,2,4,5-tetracyanobenzene and Cu(I) [34] but also by more complex assemblies like the binuclear zinc carboxylate (square planar) and methyltetrabensoate, MTB *secondary building units, SBUs* see Figure 5.17 [35].

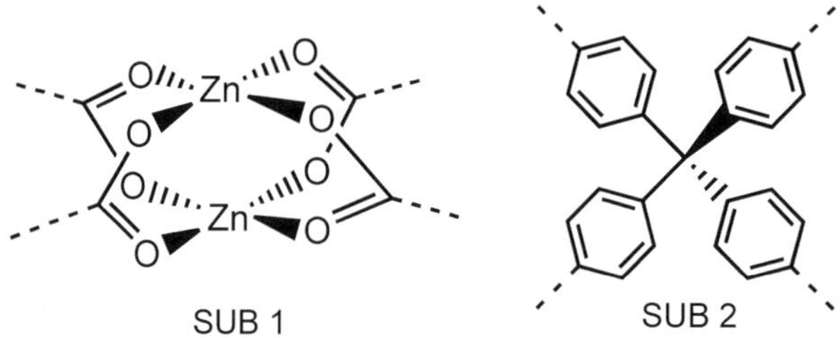

Figure 5.17 The secondary building unit approach: The binuclear zinc(II)tetracarboxylate unit will give square planar propagation vectors and the methyltetraphenyl tetrahedral geometry of the net. Together they give the **pts** net. Note that SUBs and synthons are not synonyms and refer to different ways of disassembling the net in order to plan a synthesis. If you do not feel helped by these concepts you can chose to ignor them for the moment.

Other types of building blocks are illustrated in Figure 5.18 where we see that the "see-saw" geometry of the Cr-centre blocked by two *cis* pyridine ligands also can function as the "tetrahedral" node. Note also the role of the potassium ion; far from being the boring spectator ions of yesterday "…it is now recognized that the alkali metals have a rich and extremely varied coordination chemistry…" [36]

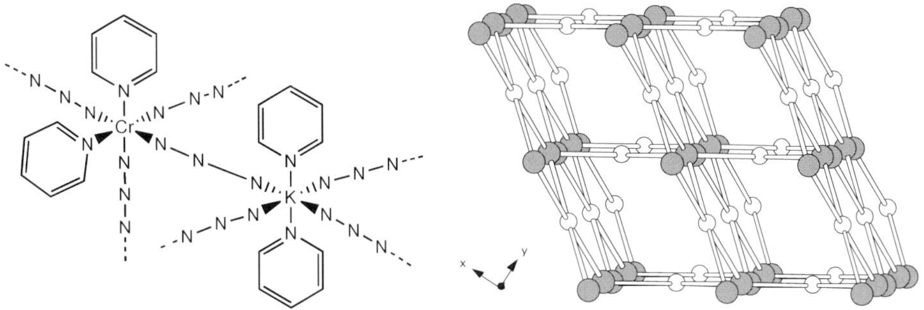

Figure 5.18 The azide ion (N_3^-) is not only potentially explosive but also a very versatile bridging ligand as for example in this catena[tetrakis(μ^2-Azido-N,N')-dipyridyl-chromium-dipyridyl-potassium] compound that contain one single **pts** or platinum sulphide net [37].

5.2.7. The $4^6.6^4$-**bnn**, or boron nitride net

As is evident from Figure 5.19 the **bnn** net is simply derived by cross-linking hexagonal layers of the 2D (6,3)-net.[5]

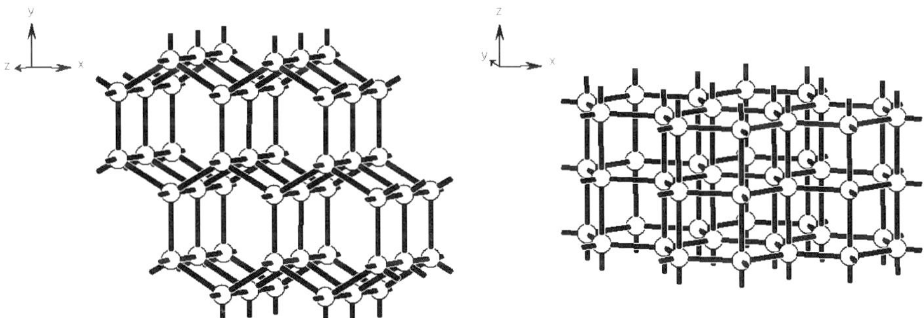

Figure 5.19 The boron nitride or **bnn** net.

A single distorted version of this net is present in [Cu(piperazine)(MoO$_4$)] [38], see Figure 5.20. The vertex symbol of this net is 4·4·4·4·4·4·6·6·6·* (Table 5.1) as can be easily verified by counting the rings; every 90° angle contains a four-ring and there are six such angles. The 120° angles are parts of a six-membered ring and there are three of those. The final * arises since there is no way to go

---

[5] Note that such "bonds" between layers do not exist in boron nitride, the net code refers only to the atom positions in this compound [36].

arises since there is no way to go between the two axial links without passing through one of the already counted four-rings.

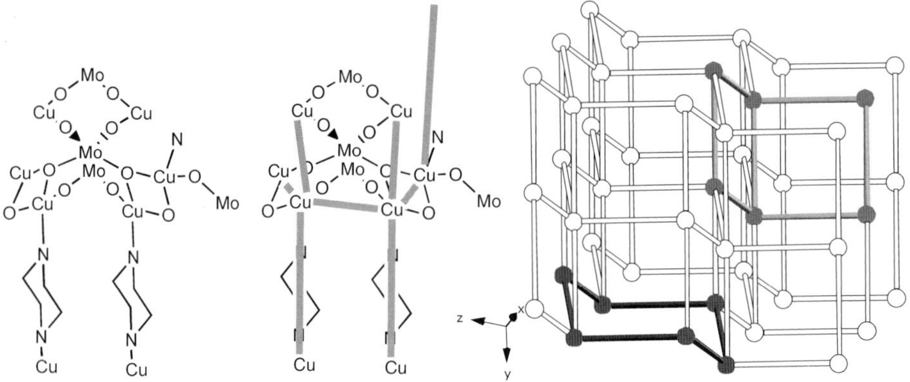

Figure 5.20 An example of the **bnn** net, [Cu(piperazine)(MoO$_4$)] [38] Node assignment is indicated in grey (left) and six- and four-rings are highlighted in the right picture to help comparison with Figure 5.19.

A recent more symmetric form of this net is found in (tris(η^4-adamantane-1,3-dicarboxylato)-di-europium(III)) [39].

5.2.8. The $4^{12}.6^3$-**pcu** net with octahedral nodes

The **pcu** net is the highest connected net in this chapter, and **pcu** stands for primitive cubic packing, *i.e.* a cube with an atom (or other object such as a globular protein) in each corner. This is not one of the close packings (called *hexagonal close packing (hcp)* and *cubic close packing (ccp)*) but nevertheless a common way for spherical objects to arranged themselves in a crystal. Just as the two close packings it has six nearest neighbours arranged in an octahedral fashion, see Figure 5.21. This net is also known as the α-Po, α-polonium or ReO$_3$ net.

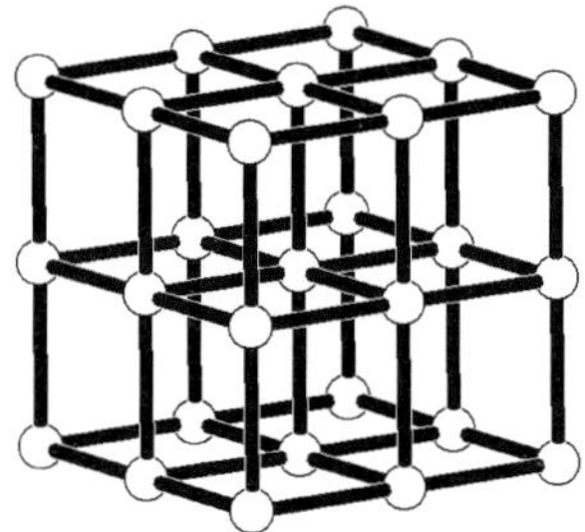

Figure 5.21 The **pcu** net also known as the α-Po or α-polonium net. The nodal positions are the same as for Na$^+$ and Cl$^-$ in NaCl(s).

Prominent examples of the pcu nets are a series of metal-organic frameworks prepared by Yaghi et al., of which an example is shown in Figure 5.22 [40]. These show remarkable porous properties such as methane and hydrogen storage capabilities, see also Figure 1.7 We also want to mention a porphyrine based pcu net [41].

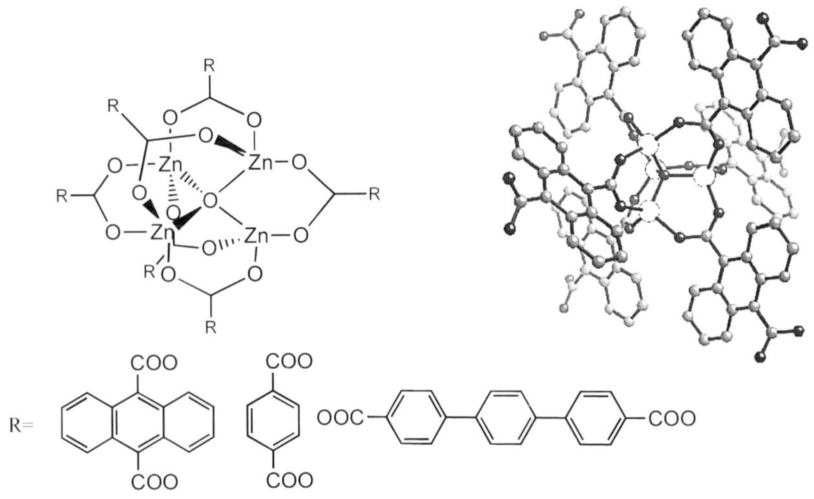

Figure 5.22 The catena-(tris(μ^4-naphthalene-1,4-dicarboxylato(μ^4-oxo)-tetra-zinc heptakis(N,N-diethylformamide) trihydrate clathrate) a non-interpenetrated **pcu** net [40]. Several, but not all, of the bridging units exploited by Yaghi's group are shown to the left, bottom. A zinc (plain white spheres) cluster and the naphtalenecarboxylates are shown to the right.

Another series of important compounds that are based on **pcu** nets are the Prussian blue[6] family, [M(CN)$_6$M'], many of them with magnetic properties [44-46]. Note however that in Prussian Blue itself one out of every four hexacyanide-units is missing from the structure.

This concludes our first excursion into net-land, and we want to emphasise two important features: first, it is important to realise that a real net may be very much distorted from an ideal net, and secondly, that among the infinite possibilities of hypothetical nets, the large majority of such structures belong to only a few different topologies presented in this chapter. Of these, the diamond, **dia**, net is by far the most common [3].

We will now move on to more systematic investigations of a number of other nets, most of them present in real compounds, but also some hypothetical ones that seem relevant to us.

---

[6] The formation of Prussian blue is also notable as being one of the few reactions depicted in a painting; that of a young Michael Faraday watching as his teacher, Professor W.T. Brande, adds colourless cyanide solution to a test tube containing the pale yellow Fe(II). Naturally, neither of them are wearing safety goggles [42,43].

References

[1] O. Delgado-Friedrichs, M. O'Keeffe, O. M. Yaghi, Acta Cryst. A 59 (2003) 22.

[2] C. Bonneau, O. Delgado-Friedrichs, M. O'Keeffe, O. M. Yaghi, Acta Cryst. A 60 (2004) 517.

[3] N. W. Ockwig, O. Delgado-Friedrichs, M. O'Keeffe, O. M. Yaghi, Acc. Chem. Res. 38 (2005) 176.

[4] M. O'Keeffe, O. M. Yaghi, Reticular Chemistry Structure Resource, Tucson, Arizona State University, 2005, http://okeeffe-ws1.la.asu.edu/RCSR/home.htm

[5] V. A. Blatov, L. Carlucci, G. Ciani, D. M. Proserpio, Crystengcomm 6 (2004) 377.

[6] O. M. Yaghi, M. O'Keeffe, N. W. Ockwig, H. K. Chae, M. Eddaoudi, J. Kim, Nature 423 (2003) 705.

[7] O. Delgado-Friedrichs, A. W. M. Dress, D. H. Huson, J. Klinowski, A. L. Mackay, Nature 400 (1999) 644.

[8] E. Koch, W. Fischer, Z. Kristallogr 210 (1995) 407

[9] M. M. J. Treacy, I. Rivin, E. Balkovsky, K. H. Randall, M. D. Foster, Microporous Mesoporous Mater. 74 (2004) 121.

[10] M. Eddaoudi, H. Li, O. M. Yaghi, J. Am. Chem. Soc. 122 (2000) 1391.

[11] J. F. Eubank, R. D. Walsh, M. Eddaoudi, Chem. Commun. (2005) 2095.

[12] C. J. Kepert, M. J. Rosseinsky, Chem. Commun. (1998) 31.

[13] C. J. Kepert, T. J. Prior, M. J. Rosseinsky, J. Am. Chem. Soc. 122 (2000) 5158.

[14] B. E. Abrahams, P. A. Jackson, R. Robson, Angew. Chem. Int. Ed. 37 (1998) 2656.

[15] L. Li, J. Fan, T. A. Okamura, Y. Z. Li, W. Y. Sun, N. Ueyama, Supramol. Chem. 16 (2004) 361.

[16] X. H. Bu, W. Chen, M. Du, K. Biradha, W. Z. Wang, R. H. Zhang, Inorg. Chem. 41 (2002) 437.

[17] S. Decurtins, H. W. Schmalle, P. Schneuwly, J. Ensling, P. Gütlich, J. Am. Chem. Soc. 116 (1994) 9521.

[18] S. Decurtins, H. W. Schmalle, R. Pellaux, P. Schneuwly, A. Hauser, Inorg. Chem. 35 (1996) 1451.

[19] L. Öhrström, K. Larsson, Dalton Trans. (2004) 347.

[20] C. S. Abrahams, R. L. Collin, W. N. Lipscomb, Acta Cryst. 4 (1951) 15.

[21] K. Biradha, M. Fujita, Angew. Chem. Int. Ed. 41 (2002) 3392.

[22] M. R. Sundberg, R. Kivekas, J. K. Koskimies, J. Chem. Soc., Chem. Commun. (1991) 526.

[23] R. Vaidhyanathan, S. Natarajan, A. K. Cheetham, C. N. R. Rao, Chem. Mat. 11 (1999) 3636.

[24] H. Y. Shen, W. M. Bu, D. Z. Liao, Z. H. Jiang, S. P. Yan, G. L. Wang, Inorg. Chem. 39 (2000) 2239.

[25] M. J. Zaworotko, Chem. Soc. Rev. 23 (1994) 283.

[26] Z. F. Chen, R. G. Xiong, B. F. Abrahams, X. Z. You, C. M. Che, J. Chem. Soc., Dalton Trans. (2001) 2453.

[27] X. H. Bu, M. L. Tong, H. C. Chang, S. Kitagawa, S. R. Batten, Angew. Chem. Int. Ed. 43 (2004) 192.

[28] B. L. Chen, F. R. Fronczek, A. W. Maverick, Chem. Commun. (2003) 2166.

[29] M. Eddaoudi, J. Kim, M. O'Keeffe, O. M. Yaghi, J. Am. Chem. Soc. 124 (2002) 376.

[30] K. N. Power, T. L. Hennigar, M. J. Zaworotko, Chem. Commun. (1998) 595.

[31] P. V. Ganesan, C. J. Kepert, Chem. Commun. (2004) 2168.

[32] I. Boldog, E. B. Rusanov, A. N. Chernega, J. Sieler, K. V. Domasevitch, Angew. Chem. Int. Ed. 40 (2001) 3435.

[33] V. Niel, A. L. Thompson, M. C. Munoz, A. Galet, A. S. E. Goeta, J. A. Real, Angew. Chem. Int. Ed. 42 (2003) 3760.

[34] L. Carlucci, G. Ciani, D. W. von Gudenberg, D. M. Proserpio, New J. Chem. 23 (1999) 397.

[35] J. Kim, B. L. Chen, T. M. Reineke, H. L. Li, M. Eddaoudi, D. B. Moler, M. O'Keeffe, O. M. Yaghi, J. Am. Chem. Soc. 123 (2001) 8239.

[36] N. N. Greenwood, A. Earnshaw, Chemistry of the Elements, 2nd ed. Pergamon Press, Oxford, 1997.

[37] M. A. S. Goher, M. A. M. Abu-Youssef, F. A. Mautner, H. P. Fritzer, Z. Naturforsch., B: Chem. Sci. 47 (1992) 1754.

[38] Y. Xu, J. Lu, N.K.Goh, J. Mat. Chem. 9 (1999) 1599.

[39] F. Millange, C. Serre, J. Marrot, N. Gardant, F. Pelle, G. Ferey, J. Mat. Chem. 14 (2004) 642.

[40] M. Eddaoudi, J. Kim, N. Rosi, D. Vodak, J. Wachter, M. O'Keefe, O. M. Yaghi, Science 295 (2002) 469.

[41] D. Hagrman, R. P. Hammond, J. Zubieta, Angew. Chem. Int. Ed. 38 (1999) 3165.

[42] T. Phillips, Prussian blue, Alfred Bader Collection, Milwaukee, Wisconsin, USA, 1816.

[43] Aldrich Chimica Acta 34 (2001).

[44] T. Mallah, S. Thiébaut, M. Verdaguer, P. Veillet, Science 262 (1993) 1554.

[45] J. S. Miller, MRS Bull. 25 (2000) 60.

[46] M. V. Bennett, L. G. Beauvais, M. P. Shores, J. R. Long, J. Am. Chem. Soc. 123 (2001) 8022.

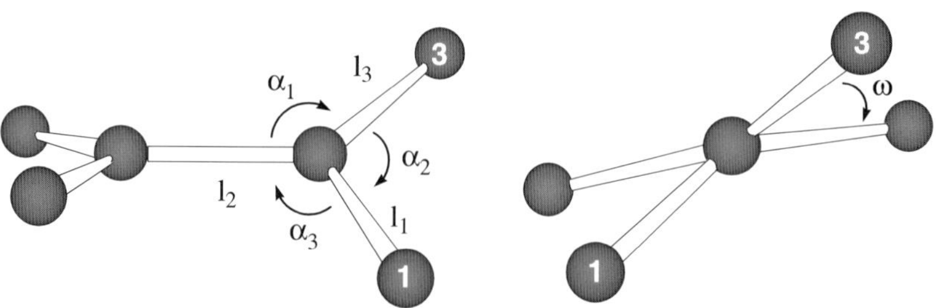

Figure 6.1 Relevant geometric parameters for three-connected nets. To a large degree these can be controlled by the choice of building blocks although in each case there is also a span of allowed values for l_i, α_i and ω.

A few examples of three-connected nodes are displayed in Figure 6.2 and we will discuss them briefly below.

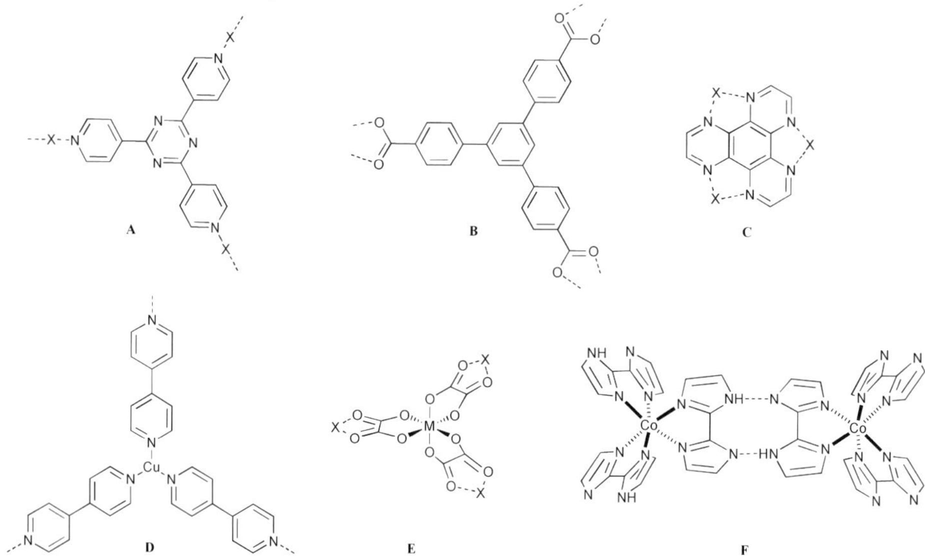

Figure 6.2 Some different building blocks that form three-connected nets. The possible modes of bridging interaction are hinted by dashed lines, but can vary from case to case.

There are essentially two different types of building blocks, organic molecules with trigonal bonding possibilities (hydrogen bonding or metal coordination) as **A**, **B** and **C**, and trigonal nodes formed by a metal ion and its ligands (**D**, **E** and **F**). In the latter case we can also distinguish between complexes formed *in situ* as in **D**, and pre-prepared complexes as in **E** and **F** (although this an ambiguous division).

The preferred geometry of a building block can be analysed by searching the Cambridge Structural Database, as shown in Figure 6.3 covering compounds of type **A** and **B**. These results indicate that these nodes will essentially have to

remain planar, but will tolerate quite large deviations in from the ideal 120°
link-link angle.

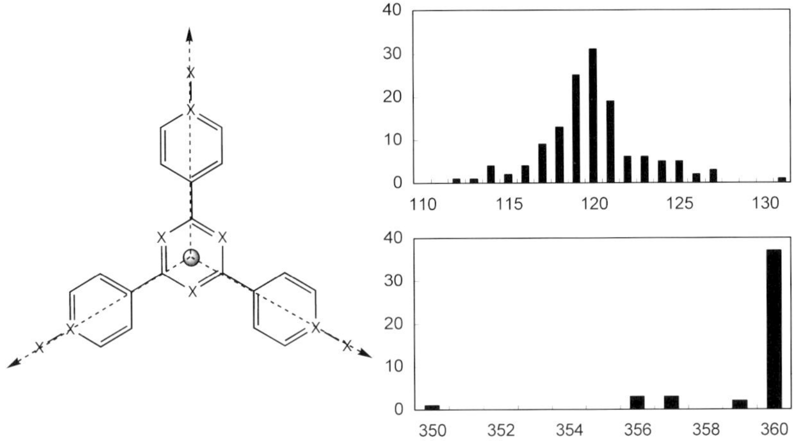

Figure 6.3 A simple CSD search[1] for the geometry preferences of buildings blocks of type **A**
and **B** in Figure 6.2 using the substructure to the left. (A centroid has to be added to enable the
determination of the angles, X are variable non-hydrogen attachments. Measurements were made
to the outermost X) The results are displayed in the diagrams showing large tolerances for the
link-link angles (top) but very little deviation from planarity (bottom).

However, often such detailed investigation is not necessary, you get a long
way with general chemical knowledge and some good reference literature, for
example concerning the different expected coordination geometries of the
transition metal ions [2,3].

All the preceding building blocks have preferred angles of 120° although for
some metal ions this may not be strictly true. In fact some of them, like Ag(I),
can easily adapt to a more T-shaped geometry (see Figure 3.13, [Ag(4,4′-
bipyridine)]NO$_3$ [4,5]). T-shaped nodes with focus on nets based on main group
elements are discussed by R. Hoffmann et al. [6] and a review of coordination
polymers based on M(NO$_3$)$_2$ by Barnett and Champness contains an extensive
analysis of T-shaped nodes [7].

6.1.2. Torsion angles, ω, between consecutive nodes

While the length of the connectors and the angles between them thus are to a
certain degree determined by our starting materials, the torsion angle ω between
neighbouring nodes is more difficult to control. Referring again to Figure 6.2
we see that in **A** and **D** the torsion angle has complete freedom, whereas in **F** it
is governed by the tolerance of the $R^2_2(10)$ hydrogen bond pattern (which may
be substantial [8]). The other cases are somewhere in-between and a detailed

survey of the CSD may be necessary to establish if there are any torsion angle preferences. Note also that the chirality in **E** and **F** has an important influence on the torsion angle (see also the discussion of the **srs** and **ths** nets in Chapter 6) [9].

6.1.3. The net and the tecton

Finally it is important to keep in mind the adaptability of the nets. The substantial deviations from ideal geometry that are possible make the relationship between building block and the type of net formed a very complex issue.

6.2. Uninodal three-connected nets

There are many possible ways to organise this type of survey but our choice is to start with the uninodal nets, that is, nets containing only one type of node, and to order these according to decreasing ring size. We start with those containing only one size of smallest rings, that is; all link pairs emerging from the nodes have shortest rings of the same size. They are also called *uniform nets*.

For easy references, we will note the Schläfli (short) symbol (see Chapter 4) together with the three-letter code and any other names in circulation.

6.2.1. The 12^3-*twt* net

One of the more spectacular predictions of Wells occurs already in the first paper in the series "The geometrical basis of crystal chemistry". In this 1954 paper he describes a chiral three connected net based on 12-rings, see Figure 6.4 [10]. Another 45 years were to pass before a compound with this net was obtained by Robson and co-workers [11].

In this net the shortest rings are interlocked and this is thus an example of self-penetrated net.

This net has vertex symbol $12_4 \cdot 12_7 \cdot 12_7$, and genus 4. Note that in the ideal form all nodes are equal, (angles 136.7° 111.6°, and 111.6°, thus outside the possible values for a network containing only tri-pyridyl-pyrazine nodes, see Figure 6.3) while in Robsons net there are two different type of nodes, i.e. T-shaped Ni(II) centred vertices and the pyrazine nodes that are trigonal planar. (There is no contradiction here; the structure can contain two chemically different types of nodes while the topology of the net is still uninodal.)

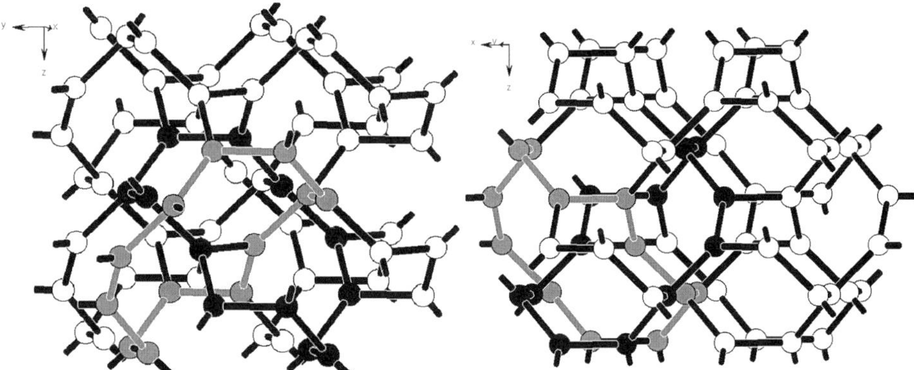

Figure 6.4 Two different views of the **twt** net, a (12,3) net derived by A.F. Wells in 1954 [10] and synthesised by Robson and co-workers 45 years later [11]. Note the entanglement of two 12-rings giving a self-penetrated net.

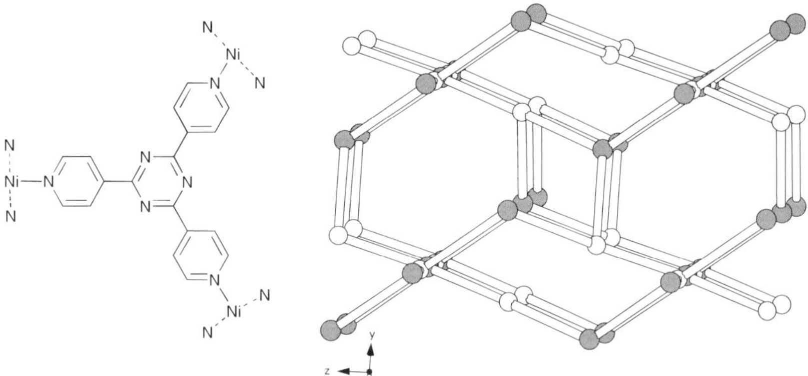

Figure 6.5 The Ni(II)-tris(2,4,6-pyridyl-triazine) complex forming the (12,3) or **twt** net [11]. Nickel nodes are light grey and triazine centroid nodes are dark grey. The coordination sphere of nickel is completed by nitrate ions and the structure also contains chloroform.

As there appears to be no three-connected nets based on 11-rings we move on the more numerous (10,3)-nets.

6.2.2. The (10,3)-c 10^3-**bto** net

This chiral net is named after the B_2O_3 structure [12] and is closely related to the much more common **ths** net ((10,3)-b). Like this net, the **bto** net has zigzag strips although not in layers, every second strip is turned 60° making the 10-gons twisted rather than chair-formed. Also note the difference this makes in comparison with **ths** when the nets are turned, see Figure 6.6. The **bto** net has vertex symbol $10 \cdot 10_2 \cdot 10_2$, and genus 4.

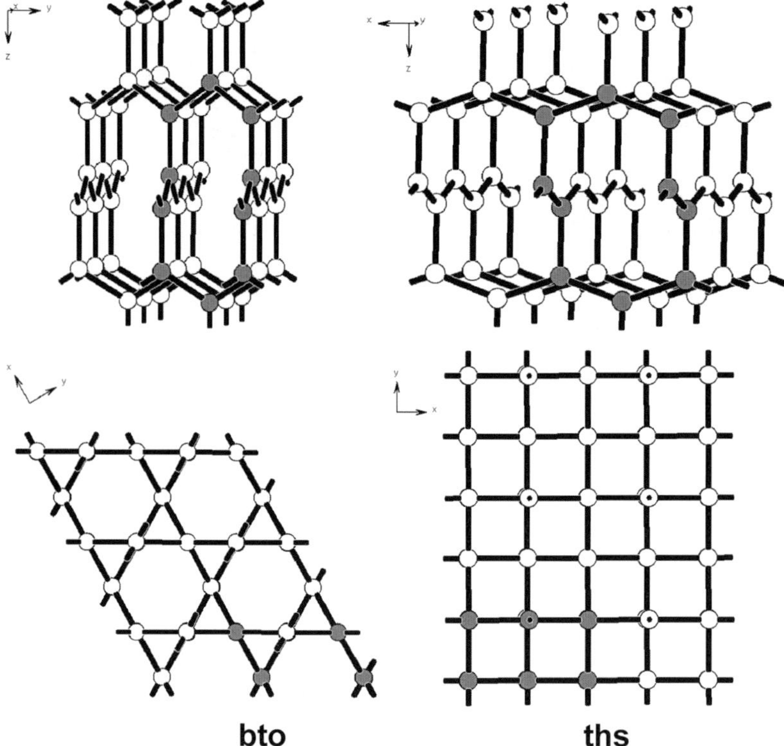

Figure 6.6 The chiral **bto** net (left) compared to the much more common **ths** net (right). Note the different conformations of the 10-gons highlighted in the picture.

This net is found in the hexagonal polymorph of 3,3′5,5′-tetramethyl-4,4′-bipyrazole, a self-complimentary hydrogen bonding molecule, see Figure 6.7. The **srs** ((10,3)-a), the **ths** ((10,3)-b) and the **bto** ((10,3)-c) nets may be assembled in different polymorphs using different solvents in the crystallisation process, whereas the unsubstituted 4,4′-bipyrazole gives the familiar 2D honeycomb (6,3) net [13,14]. The torsion angle ω plays a critical part in the net formation, forcing the two pyrazole pentagons out-of-plane, but the exact role of the different solvents in controlling this is not known.

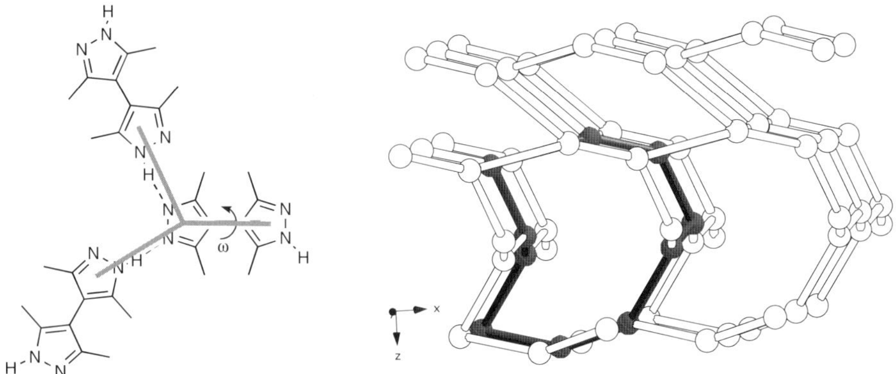

Figure 6.7 The hexagonal polymorph of 3,3'5,5'-tetramethyl-4,4'-bipyrazole crystallises to give the **bto** net. Nodes are inserted at the centroids of the pentagons. The torsion angle ω plays a critical part in the net formation, but the exact role of the different solvents in controlling this is not known [13,14].

6.2.3. The (10,3)-d or 10³-***utp*** net

This net has vertex symbol $10_2 \cdot 10_4 \cdot 10_4$, $Z_t = 8$ and genus 5. A casual glance may give the impression of a **srs** net, see Figure 6.8, but if viewed from another direction, or if we check the chirality of the helices, it is apparent that this is another net. As we shall see, interconnected four-fold helices is a reoccurring theme in 3-connected nets. The **utp** net is non-chiral and the helices are turning in opposite directions.

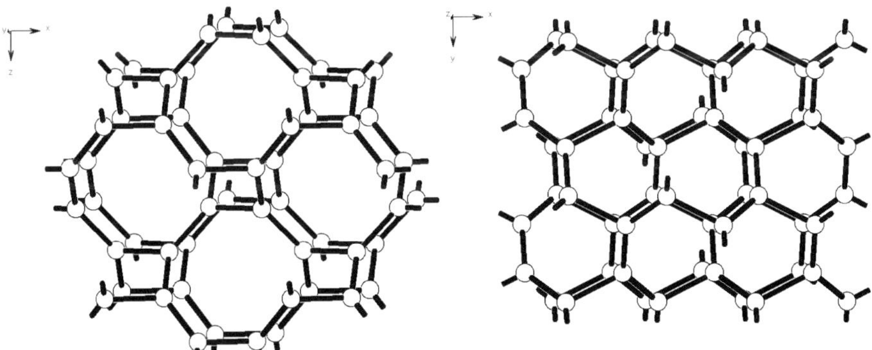

Figure 6.8 The **utp** net is non-chiral and the helices are turning in opposite directions along the diagonal.

There are several known examples, hydrogen bonded or coordination polymers, of this net in the literature, interpenetrating [8,15,16] and single nets [17,18] although only a few of them fully assigned. A recent example is a compound based on iron(III) or Co(III)*tris*(dipyrrinato) complexes bridged by silver ions [19,20], see Figure 6.9.

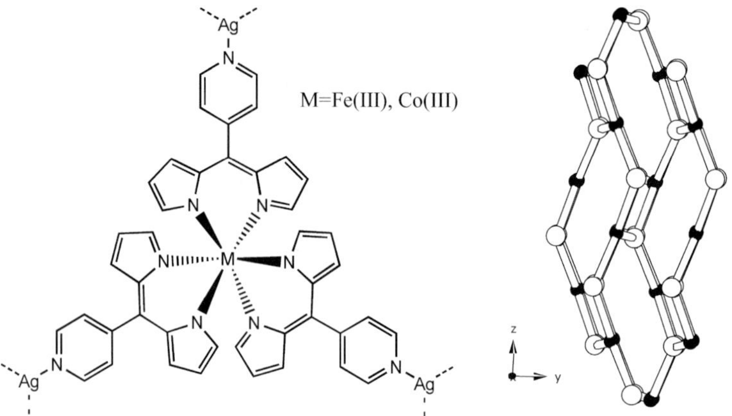

Figure 6.9 [[M(5-(4-pyridyl)-4,6-dipyrrinato))₃]Ag]SO₃CF₃ [19] (M=Co(III) or Fe(III) form isostructural compounds containing doubly interpenetrated **utp** or (10,3)-d nets. Ag nodes are black and Fe nodes are white.

6.2.4. Other uninodal (10,3)-nets

There are a number of other uninodal (10,3) nets derived by Wells [21] and by Fischer and Koch [22]. We give a summary of these nets in Table 6.1 and show the **utk, utj** and **utm** nets in Figure 6.10.

Table 6.1 Other known uninodal (10,3)-nets [21-23]. Measurements refer to the most symmetric case and may deviate substantially if the net is distorted.

Net	Fischer symbol	Vertex symbol	Genus	Planarity[a]	Angles[b]	Chiral
utk	3/10/t3	$10 \cdot 10 \cdot 10_3$	5	179.6°	95°-144°	yes
utj	3/10/t2	$10 \cdot 10 \cdot 10_3$	9	179.8°	99°-133°	no
utm	3/10/t5	$10_2 \cdot 10_4 \cdot 10_4$	9	179.5°	100°-149°	no
utn	3/10/t6	$10 \cdot 10 \cdot 10_3$	9	179.8°	80°-142°	no
uto	3/10/t7	$10 \cdot 10 \cdot 10_3$	9	179.8°	91°-135°	no

[a] Dihedral angle of a node and its three neighbours [b] Smallest and largest angles

None of these nets seem to have been observed to date, and this also applies to the other binodal (10,3)-e, -f and –g nets derived by Wells [21].

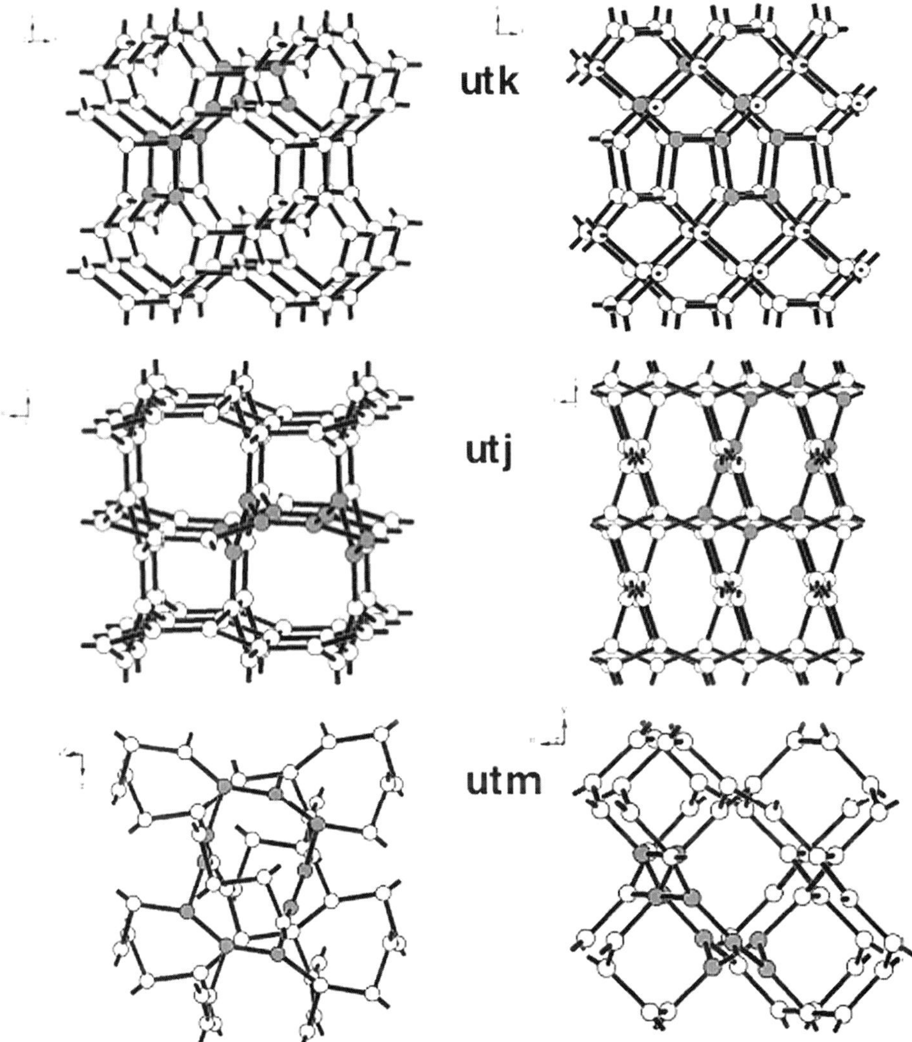

Figure 6.10 Two views each of the uninodal nets **utk** (10/3/t3), **utj** (10/3/t2) and **utm** (10/3/t5) Note that the **utk** net is chiral.

The next size is 9, but although there are numerous nets of this type, none of them are uninodal so for the moment we will move on to the nets based on octagons.

6.2.5. The (8,3)-a and (8,3)-b nets (8^3-eta and 8^3-etb)

The (8,3)-a and (8,3)-b nets (8^3-**eta** and 8^3-**etb**) are the two uninodal (8,3) nets and they can both be constructed with perfect symmetry around the vertex, that is angles of 120° and equal lengths of the connectors [21]. Wells labelled the first (8,3)-a and the second (8,3)-b. The corresponding three letter codes are **eta** and **etb** (Fischer symbol 3/8/h2 and 3/8/h1). Unfortunately, they have both the same vertex symbol; 8·8·8$_2$, and genus 4, which could make it difficult to distinguish between them. However, **eta** is chiral, and **etb** not; this should make the identification easier. The nets are shown in Figure 6.11 and Figure 6.12.

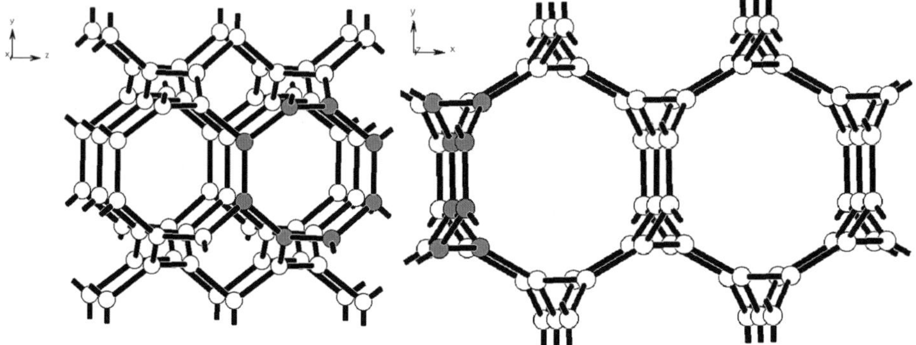

Figure 6.11 The (8,3)-a or **eta** net. Another example of a three-connected net comprising four-fold helices, this time running parallel to the x and y axes. Note the chirality.

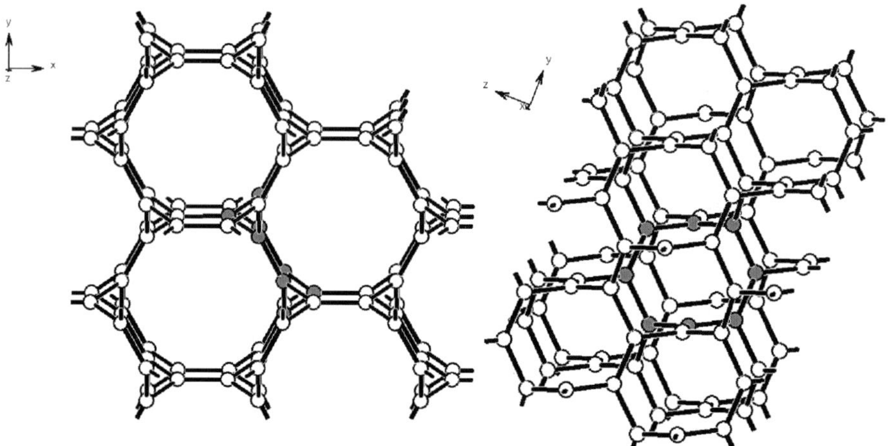

Figure 6.12 The (8,3)-b or **etb** net. This net is not chiral.

While there are numerous structures based on the (10,3)-nets, only a few based on octagons are known. This may seem quite surprising, especially if we compare the **srs** and **eta** nets that can both be built from identical nodes, but that also have identical torsion angles between consecutive sets of nodes.

It is not possible to give a straight answer as to why this is so, but both the symmetry of the net (The connectors in **eta** are all of equal length, but of two different types, those within a four-fold helix and those connecting the helices. In **srs** there is only one type) and the possibilities of interpenetration have to be considered.[1] (A majority of the **srs** structures are interpenetrated.)

Either way, in Figure 6.13 we present one of the known structures with **etb** nets [24]. In this compound Cd(II) ions are doubly bridged by 2,4'-(1,4-phenylene)bis(pyridine) ligands and further linked via nitrate ions to give a neutral, porous compound that can host a variety of guests in its channels [25].

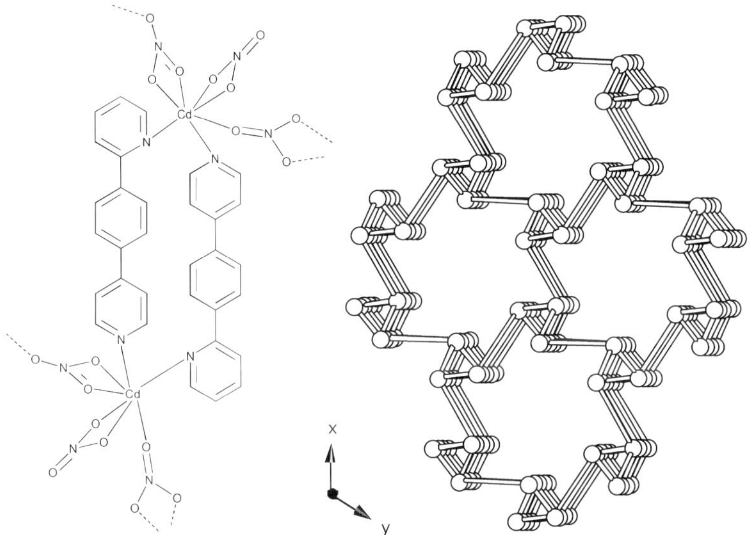

Figure 6.13 One of the very few examples of the (8,3)-b or **etb** net found in a structure with Cd(II) ions doubly bridged by 2,4'-(1,4-phenylene)bis(pyridine) ligands and further linked via nitrate ions to give a non-interpenetrated, neutral, porous compound that can host a variety of guests in its channels [25].

Another example is the recently synthesised neutral, non-interpenetrated net formed by [Cu(5-methyltetrazolate)]·0.17H$_2$O (tetrazolate = N$_4$CH$^-$) [26].

These seem to be the most of three-connected nets containing only one kind of node and one size of ring. We will now move on to some nets with several kinds of rings.

---

[1] These factors are probably related, see Chapter 10 and Chapter 11.

6.2.6. The (8²10)-a, LiGe or 8².10-**lig** net

As pointed out by Wells, many three connected nets can be constructed by joining four-fold helices in different ways, and one of the nets he derived in this way was the (8²10)-a net, see Figure 6.14. This net has vertex symbol $8 \cdot 8 \cdot 10_3$ and genus 5.

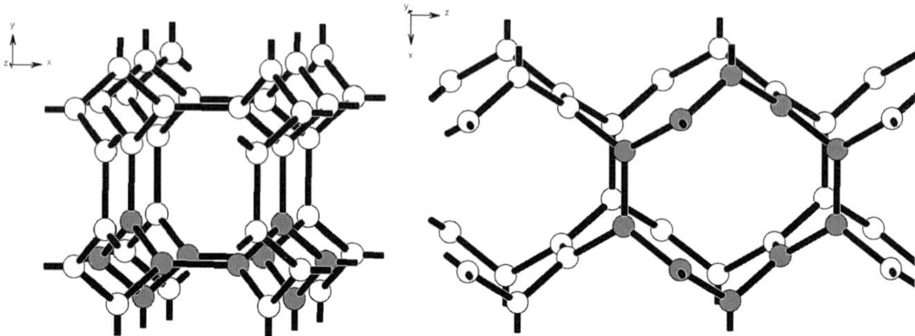

Figure 6.14 The (8²10)-a LiGe or **lig** net. This net is not chiral.

We are aware of two examples of this net, the cadmium(II)–terephthalate coordination polymer ([Ph₃PCH₂Ph][Cd(terephthalate)Cl]·2H₂O) where two anionic **lig** nets interpenetrated by forming double helices, leaving large rhombic channels for the cations, see Figure 6.15, [27] and a pseudo-polymorph of [Cu(5-methyltetrazolate)], (tetrazolate = N_4CH^-) [26].

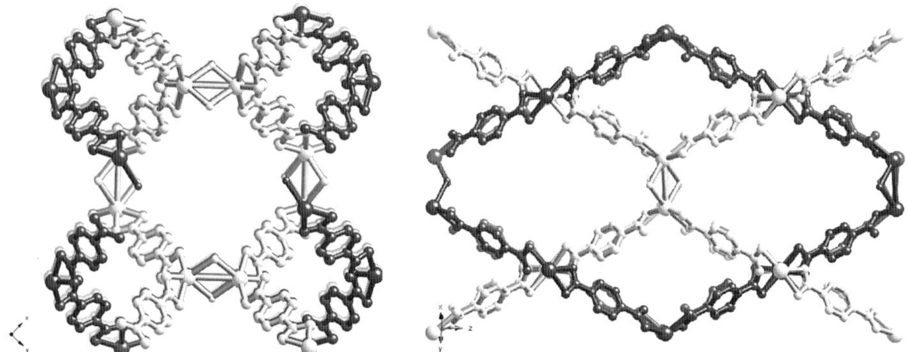

Figure 6.15 In [Ph₃PCH₂Ph][Cd(terephthalate)Cl]·2H₂O two anionic **lig** nets(white and black) are interwoven giving double helices and large channels containing the cations [27].

6.2.7. The 4.14²-**dia-f** net

Another strategy for deriving nets is to take a basic net and replace the nodes by a set of nodes with equivalent total connectivity. For example, in the diamond (**dia**) net one four-connected node can be replaced by two three-connected nodes. One possible outcome of such an operation is the **dia-f** net with vertex symbol $4 \cdot 14_{12} \cdot 14_{12}$, genus 5 and C10 value of 352.

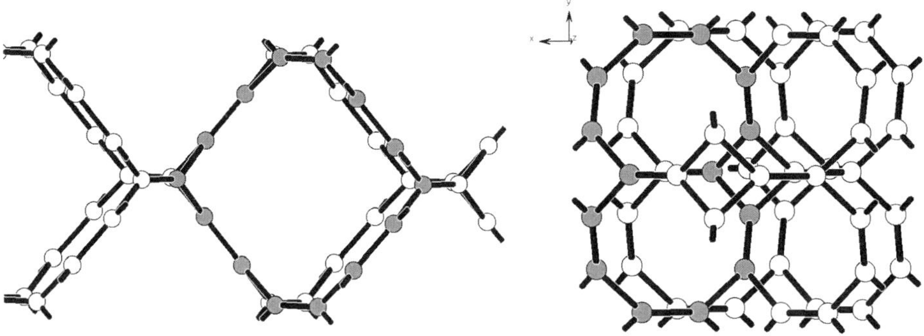

Figure 6.16 The **dia-f** net has vertex symbol $4 \cdot 14_{12} \cdot 14_{12}$ and a 14-ring is highlighted in both views. It is related to the four-connected **dia** net by replacing the tetrahedral nodes with two three-connected nodes.

An example seems to be a $CuBr_2$ complex with 2,2'-dipyridyldiselenide, see Figure 6.17, that forms doubly interpenetrated **dia-f** nets [28].

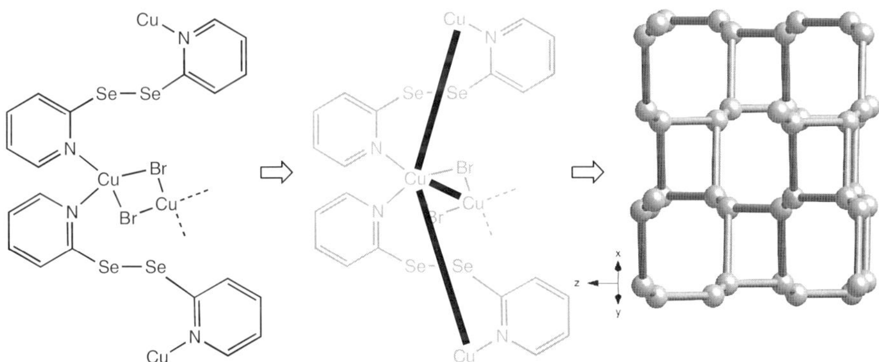

Figure 6.17 The compound $(bis((\eta^2\text{-Bromo})\text{-}(\eta^2\text{-}2,2'\text{-dipyridyldiselenide})\text{-copper(II)}))$ forms doubly interpenetrated **dia-f** nets with dichloromethane in the voids (only one net shown) [28].

6.2.8. The 4.14^2-**dia-g** net

The 4.14^2 **dia-g** net has vertex symbol $4 \cdot 14_{12} \cdot 14_{12}$, genus 5 and a C10 value of 350, thus in this respect almost indistinguishable from **dia-f**. However, this net is chiral.

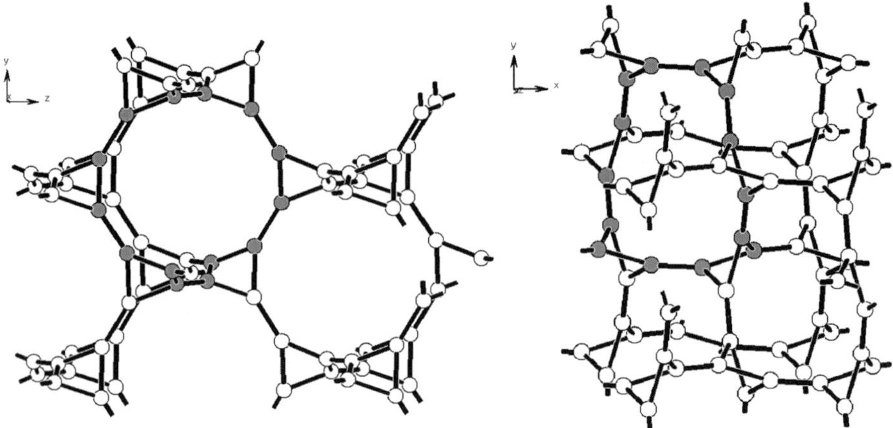

Figure 6.18 The **dia-g** net has vertex symbol $4 \cdot 14_{12} \cdot 14_{12}$, genus 5 and is chiral.

6.2.9. The 4.8.10-*lvt-a* net

The **lvt-a** net has vertex symbol $4 \cdot 8 \cdot 16_3$ and genus 9, see Figure 6.19. Note the difference in the Schläfli symbol and the vertex symbol. The 10-gons in the Schläfli symbol are circuits containing shortcuts while the 16-gons are the smallest fundamental rings.

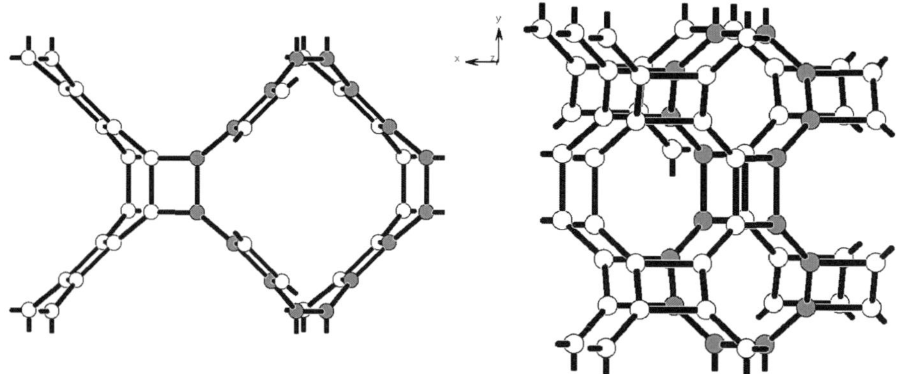

Figure 6.19 The **lvt-a** net has vertex symbol $4 \cdot 8 \cdot 16_3$ and a 16-ring is highlighted in both views. It is related to the 4-connected **lvt** net by replacing the square planar nodes with a square having a node in each corner.

It is found in the rather complicated compound [*bis*(η^2-1,2-*bis*(4-pyridyl)ethane))-(*hexakis*(η^2-*N*-acetylsalicylhydrazidato)-*hexa*-manganese(III)) dimethylformamide solvate pentahydrate] [29]. The main molecular motif in this structure is hexanuclear rings of bridged manganese complexes, two of the complexes being capped by coordinated DMF solvent molecules, but four bridging to other hexanuclear rings by the 1,2-bis(4-pyridyl)ethane ligands, see Figure 6.20. This structure is doubly interpenetrated.

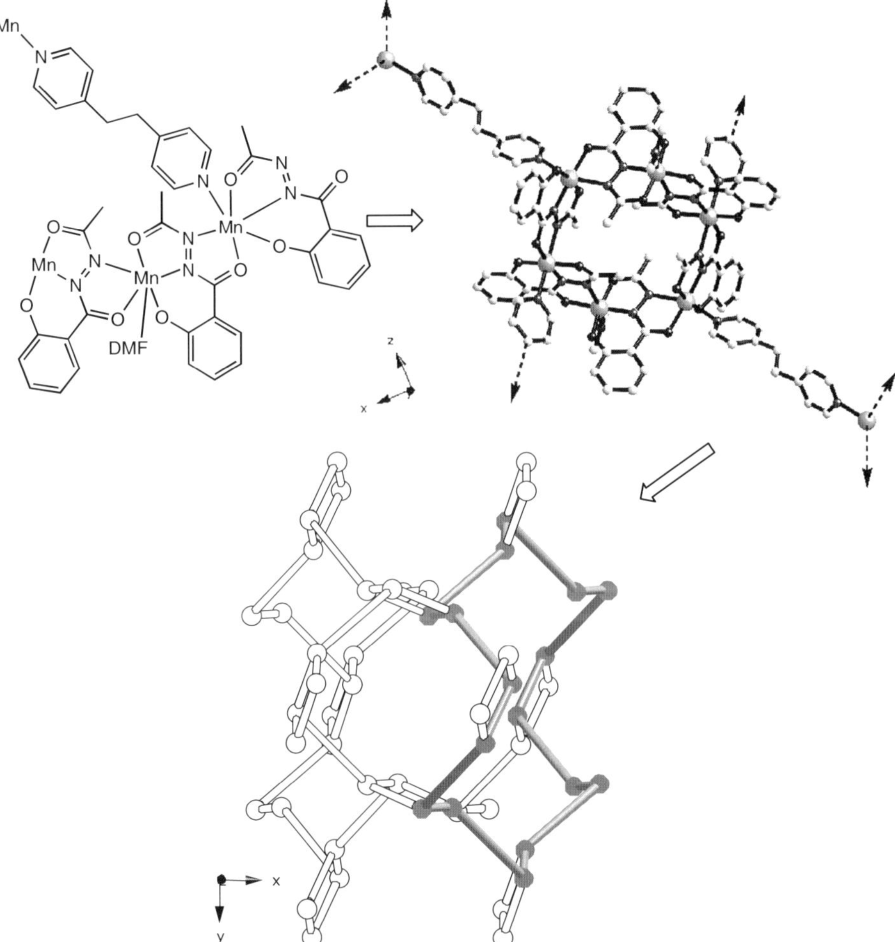

Figure 6.20 Molecular constituents of [*bis*(η^2-1,2-*bis*(4-pyridyl)ethane))-(*hexakis*(η^2-*N*-acetylsalicylhydrazidato)-*hexa*-manganese(III)) dimethylformamide solvate pentahydrate] [29], the hexanuclear rings and the resulting **lvt-a** net (doubly interpenetrated only one net shown).

This survey of the uninodal nets is not exhaustive, however, our point is not to cover all possibilities but rather to show that even with the restriction of all nodes being the same, the number of plausible structure alternatives for a given building block is large. And of course, with binodal nets it gets even larger. We will now move on and have a look at some of those nets.

6.3. Binodal three-connected nets

6.3.1. The (6.10²)(6².10)-**nof** net

The **nof** net has vertex symbols 6·6·10 and 6·10·10₂ (since there are two nodes there will be two vertex symbols) and genus is 5. The main theme is parallel chains of chair-formed hexagons interconnected at the apexes, see Figure 6.21

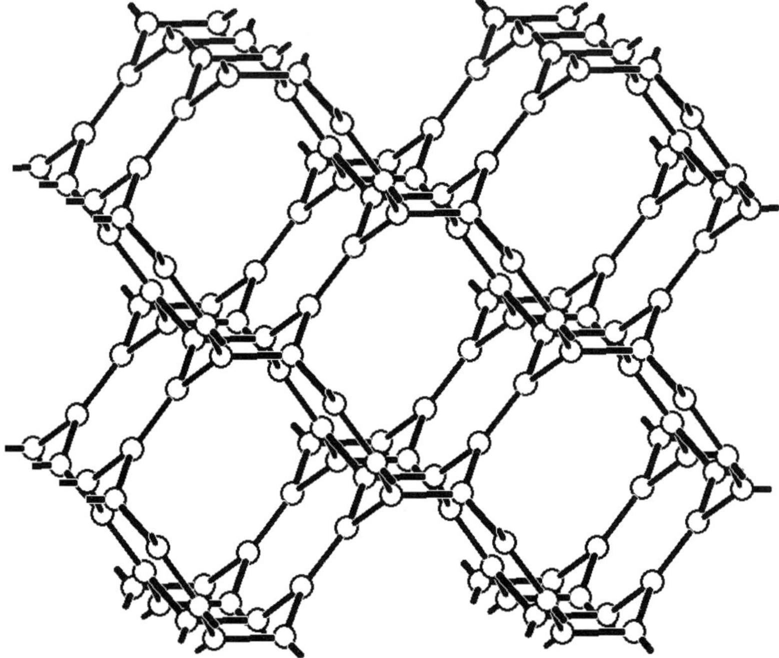

Figure 6.21 The **nof** net has vertex symbols 6·6·10 and 6·10·10₂. Note the chains of chair formed hexagons.

An example is found in the compound [3,3':5',3'':5'',3'''-quaterpyridine(Cu₂I₂)]·2G, where G is the guest molecule, either nitrobenzene or cyanobenzene, see Figure 6.22. Apparently the 3D net is templated by these inclusion molecules as a "guest free" preparation results in a 2D-net [30].

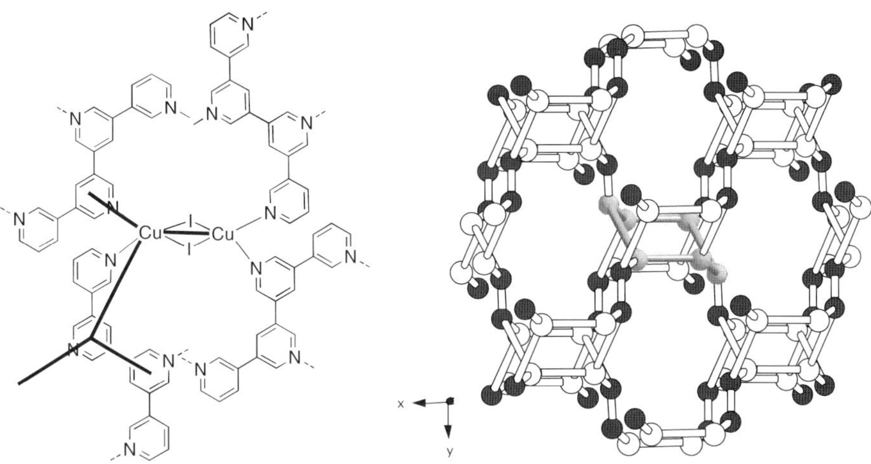

Figure 6.22 [3,3':5',3":5",3'''-quaterpyridine(Cu$_2$I$_2$)]·2nitrobenzene forms a distorted **nof** net templated by the guest molecules. These guests are found in between the chains of chair-formed hexagons while the apparent larger cavities are occupied by iodides and the quaterpyridines from the net [30]. (White nodes are centroids of the middle pyridyl groups and black nodes are Cu(I))

6.3.2. The $(8^3)(8^3)$-*noj* net

This net has vertex symbols 8·8·8 and 8·8·8$_3$ and is thus a uniform net (having only one kind of shortest rings) with genus 5. It is formed by four-fold helices, well known by now, this time interconnected by zigzag chains, see Figure 6.24 It is also chiral. An example can be found in the cyano-bridged iridium cyclopentadienyl silver(I)pyridine compound in Figure 6.24 [31].

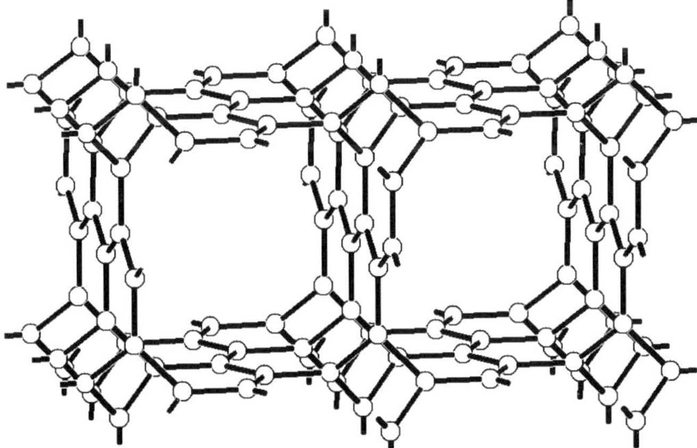

Figure 6.23 The **noj** net. Note the four-fold helices, a reoccurring theme in this chapter.

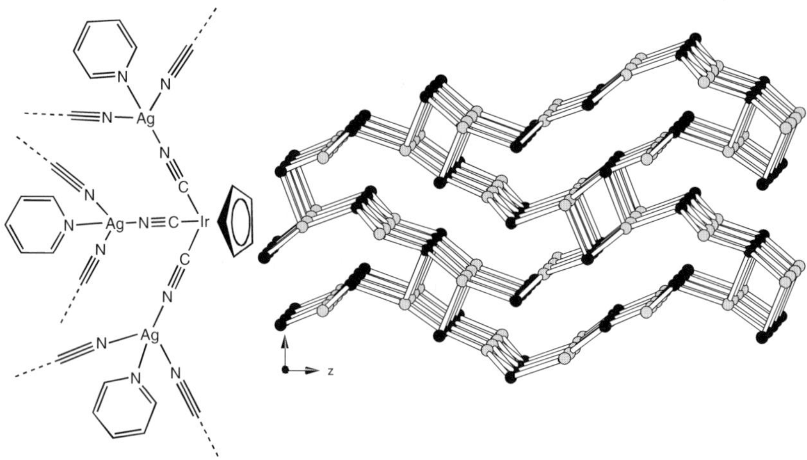

Figure 6.24 The **noj** net in hexakis(η^2-cyano)-bis((η^5-pentamethyl-cyclopentadienyl)-(pyridine)-di-iridium-di-silver [31]. Iridium atoms are black and silver atoms grey.

6.3.3. The $(10^3)(6.10^2)_3$-**noh** net

The $(10^3)(6.10^2)_3$-**noh** net has vertex symbols $10_3 \cdot 10_3 \cdot 10_3$ and $6 \cdot 10_3 \cdot 10_3$ with genus 17. As can be expected for such a high genus this net is complicated and difficult to visualise, but still it contains only two types of nodes, although with different stoichiometry, 1:3, see Figure 6.25.

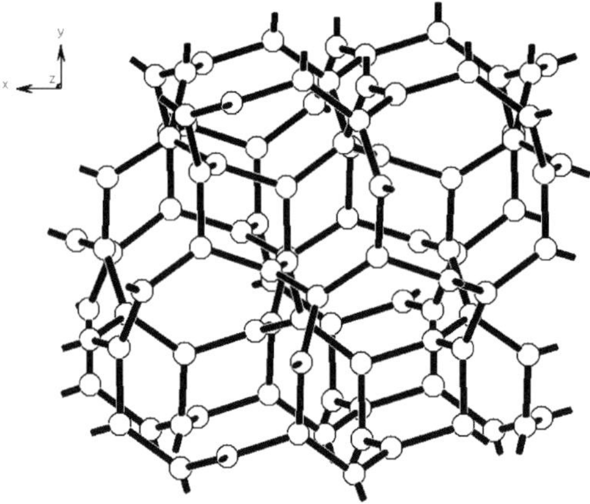

Figure 6.25 The **noh** net has vertex symbols $10_3 \cdot 10_3 \cdot 10_3$ and $6 \cdot 10_3 \cdot 10_3$ and genus 17

An example of the **noh** net is found with the adamantane-like building block hexamethylenetetramine acting as a three connected node, and Ag(I) ions, acting as connectors (3/4) or additional nodes (1/4), see Figure 6.26 [32].

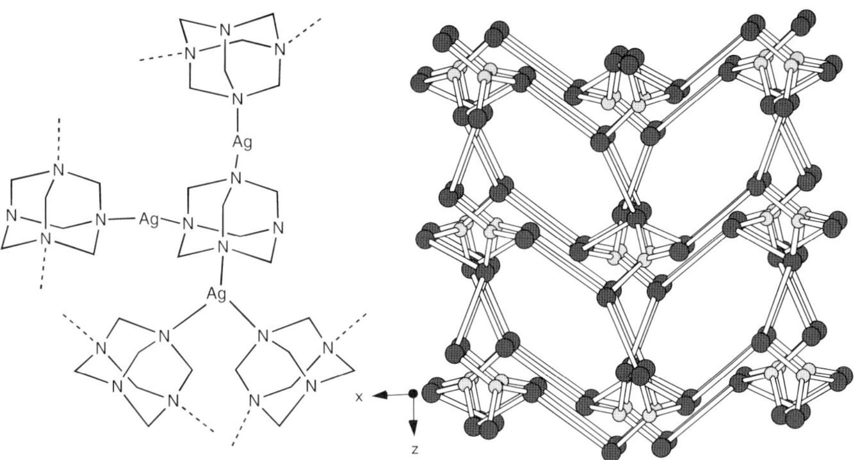

Figure 6.26 Non-interpenetrated **noh** nets in [Ag$_4$(hexamethylenetetramine)$_3$(H$_2$O)](PF$_6$)$_4$·3EtOH [32].

This example (and also the following **nod** net) also illustrates the problem of controlling the synthesised network. While hexamethylenetetramine could be expected to be a four-connected node, and silver(I) to act simply as a connector to give diamond like nets, the actual outcome may be different and controlled by more subtle intermolecular forces. Thus, using only slightly different conditions the same group also obtained an **srs** net with the same reagents and solvents [33].

6.3.4. The $(8^2.10)(8^2.10)$-**nod** or $(8^2 10)$-b net

The $8^2 10$ **nod** net is also built from interconnected four-fold helices and has vertex symbols 8·8·10$_3$ and 8·8·10$_3$ with genus 5, see Figure 6.27.

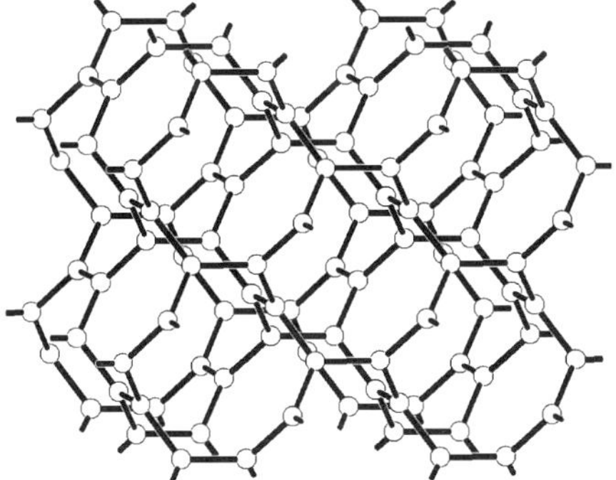

Figure 6.27 The **nod** net, vertex symbols 8·8·10$_3$ and 8·8·10$_3$, genus 5

6.4. Trinodal three-connected nets

Some trinodal nets have also been observed. As was briefly noted in Chapter 4 it is very difficult to search for nets in the literature for a variety of reasons, thus, no doubt, there are other nets of this type to be uncovered even though the two systematic studies published so far seem to be fairly comprehensive [24,36].

6.4.1. The $(4.12^2)(4.12^2)(4.12^2)$-**mot-a** net

The **mot-a** net has vertex symbols $4 \cdot 12_2 \cdot 12_2$, $4 \cdot 12_2 \cdot 12_2$ and $4 \cdot 12_2 \cdot 12_2$ with genus 7.

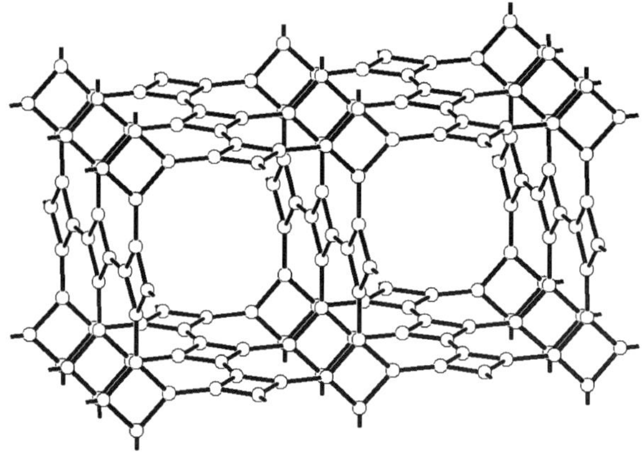

Figure 6.32 The **mot-a** net has vertex symbols $4 \cdot 12_2 \cdot 12_2$, $4 \cdot 12_2 \cdot 12_2$ and $4 \cdot 12_2 \cdot 12_2$

It can be found in $[Cu(II)_2(4\text{-bromobenzene-1,3-dicarboxylato})_2(DMF)_2]$ ·2DMF, see Figure 6.33 [37]

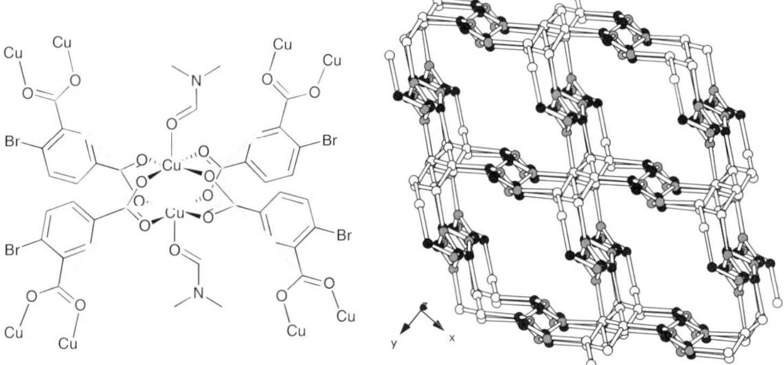

Figure 6.33 The **mot-a** net in $[Cu(II)_2(4\text{-bromobenzene-1,3-dicarboxylato})_2(DMF)_2]\cdot 2DMF$ [37]. The node assignment is shown in grey.

This compound also demonstrate the difficulty of the node assignment in certain cases (see section 3.3). We might equally well put the nodes in the centre of the O_8-polygon, at the average position of the two copper(II) ions. This will create a four connected binodal $(6^6)(6^4.8^2)_2$-**mot** net. (**mot-a** (a=augmented) is derived from **mot** by replacing the four-connected node with four three-connected nodes, see Figure 6.34.) While this net is conceptually simpler with higher symmetry it has the drawback of leaving the entire benzene rings outside the net, thus being slightly misleading as to the voids in the structure. However, it may be regarded as somewhat disturbing that the same connectivity and basic structural units should give different node assignments, compare for example with the **pts** net in Figure 1.6.

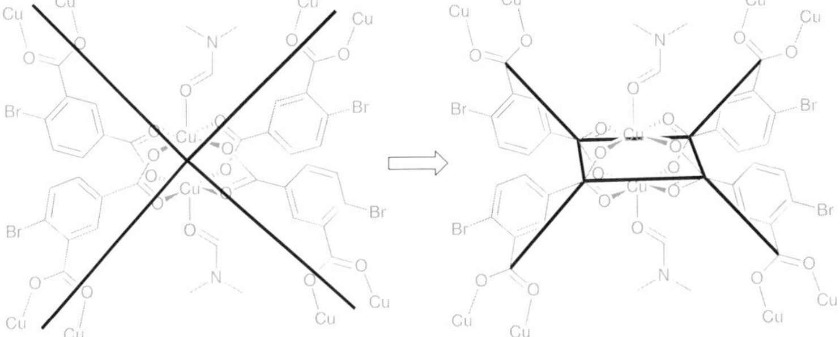

Figure 6.34 The **mot-a** (a=augmented) is derived form **mot** by replacing the four-connected node with four three-connected nodes.

6.4.2. The $(7^2.8)_2(7^2.8)(7.12^3)$-**noe** net

Seven member rings are uncommon but we encounter them in the **noe** net. This net has vertex symbols $7 \cdot 7_8 \cdot 7_8$, $7 \cdot 7_8 \cdot 7_8$, and $7 \cdot 12_3 \cdot 12_3$ with genus 5.

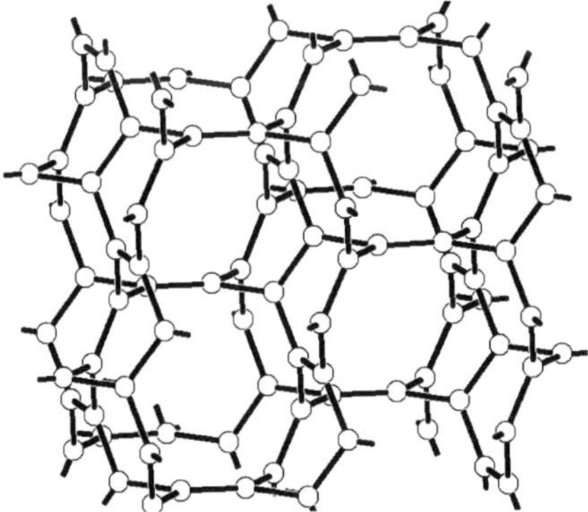

Figure 6.35 The **noe** net has vertex symbols $7 \cdot 7_8 \cdot 7_8$, $7 \cdot 7_8 \cdot 7_8$, and $7 \cdot 12_3 \cdot 12_3$ with genus 5.

It has been found as the anionic array of trigonal Cu(I) coordinated by cyanide ions in bis(tetrabutylammonium)hexakis(η^2-cyano)-tetra-copper(I), see Figure 6.36 [38].

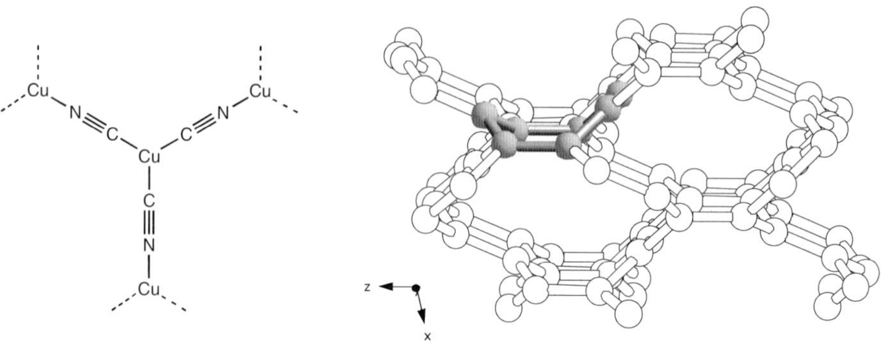

Figure 6.36 The **noe** net in bis(tetrabutylammonium)hexakis(η^2-cyano)-tetra-copper(I) [38]. A seven ring is emphasised.

6.4.3. $(4.12^2)(4.12^2)_2(12^3)_2$-"net 10"

In his very first paper on three dimensional nets Wells also derives a net he calls simply "net 10" [10]. Probably, since it is a trinodal net, it does not appear in any of his subsequent books, [21,39,40] and its next appearance in the literature seems to be the report of a five fold interpenetrated hydrogen bonded net in [Co(Hbiim)$_2$(H$_2$biim)]$_2$(*para*-OOCC$_6$H$_4$COOH)$_2$·H$_2$O in 2004 [41]. It has vertex symbols $4·12_2·12_2$, $4·12_6·12_6$, $12_4·12_6·12_6$, and has already been shown in Figure 1.14, Wells' model of this net is shown in Figure 6.37

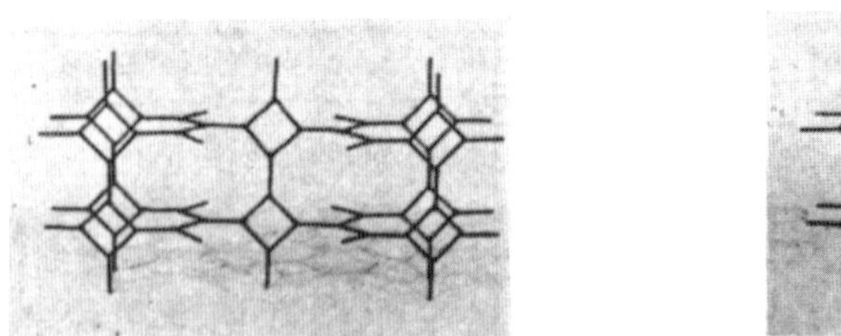

Figure 6.37 Wells' original stereo photograph of his model of the "net 10". Reproduced with permission [10].

6.5. Multinodal three-connected nets

At last we will see a rare example of a hexanodal net.

6.5.1. The $(8^3)(8^3)(8^2.10)(8^2.10)(8^3)(8^3)$-nos net, a hexanodal net

The **nos** net has six different nodes and vertex symbols 8·8·8, 8·8·8$_2$, 8·8$_2$·10$_2$, 8·8$_2$·10$_2$, 8·8·8, 8·8·8$_2$ and genus 13. It has been found as the anionic array of trigonal Cu(I) coordinated by cyanide ions in [(Ln(dmf)$_8$Cu$_6$(CN)$_9$)]·2DMF (Ln = Gd, Eu, Er) see Figure 6.38 [42,43]. The voids in the structure are filled with the Gd(III) complex and DMF molecules.

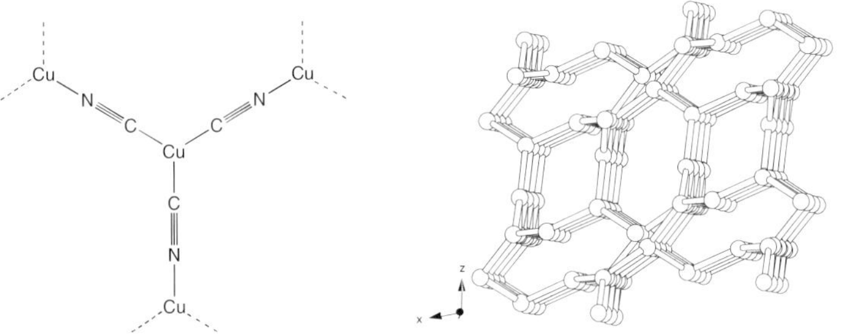

Figure 6.38 Trigonal Cu(I) coordinated by cyanide ions in [(Gd(dmf)$_8$Cu$_6$(CN)$_9$)·2DMF] [42] form the **nos** net.

6.6. Summary of three-connected nets

Table 6.2 Summary of the three-connected nets discussed in this chapter and in chapter 5

Net	Other names	Vertex Symb.	Short symb.	Genus	Node[b]
twt	(12,3)	12$_4$·12$_7$·12$_7$,	12^3	4	1
srs	(10,3)-a, SrSi$_2$	10$_5$·10$_5$·10$_5$	10^3	3	1
ths	(10,3)-b, ThSi$_2$	10$_2$·10$_4$·10$_4$	10^3	3	1
bto	3/10/h1, (10,3)-c	10·10$_2$·10$_2$	10^3	4	1
utp	3/10/o1, (10,3)-d	10$_2$·10$_4$·10$_4$	10^3	5	1
utk	3/10/t3	10·10·10$_3$	10^3	5	1
utj	3/10/t2	10·10·10$_3$	10^3	9	1
utm	3/10/t5	10$_2$·10$_4$·10$_4$	10^3	9	1
utn	3/10/t6	10·10·10$_3$	10^3	9	1

Table continued ...

Net	Other names	Vertex symb.	Short symb.	Genus	Node[b]
uto	3/10/t7	$10 \cdot 10 \cdot 10_3$	10^3	9	1
eta	3/8/h2, (8,3)-a	$8 \cdot 8 \cdot 8_2$	8^3	4	1
etb	3/8/h1, (8,3)-b	$8 \cdot 8 \cdot 8_2$	8^3	4	1
lig	$(8^2 10)$-a	$8 \cdot 8 \cdot 10_3$	$8^2.10$	5	1
dia-f		$4 \cdot 14_{12} \cdot 14_{12}$	4.14^2	5	1
dia-g		$4 \cdot 14_{12} \cdot 14_{12}$	4.14^2	5	1
lvt-a		$4 \cdot 8 \cdot 16_3$	$4.8.10$	9	1
nof		$6 \cdot 6 \cdot 10$ $6 \cdot 10 \cdot 10_2$	$6^2.10$ 6.10^2	5	2
noj	(8,3)-d	$8 \cdot 8 \cdot 8$ $8 \cdot 8 \cdot 8_3$	8^3 8^3	5	2
noh		$10_3 \cdot 10_3 \cdot 10_3$ $6 \cdot 10_3 \cdot 10_3$	10^3 6.10^2	17	2
nod	$(8^2 10)$-b	$8 \cdot 8 \cdot 10$ $8 \cdot 8 \cdot 10_3$	$8^2.10$ $8^2.10$	5	2
nob		$6 \cdot 12_2 \cdot 12$ $6 \cdot 10 \cdot 10$	6.12^2 6.10^2	7	2
nta	(9,3)-a	$9_2 \cdot 9_2 \cdot 9_2$ $9_2 \cdot 9_2 \cdot 9_2$	9^3 9^3		2
ntb	(9,3)-b	$9_2 \cdot 9_2 \cdot 9_2$ $9_2 \cdot 9_2 \cdot 9_2$	9^3 9^3		2
mot-a		$4 \cdot 12_2 \cdot 12_2$ $4 \cdot 12_2 \cdot 12_2$ $4 \cdot 12_2 \cdot 12_2$	4.12^2 4.12^2 4.12^2	7	3
noe		$7 \cdot 7_8 \cdot 7_8$ $7 \cdot 7_8 \cdot 7_8$ $7 \cdot 12_3 \cdot 12_3$	$7^2.8$ $7^2.8$ 7.12^2	5	3
-	Net 10	$4 \cdot 12_2 \, 12_2$ $4 \cdot 12_6 \cdot 12_6$ $12_4 \cdot 12_6 \cdot 12_6$	4.12^2 4.12^2 12^3	un	3
nos		$8 \cdot 8 \cdot 8$ $8 \cdot 8 \cdot 8_2$ $8 \cdot 8_2 \cdot 10_2$ $8 \cdot 8_2 \cdot 10_2$ $8 \cdot 8 \cdot 8$ $8 \cdot 8 \cdot 8_2$	8^3 8^3 $8^2.10$ $8^2.10$ 8^3 8^3	13	6

References

[1] F. H. Allen, O. Kennard, Chem. Design Auto. News 8 (1993) 31.

[2] F. A. Cotton, G. Wilkinson, Advanced Inorganic Chemistry, 4th ed. Wiley, New York, 1989.

[3] N. N. Greenwood, A. Earnshaw, Chemistry of the Elements, 2nd ed. Pergamon Press, Oxford, 1997.

[4] O. M. Yaghi, H. Li, J. Am. Chem. Soc. 118 (1996) 295.

[5] F. Robinson, M. J. Zaworotko, J. Chem. Soc., Chem. Commun. (1995) 2413.

[6] A. Ienco, D. M. Proserpio, R. Hoffmann, Inorg. Chem. 43 (2004) 2526.

[7] S. A. Barnett, N. R. Champness, Coord. Chem. Rev. 246 (2003) 145.

[8] L. Öhrström, K. Larsson, S. Borg, S. T. Norberg, Chem. Eur. J. 7 (2001) 4805.

[9] L. Öhrström, K. Larsson, Dalton Trans. (2004) 347.

[10] A. F. Wells, Acta Cryst. 7 (1954) 535.

[11] B. F. Abrahams, S. R. Batten, M. J. Grannas, H. Hamit, B. F. Hoskins, R. Robson, Angew. Chem. Int. Ed. 38 (1999) 1475.

[12] S. L. Strong, R. Kaplow, Acta Cryst. B24 (1968) 1032.

[13] I. Boldog, E. B. Rusanov, A. N. Chernega, J. Sieler, K. V. Domasevitch, Angew. Chem. Int. Ed. 40 (2001) 3435.

[14] I. Boldog, E. B. Rusanov, J. Sieler, S. Blaurock, K. V. Domasevitch, Chem. Commun. (2003) 740.

[15] J. M. Robertson, Proc. Roy. Soc (A) 157 (1936) 79.

[16] Y. B. Dong, M. D. Smith, H. C. zur Loye, Inorg. Chem. 39 (2000) 4927.

[17] S. B. Qin, S. M. Lu, Y. X. Ke, H. M. Li, X. T. Wu, W. X. Du, Solid State Sc. 6 (2004) 753.

[18] R. Kuhlman, G. L. Schimek, J. W. Kolis, Inorg. Chem. 38 (1999) 194.

[19] S. R. Halper, S. M. Cohen, Inorg. Chem. 44 (2005) 486.

[20] S. R. Halper, S. M. Cohen, work in progress (2005).

[21] A. F. Wells, Three-dimensional nets and polyhedra, John Wiley & Sons, New York, 1977.

[22] E. Koch, W. Fischer, Z. Kristallogr 210 (1995) 407

[23] C. Bonneau, O. Delgado-Friedrichs, M. O'Keeffe, O. M. Yaghi, Acta Cryst. A 60 (2004) 517.

[24] N. W. Ockwig, O. Delgado-Friedrichs, M. O'Keeffe, O. M. Yaghi, Acc. Chem. Res. 38 (2005) 176.

[25] A. J. Blake, N. R. Champness, A. N. Khlobystov, S. Parsons, M. Schröder, Angew. Chem. Int. Ed. 39 (2000) 2317.

[26] D. Li, W. J. Shi, L. Hou, Inorg. Chem. 44 (2005) 3907.

[27] J. C. Dai, X. T. Wu, Z. Y. Fu, S. M. Hu, W. X. Du, C. P. Cui, L. M. Wu, H. H. Zhang, R. Q. Sun, Chem. Commun. (2002) 12.

[28] C. O. Kienitz, C. Thone, P. G. Jones, Z. Naturforsch., B: Chem. Sci. 55 (2000) 587.

[29] M.Moon, I.Kim, M.S.Lah, Inorg. Chem. 39 (200) 2710.

[30] K. Biradha, M. Aoyagi, M. Fujita, J. Am. Chem. Soc. 122 (2000) 2397.

[31] S. M. Contakes, K. K. Klausmeyer, T. B. Rauchfuss, Inorg. Chem. 39 (2000) 2069.

[32] L. Carlucci, G. Ciani, D. M. Proserpio, A. Sironi, Inorg. Chem. 36 (1997) 1736.

[33] L. Carlucci, G. Ciani, D. M. Proserpio, A. Sironi, J. Am. Chem. Soc. 117 (1995) 12861.

[34] L. Carlucci, G. Ciani, D. W. von Gudenberg, D. M. Proserpio, New J. Chem. 23 (1999) 397.

[35] B. Rossenbeck, W. S. Sheldrick, Z. Naturforsch., B: Chem. Sci. 54 (1999) 1510.

[36] V. A. Blatov, L. Carlucci, G. Ciani, D. M. Proserpio, Crystengcomm 6 (2004) 377.

[37] M. Eddaoudi, J. Kim, D. Vodak, A. Sudik, J. Wachter, M. O'Keeffe, O. M. Yaghi, Proc. Nat. Acad. Sci. USA 99 (2002) 4900.

[38] E. Siebel, P. Schwarz, R. D. Fischer, Solid State Ionics 101 (1997) 285.

[39] A. F. Wells, Further Studies of Three-Dimensional Nets, Polycrystal book service, Pittsburgh, 1979.

[40] A. F. Wells, Structural Inorganic Chemistry, 5th ed. Clarendon Press, Oxford, 1984.

[41] K. Larsson, L. Öhrström, Crystengcomm 6 (2004) 354.

[42] S. M. Liu, E. A. Meyers, S. G. Shore, Angew. Chem. Int. Ed. 41 (2002) 3609.

[43] S. M. Liu, C. E. Plecnik, E. A. Meyers, S. G. Shore, Inorg. Chem. 44 (2005) 282.

Chapter 7

Four-connected nets

Four-connected nets with tetrahedral building blocks are of particular interest since they form the basis of zeolites and related materials. Zeolites are important in many areas, from large scale cracking of crude oils to the performance of washing powders in areas with hard water [1]. Thus, there is an abundant literature on the subject in contrary to the three-connected nets, and this chapter therefore will be briefer than it otherwise would have merited.

The overwhelming richness of structural types can be seen from the fact that there are over 140 recognized zeolite networks (thermally stable and microporous), this number growing by about six every year [2,3]. On the theoretical side there are estimates that the number of plausible regular tetrahedral frameworks exceeds 100 000, [2] and the uninodal nets alone number more than 150 [4]. The IUPAC approved Structure Commission of the International Zeolite Association are in charge of the assignment of three letter codes to each unique zeolite structure and publish them in a web-based database [5].

Luckily, we do not have to deal with all those; the number of known molecular based four-connected nets is much lower. In the study by Ockwig et al. 28 nets were found containing tetrahedral or approximately tetrahedral nodes, three containing tetrahedral and square planar nodes, and four containing square planar nodes, [6] and in their investigation of interpenetration Blatov et al. found 14 different four-connected nets [7].

It is useful to distinguish between nets based on tetrahedral nodes, those based on squares planar nodes, and those that are a mixture of the two, although intermediate configurations also exists, and this chapter will be divided accordingly.

7.1. Uninodal tetrahedral nets

7.1.1. The SrAl$_2$ or 4^2.6^3.8-*sra* net

The **sra** net has vertex symbol 4·6·4·6·6·8$_2$ and genus 5 and it is shown in Figure 7.1. Since it contains four-rings, it is obviously very distorted from the ideal tetrahedral symmetry, but is nevertheless the second most common of those nets in molecular chemistry. Look for the characteristic zigzag ladder motif.

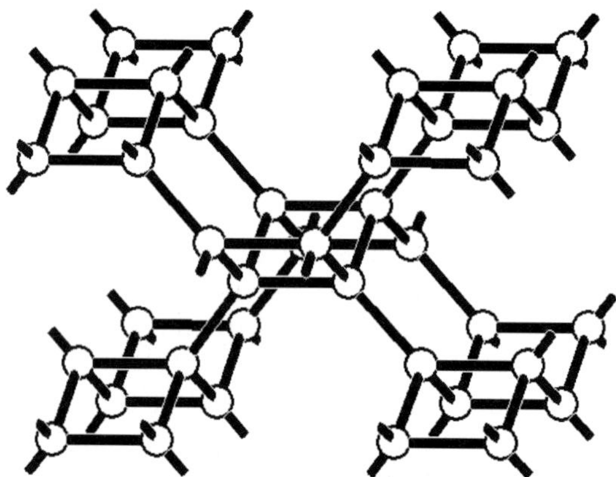

Figure 7.1 The **sra** net contains very distorted tetrahedrons since it is based on four-rings, but it is still the second most common tetrahedral type four connected net. It is formed by interconnected ladders.

An example of this net can be found in the triply interpenetrated net structure of 1,3,5-tris[4-pyridyl(ethenyl)]benzene [8].

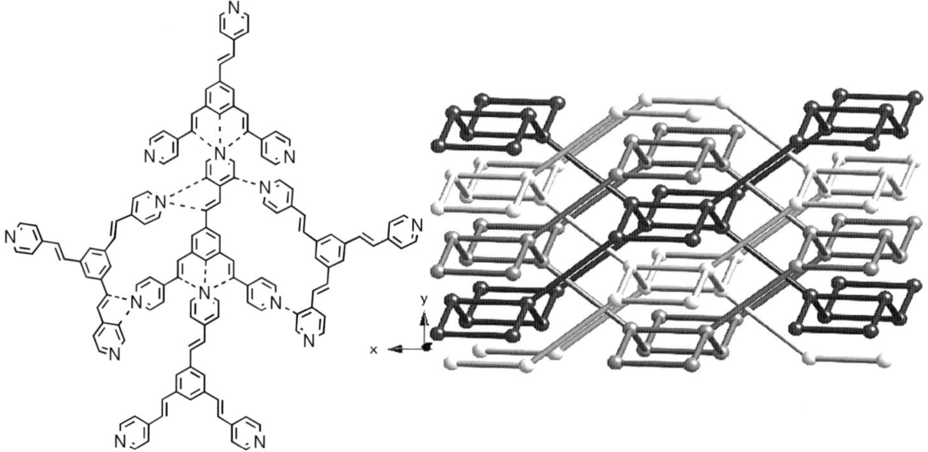

Figure 7.2 Weak hydrogen bonds (dotted lines) are the strongest forces in the triply interpenetrated **sra** net structure of 1,3,5-tris[4-pyridyl(ethenyl)]benzene [8].

Another recent, non-interpenetrated, example can be found in [Zn(II)((S)-5-(3-tetrazoyl)phenylalaninato)] and [Cd(II)(S)-5-(3-tetrazoyl)-phenylalaninato)]·H$_2$O [9].

7.1.2. The sodalite or $4^2.6^4$-**sod** net

The sodalite or $4^2.6^4$ **sod** net is a zeolite net with vertex symbol 4·4·6·6·6·6 and genus 7. The typical feature of this net is that it can be seen as the packing of octahedrons whose six corners have been chopped of to yield truncated octahedrons, see Figure 7.3 right. Just as in the previous net the four rings prevent perfect tetrahedral symmetry at the nodes. A characteristic feature are the chains of four rings, every second ring rotated 90° vis-à-vis its neighbours.

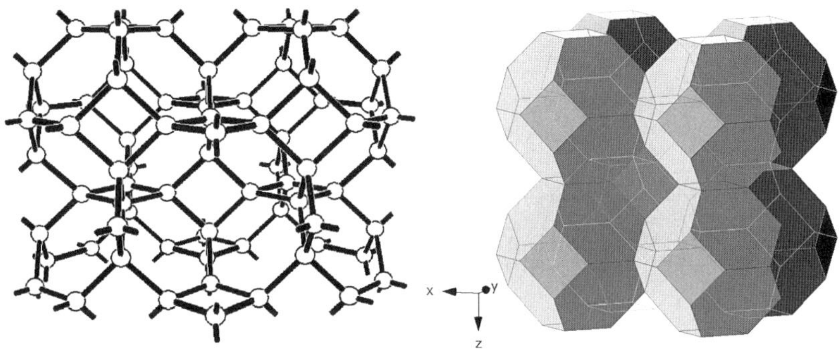

Figure 7.3 The sodalite or $4^2.6^4$ **sod** net is a zeolite net with vertex symbol 4·4·6·6·6·6 and genus 7. The net is shown to the left, and a polyhedron drawing is shown to the right.

An example is the carbonato-bridged copper(II) compound $[Cu_6(CO_3)_{12}]$ $(C(NH_2)_3)_8 \cdot 4K^+ \cdot 8H_2O$ and several isostructural materials prepared by Abrahams et al. where also the guanidinium ion, $(C(NH_2)_3)^+$, appears to play an important role hydrogen bonding inside the hexagonal faces, see Figure 7.4 [10,11].

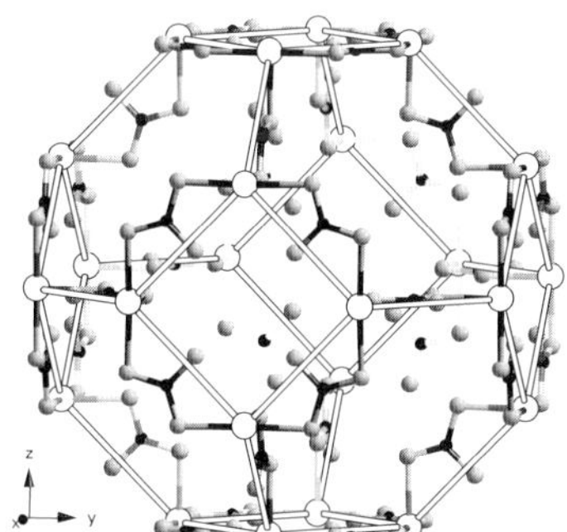

Figure 7.4 The copper(II) compound $[Cu_6(CO_3)_{12}](C(NH_2)_3)_8 \cdot 4K^+ \cdot 8H_2O$ contains a carbonato-bridged **sod** net (white) with disordered potassium ions and water molecules (not shown) [10,11].

7.1.3. The quartz or $6^6.8^2$-*qtz* net

The **qtz** net is the Si net in the quartz form of SiO_2, the second most abundant mineral in the earth's crust. It has vertex symbol $6\cdot6\cdot6_2\cdot6_2\cdot8_7\cdot8_7$, genus 4 and it is chiral. As can be seen in Figure 7.5 it can be constructed with less distorted tetrahedrons than the two preceding structures.

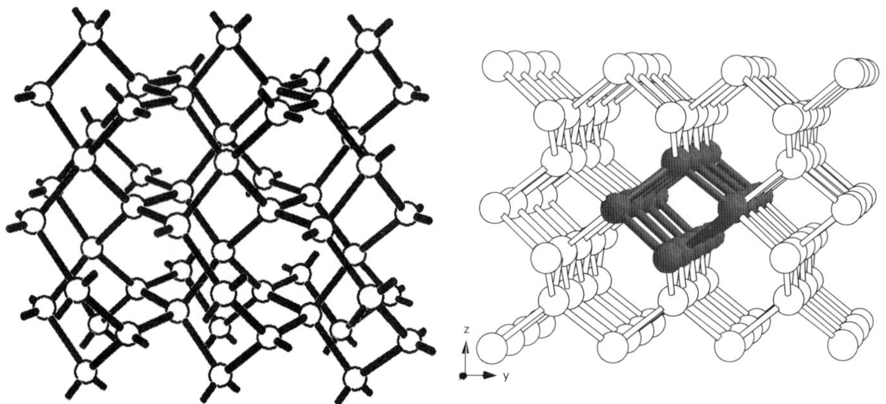

Figure 7.5 The quartz or $6^6.8^2$-**qtz** net is chiral. This can easiest be seen by inspecting the parallel four-fold helices and noting that they all have the same helicity.

Just as *tris*-oxalato complexes are common nodes in three-connected nets and can form the chiral (10,3)-a or **srs** nets, so are *tetrakis*-oxalato complexes possible building blocks for tetrahedral type nets. They normally require larger metal ions that do not have any preferences for octahedral geometries due to crystal field stabilising energies, such as the larger d^{10} or d^0 ions. An example of the **qtz** net is found in $CdZr(C_2O_4)_4(NH_4)_2\cdot3.9H_2O$ and in the closely related $CdZr(C_2O_4)_4(H_3NCH_2CH_2NH_3)_2\cdot4.4H_2O$ see Figure 7.6 [12]. Many other analogous compounds have also been prepared [13-15].

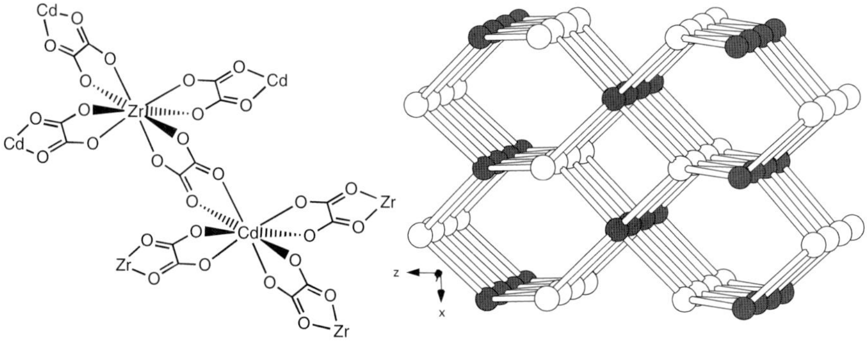

Figure 7.6 $[CdZr(C_2O_4)_4(H_3NCH_2CH_2NH_3)_2]\cdot4.4H_2O$ have the quartz or **qtz** net [12].

While the M(oxalato)$_3^{n+}$ ions almost always have octahedral coordination, and the resulting complexes are therefore chiral, the same is not true for the eight-coordinated M(oxalato)$_4^{n+}$ species. In their highest symmetry they are achiral and the net propagation vectors passing through the centres of the oxalate ligands are all in the same plane, thus making a square planar connector. Even when the complex is distorted to give a propeller like arrangements of the ligands, it will be a square planar connector, see Figure 7.7, so a further lowering of the symmetry is needed to get closer to a tetrahedral node. Note, however, that in [CdZr(C$_2$O$_4$)$_4$(H$_3$NCH$_2$CH$_2$NH$_3$)$_2$]·4.4H$_2$O there is both a distortion out of plane and chirality, and that all centres have the same chirality.[1]

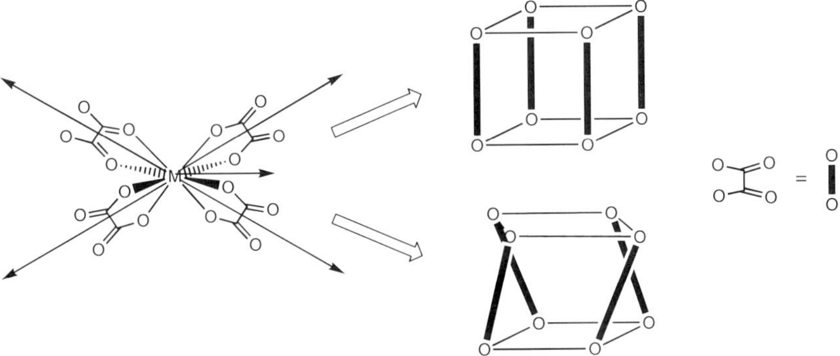

Figure 7.7 Chiral (bottom) and non chiral (top) configurations of the M(oxalato)$_4^{n+}$ ion.

Note that just as for the M(L)$_3^{n+}$ units (section 5.2.1), adding nodes of with alternating chirality will keep the propagation vectors in the same plane, while adding centres of the same chirality will turn them out of the plane, see Figure 7.8.

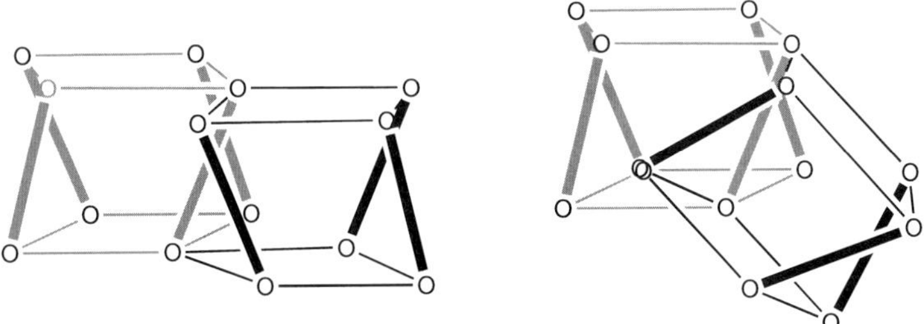

Figure 7.8 Connecting two M(oxalato)$_4^{n+}$ units (grey and black) with different chirality (left) and the same chirality (right). In the form case a 2D structure can be envisaged, while in the latter case a 3D structure is probable.

---

[1] It is not stated in the paper if the products were obtained as a mixture of crystals with different chirality, or if an enantiopure product was formed.

There are also a number of examples of $M(oxalato)_4^{n+}(M')$ structures giving the diamond (**dia**) net [16-18].

7.1.4. The CrB_4 or 4.6^5-**crb** net

The CrB_4 or **crb** net has vertex symbol $4·6_2·6·6·6·6$ and genus 5. Just as the sodalite or **sod** net it is built from interconnected four-rings, but these are no longer directly connected. Each square instead connects to eight other squares by corner-to-corner linking, see Figure 7.9.

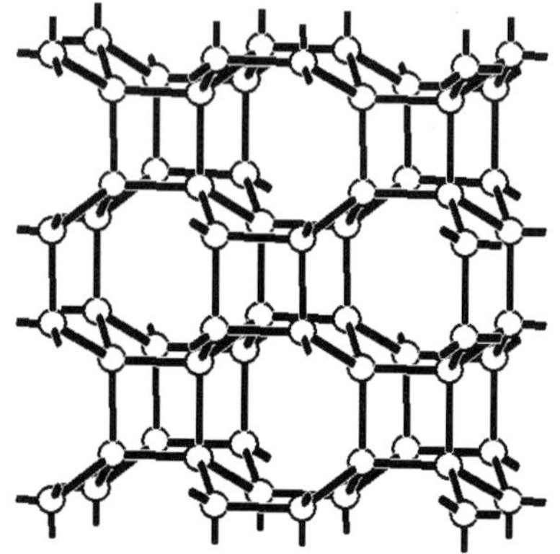

Figure 7.9 The CrB_4 or **crb** net has vertex symbol $4·6_2·6·6·6·6$ and genus 5.

An example can be found in the bromide bridged [CuBr(2-amino-5-bromopyrimidine)]·H_2O coordination polymer, see Figure 7.10 [19]. In this structure the coordinated bromide anions and the amino group all point into the smaller cube-like void where the (disordered) water molecules are found. The larger voids are entirely filled by the ligand and the bromine substituents point towards each other.

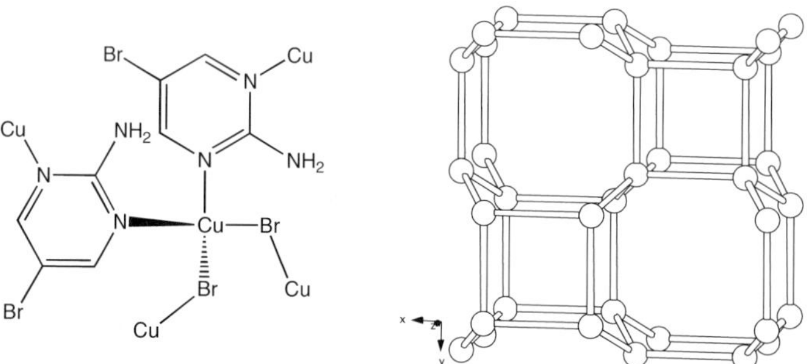

Figure 7.10 The bromide bridged [CuBr(2-amino-5-bromopyrimidine)]·H_2O coordination polymer forms the **crb** net. Coordinated bromide anions and the amino group all point into the smaller cube-like void where the (disordered) water molecules are found.

7.1.5. The gismondine or $4^3.8^3$-**gis** net

The mineral gismondine with composition $Ca_2Al_4Si_4O_{16}\cdot9(H_2O)$ has given its name to the **gis** net with vertex symbol $4\cdot4\cdot4\cdot8_2\cdot8\cdot8$ and genus 9. This is another example of a net with a ladder-like theme, just as the **sra** net.

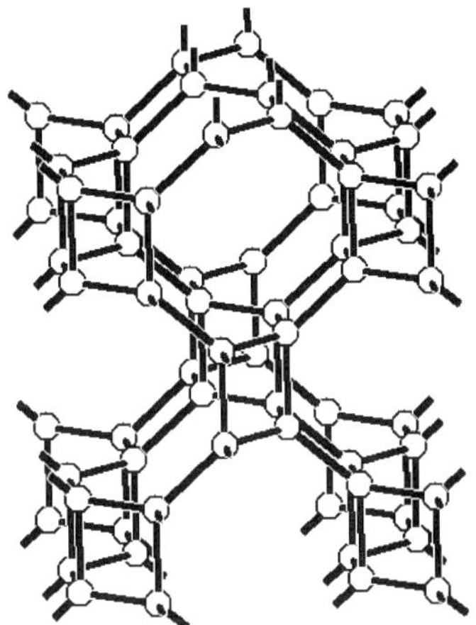

Figure 7.11 The **gis** net with vertex symbol $4\cdot4\cdot4\cdot8_2\cdot8\cdot8$ and genus 9 is another example of a net with a ladder-like theme.

This net can be found in $[Cu_4(CN)_4[\mu_4WSe_4]](N(CH_2CH_3)_4)_2$, see Figure 7.12 [20].

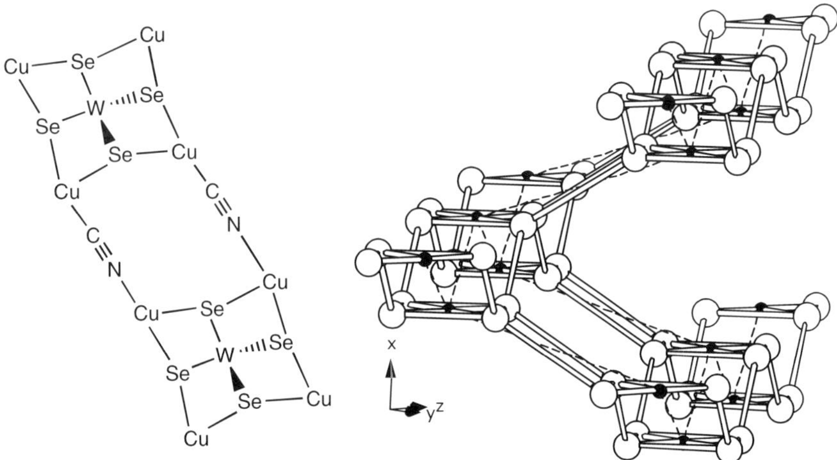

Figure 7.12 The **gis** net in $[Cu_4(CN)_4[\mu_4WSe_4]](N(CH_2CH_3)_4)_2$ [20]. An alternative net assignment using the tungsten atoms (black) giving the diamond **dia** net is shown with dashed lines.

This net assignment leaves the tungsten atoms outside the net. An alternative assignment, also given in Figure 7.12, using only the tungsten atoms as nodes yields the much simpler diamond **dia** net and seems equally valid.

7.1.6. The lonsdaleite or 6^6-**lon** net

This net is named after a rare form of carbon also called "hexagonal diamond" that in its turn was named after British crystallographer Kathleen Lonsdale (1901-1971). The **lon** net has the same vertex symbol as the diamond **dia** net, $6_2·6_2·6_2·6_2·6_2·6_2$, genus 5 (**dia** has genus 3) and can be differentiated numerically from the **dia** net by the C10 value 1027 (**dia** has 981). By inspection one should look for the boat formed six-rings that are absent in the **dia** net, and for the adamantane units that are absent in the **lon** net, see Figure 7.13.

In paranthesis we might add that these two forms of carbon correspond to the two forms of zinc sulphide, wurtzite and zinc blende.

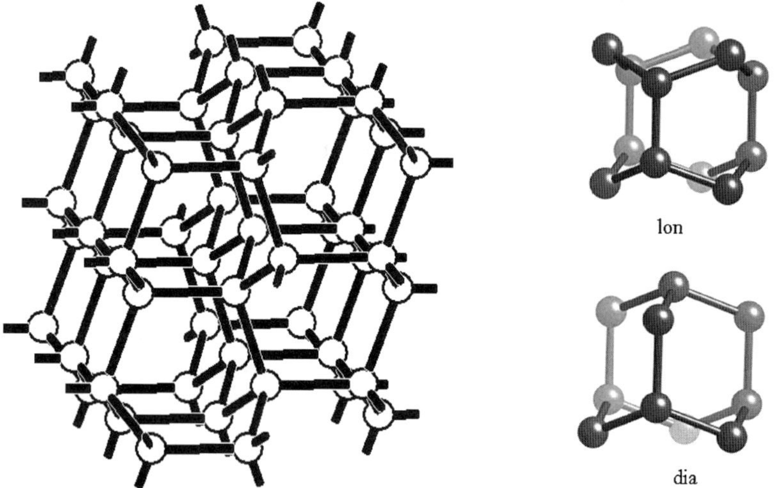

Figure 7.13 The **lon** net (left) has the same vertex symbol as the diamond **dia** net, $6_2·6_2·6_2·6_2·6_2·6_2$. To differentiate between the two, one can look for the boat-conformation of the six-rings that are absent in the **dia** net, and for the adamantane units that are absent in the **lon** net (right).

An example of this net is found in [Mn(N$_3$)$_2$(2-pyridazine)], see Figure 7.14 [21].

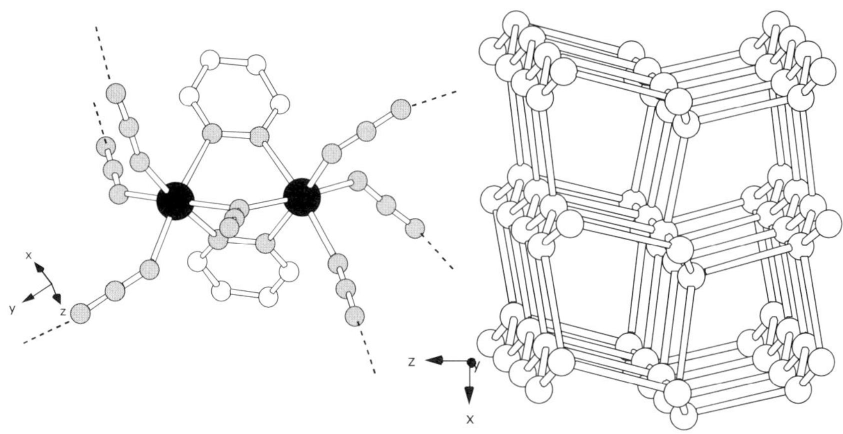

Figure 7.14 [Mn(N$_3$)$_2$(2-pyridazine)] [21] contains the **lon** net.

7.1.7. The Irish ladder or 4^2.6^3.8-***irl*** net

The **irl** or Irish ladder net looks somewhat like the **sra** net, but instead of connecting up or down pair-wise along the ladder, each step of the ladder connects both up and down, giving a slightly puckered form of the ideal net, see Figure 7.15. It has vertex symbol 4·6·4·6·6·10$_{12}$ and genus 5.

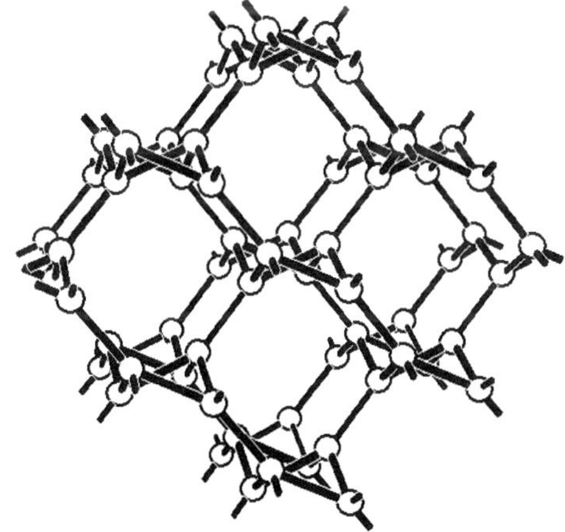

Figure 7.15 The **irl** or Irish ladder net.

An example of this net can be found in the doubly interpenetrated net structure of [Cu(SCN)(1,2-bis(4-pyridyl)ethane)] [22].

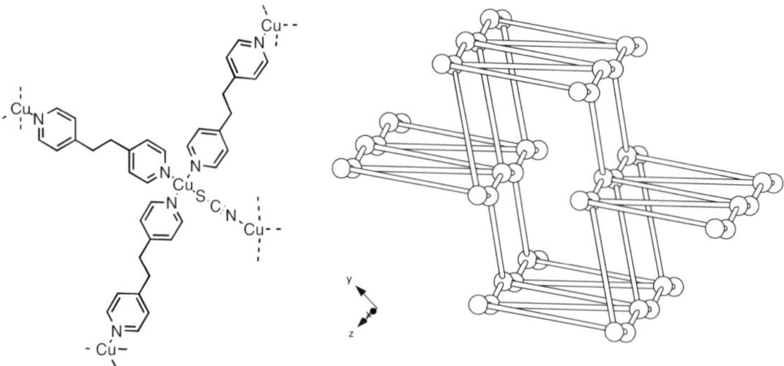

Figure 7.16 The **irl** net in the structure of [Cu(SCN)(1,2-bis(4-pyridyl)ethane)] is doubly interpenetrated, only one net is shown here [22].

7.1.8. The polycubane or $4^3.8^3$-**pcb** net

As the name suggests the polycubane or **pcb** net consists of interconnected cubes, each cube linking to eight other cubes by corner-to-corner links, see Figure 7.17. The vertex symbol for this net is $4 \cdot 8_2 \cdot 4 \cdot 6_2 \cdot 8_2 \cdot 4$, and genus is 9. Note the similarity with the **crb** net (section 7.1.4).

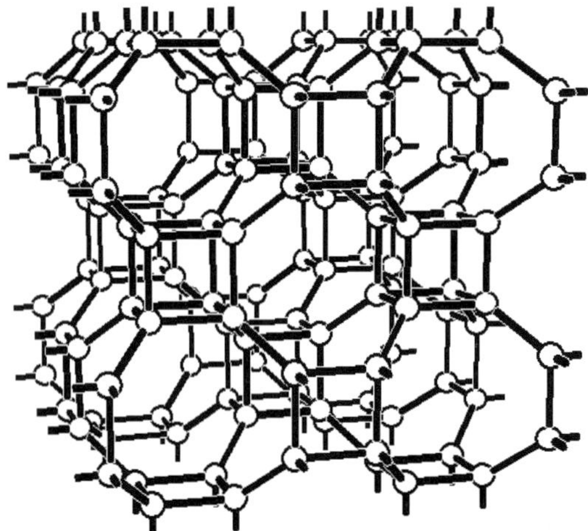

Figure 7.17 The polycubane or **pcb** net consists of interconnected cubes, each cube linking to eight other cubes by corner-to-corner links.

We can find this net in [Co$_2$(2,2'-bipyrimidine)(H$_2$O)$_2$(μ_4-CO$_3$)(μ_2-OH)]NO$_3$·4H$_2$O, see Figure 7.18 [23].

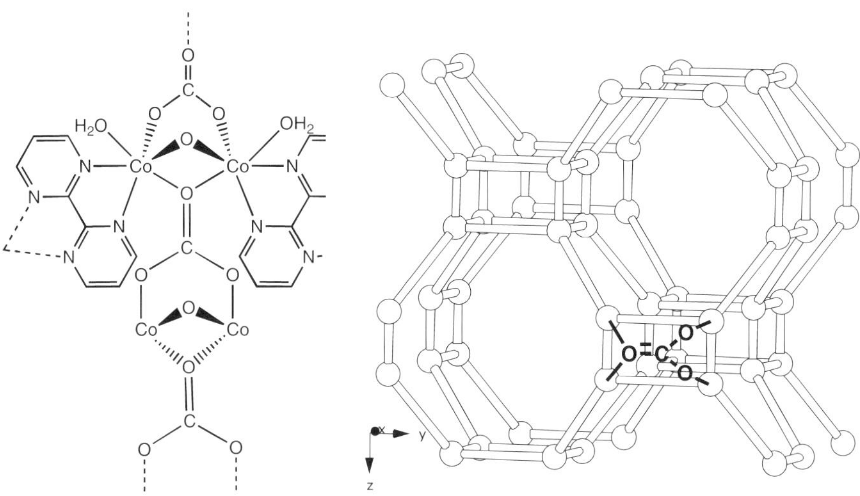

Figure 7.18 The **pcb** net (white) in [Co$_2$(2,2'-bipyrimidine)(H$_2$O)$_2$(μ_4-CO$_3$)(μ_2-OH)] NO$_3$·4H$_2$O [23]. The "boxes" are formed by eight cobalt atoms bridged by four carbonates closing the sides leaving "top" and "bottom" empty. Bipyrimidine ligands bridge the boxes.

7.1.9. The $4^2.6^3.8$-*pcl* net

The **pcl** net takes its name from the mineral Paracelsian, a form of CaAl$_2$Si$_2$O$_8$ (the name means that it is close to another mineral called celsian [24]). The vertex symbol for this net is 4·6·4·6·6·8$_3$ and genus is 9. It is another example of a net with a ladder motif, see Figure 7.19.

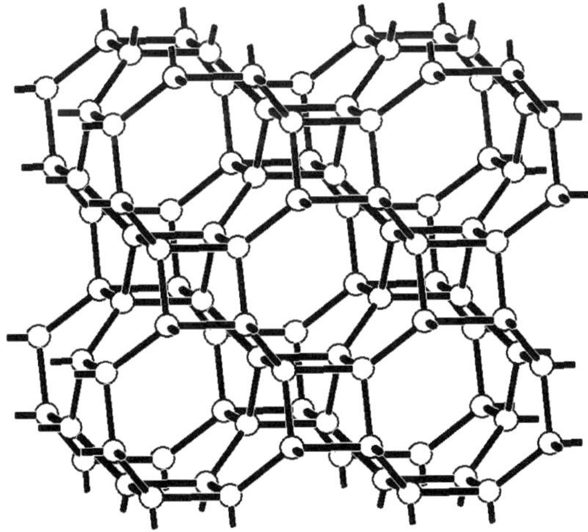

Figure 7.19 The **pcl** net has ladders as a major motif

By using a 2-pyrimidine ligand tetrahedral metal ions can be connected to form moderately strained squares, which is necessity for this net. Such a case is encountered in $[Cu(\mu_2\text{-pyrimidine})_2]BF_4$, see Figure 7.20 [25].

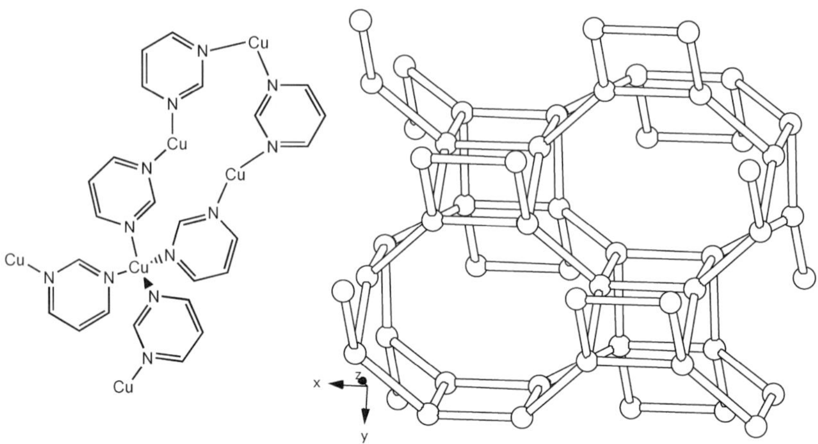

Figure 7.20 $[Cu(\mu_2\text{-pyrimidine})_2]BF_4$ contains the **pcl** net [25].

7.1.10. 6^6-*neb*, $5^4.6^2$-*unh*, 4.6^5-*cag*, 6^6-*gsi*, 6^6-*mmt*, and $4.6^4.8$-*zni*

There are a number of other "tetrahedral" nets with quoted examples in Ockwig et al. [6] but since there are only one or two examples of each net, we will not treat them in any detail. A summary of the uninodal nets is shown in Table 7.1 and in Figure 7.21.

Table 7.1 Summary of very rare four-connected uninodal nets [6]. These nets are shown and described in Figure 7.21.

Net	Vertex Symbol	Genus	Chiral	Ref.
neb	$6\cdot6\cdot6\cdot6_2\cdot6_2\cdot6_2$	5	no	[26]
unh	$5\cdot5\cdot5\cdot5\cdot12_2\cdot12$	7	yes	[27]
cag	$4\cdot6_2\cdot6\cdot6\cdot6\cdot6$	9	no	[28,29]
gsi	$6\cdot6_2\cdot6\cdot6_2\cdot6\cdot6_2$	9	no	[30]
mmt	$6\cdot6\cdot6\cdot6_2\cdot6_2\cdot6_2$	9	no	[31]
zni	$4\cdot6\cdot6\cdot6_3\cdot6_2\cdot12_{40}$	17	no	[32]

In addition eight binodal, two trinodal and one pentanodal net are mentioned by Ockwig et al. [6]

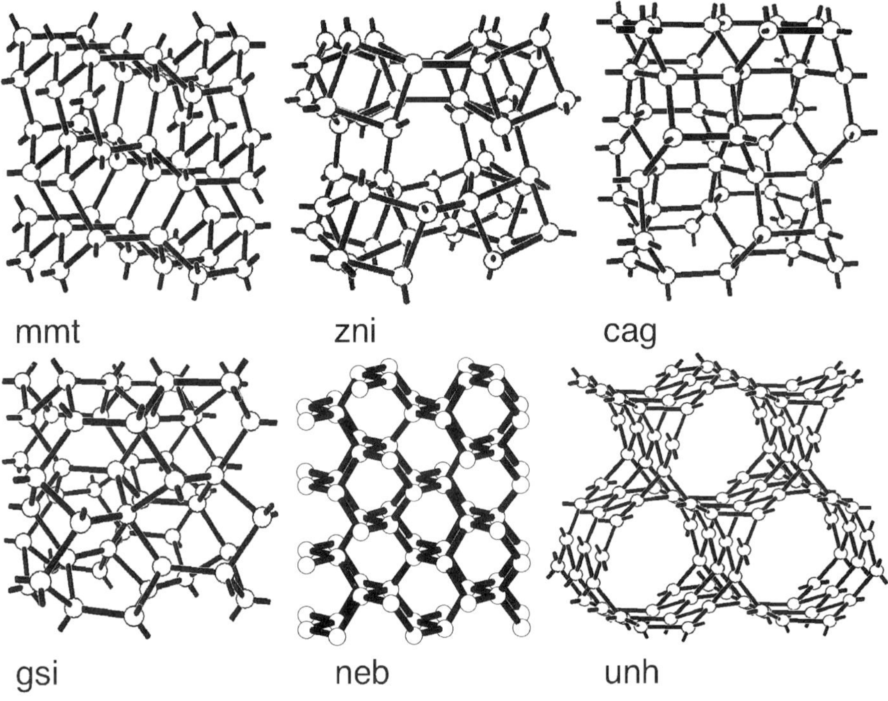

Figure 7.21 Some rare uninodal four-connected nets, see Table 7.1. Some characteristics of these nets are: **mmt**; interconnected sheets of six-rings in boat configuration, **zni**; columns of squares linked by two corners and interlinked by the remaining two, **cag**; parallel interlinked crankshafts, **gsi**; four-fold helices of opposing chirality, each helix links to six other helices, **neb**; parallel zigzag bands, **unh**, chiral, five-rings.

7.2. Uninodal nets with "square planar" nodes

We will now turn to the nets based on square planar units, and we refer the reader to Chapter 5 for the **nbo** net. Then only two nets of this type remains, the **lvt** net and the somewhat unique **tcb**.

7.2.1. The $4^2.8^4$-**lvt** net

The **lvt** net has vertex symbol $4{\cdot}4{\cdot}8_4{\cdot}8_4{\cdot}8_8{\cdot}8_8$ and genus 5, see Figure 7.22.

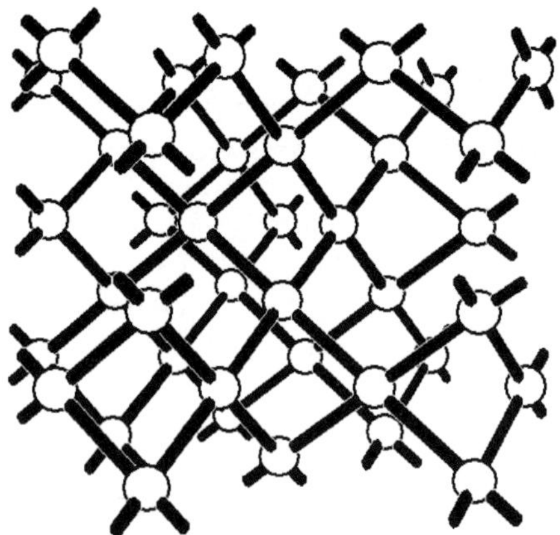

Figure 7.22 The **lvt** net has vertex symbol $4 \cdot 4 \cdot 8_4 \cdot 8_4 \cdot 8_8 \cdot 8_8$ and genus 5.

An example of this net is found in bis(3-cyano-2,4-pentanedionato)cobalt(II) shown in Figure 7.23 [33]. This structure contains two interpenetrated **lvt** nets.

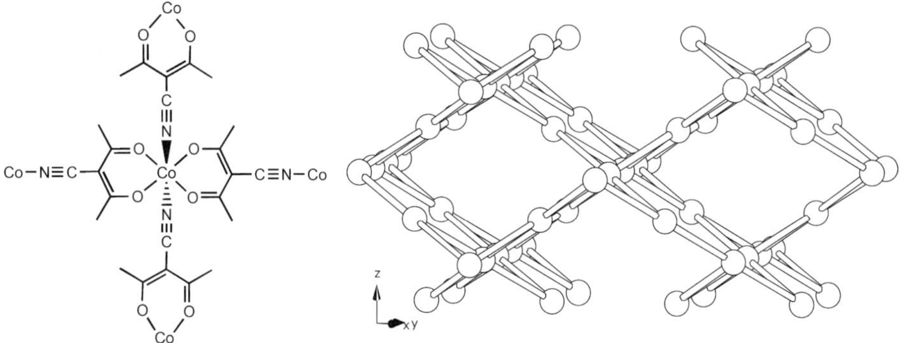

Figure 7.23 Bis(3-cyano-2,4-pentanedionato)cobalt(II) [33] contains two interpenetrated **lvt** nets.

7.2.2. The 8^6-**tcb** net

The **tcb** net has only 8-rings as smallest rings, and O'Keeffe has pointed out that "no other 4-coordinated net is known in which the smallest ring is greater than a 7-ring" see Figure 7.24 [34]. It is also self-interpenetrated, that is the smallest rings are catenated, and the distance along the links are not the shortest node-node distances in the structure. The **tcb** net has vertex symbol $8_2 \cdot 8_2 \cdot 8_5 \cdot 8_5 \cdot 8_5 \cdot 8_5$ and genus 5.

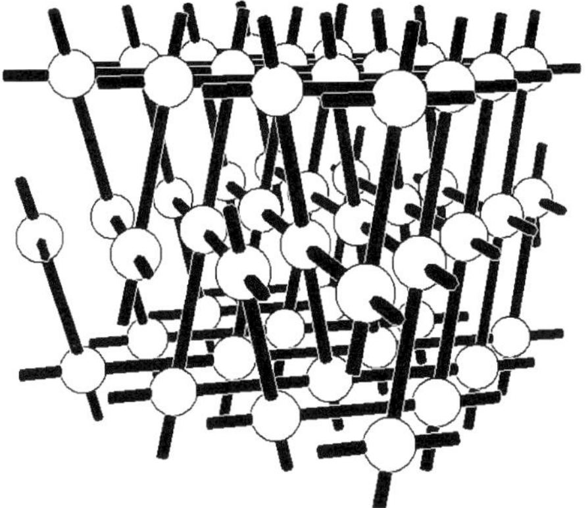

Figure 7.24 The **tcb** net is unique among the four-connected nets in that the smallest ring is greater than a 7-ring. It is also self-interpenetrating.

This net was found independently by two groups in the compound [Cu(1,1'-(1,4-butanediyl)bis(imidazole))(*m*-phthalate)(H$_2$O)]·5H$_2$O, [35] and in [Ni(3-(3-pyridyl)acrylate)$_2$(H$_2$O)$_2$] [36] see Figure 7.25. The latter structure two crystallographically unique (but chemically similar) Ni atoms.

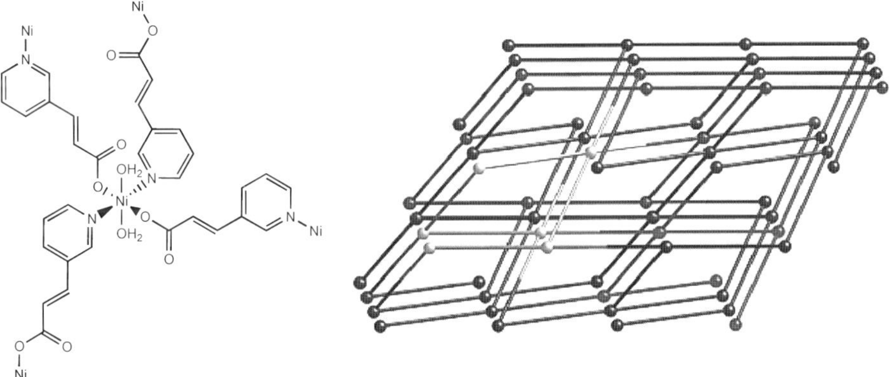

Figure 7.25 The **tcb** net in [Ni(3-(3-pyridyl)acrylate)$_2$(H$_2$O)$_2$] Reprinted with permission from [36]. Copyright 2003 American Chemical Society.

7.3. Nets with both tetrahedral and square planar nodes

These nets are by definition at least binodal, and the **pts** net was treated in Chapter 5. We will look at three more of these nets here.

                                              L. Öhrström & K. Larsson

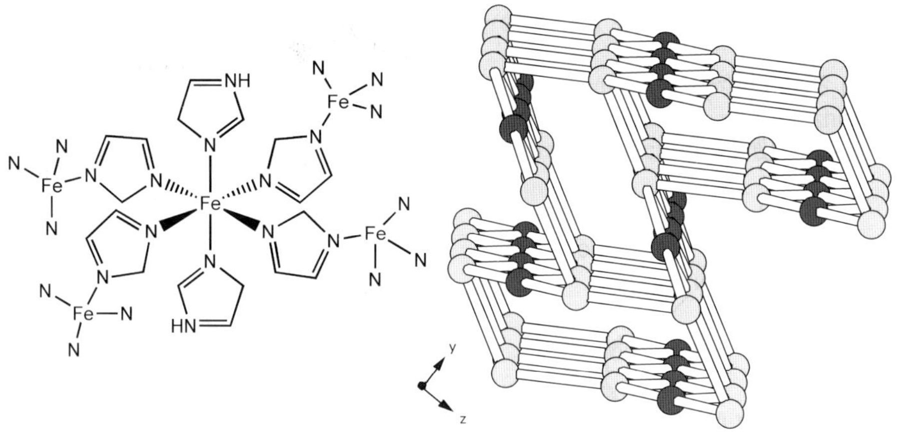

Figure 7.30 The **mog** net in [Fe[Fe(imidazolyl)$_4$(imidazole)$_2$]$_2$] [32].

7.3.3. The $(4^3.6^3)_4(6^4.10^2)$-**asv** net

In the **asv** net cubes are linked via square planar nodes, see Figure 7.31 (in the polycubane, **pcb** net in Figure 7.17 the cubes were directly linked). The vertex symbols are 4·6·4·6·4·6(tetrahederal), 6·6·6$_2$·6$_2$12$_8$12$_8$(square planar) and genus is 11.

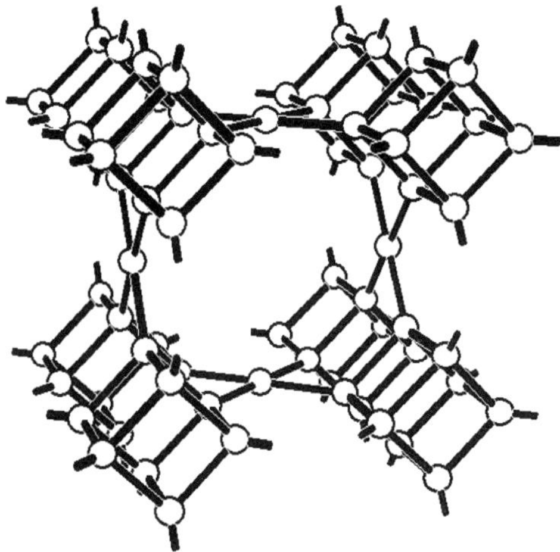

Figure 7.31 The **asv** net contains cubes linked via square planar nodes.

Such a net can be broken down into the cubic parts and their linkers. This was done when $Cd_8(SC_6H_5)_{12}^{4+}$ clusters were linked together with 1,2,4,5-tetra(4-pyridyl)benzene to give the compound [$Cd_8(SC_6H_5)_{12}$(1,2,4,5-tetra(4-

pyridyl)benzene)$_2$](SO$_4$)(HSO$_4$)$_2$(H$_2$O)$_4$, see Figure 7.32 [39]. This material can be regarded as a "hybrid inorganic-organic open framework chalcogenide" and in this context we would like to mention that zeolites are not the only porous "inorganic" materials of interest, research on porous chalcogenides and phosphates is also actively pursued [40-43].

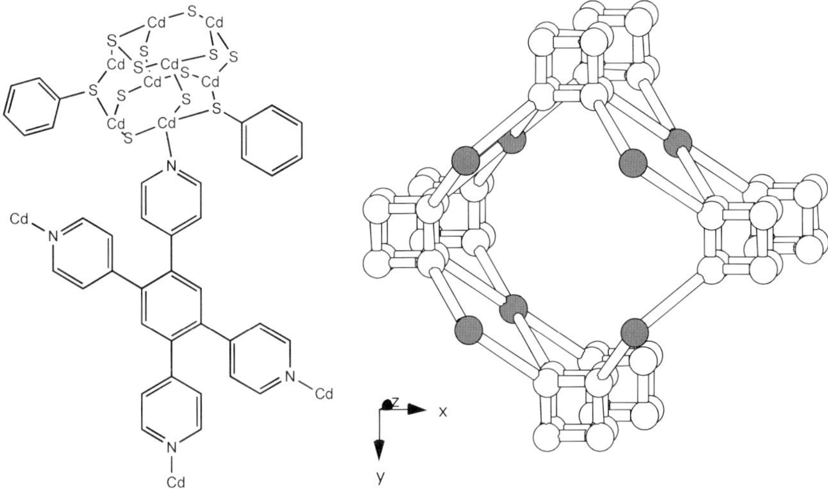

Figure 7.32 The "hybrid inorganic-organic open framework chalcogenide" [Cd$_8$(SPh)$_{12}$(1,2,4,5-tetra(4-pyridyl)benzene)$_2$](SO$_4$)(HSO$_4$)$_2$(H$_2$O)$_4$ form the **asv** net.

[30] T. Kitazawa, Mol.Cryst.Liq.Cryst.Sci.Technol.,Sect.A 276 (1996) 167.

[31] H. Kinoshita, A. Ouchi, Bull. Chem. Soc. Jpn 59 (1986) 3495.

[32] R. Lehnert, F. Seel, Z.Anorg.Allg.Chem 444 (1978) 91.

[33] O. Angelova, J. Macicek, M. Atanasov, G. Petrov, Inorg. Chem. 30 (1991) 1943.

[34] M. O'Keeffe, O. M. Yaghi, Reticular Chemistry Structure Resource, Tucson, Arizona State University, 2005, http://okeeffe-ws1.la.asu.edu/RCSR/home.htm

[35] J. F. Ma, J. Yang, G. L. Zheng, L. Li, J. F. Liu, Inorg. Chem. 42 (2003) 7531.

[36] M. L. Tong, X. M. Chen, S. R. Batten, J. Am. Chem. Soc. 125 (2003) 16170.

[37] K. M. Park, T. Iwamoto, J. Chem. Soc., Chem. Commun. (1992) 72.

[38] K. M. Park, T. Iwamoto, J. Chem. Soc., Dalton Trans. (1993) 1875.

[39] N. F. Zheng, X. H. Bu, P. Y. Feng, J. Am. Chem. Soc. 124 (2002) 9688.

[40] F. Liebau, Microporous Mesoporous Mater. 58 (2003) 15.

[41] C. N. R. Rao, S. Natarajan, R. Vaidhyanathan, Angew. Chem. Int. Ed. 43 (2004) 1466.

[42] K. Maeda, Microporous Mesoporous Mater. 73 (2004) 47.

[43] P. Y. Feng, X. H. Bu, N. F. Zheng, Acc. Chem. Res. 38 (2005) 293.

Chapter 8

Nets with both three- and four-connected nodes

Since most scientists working with network synthesis have used systems that either have forced the nodes to have the same connectivity or left this as an "option" to the chemical system to work out by itself, often yielding the same result, there are relatively few examples of nets with both three and four connected nodes.

From our experiences in Chapters 5 to 7 we would expect 3,4-connected nets from preformed organic ligands with tetrahedral linking possibilities and metal ions that can form trigonal linking complexes (either by trigonal coordination, or by blocking additional coordination sites with other ligands, see Figure 6.5) or vice versa, see Figure 8.1. Note, however, that even the best of such designs may fail, for example we have seen the hexamethylenetetramine (in B below) turn up as a trigonal node in $[Ag_4(hmt)_3(H_2O)](PF_6)_4 \cdot 3EtOH$, [1] see Figure 6.27.

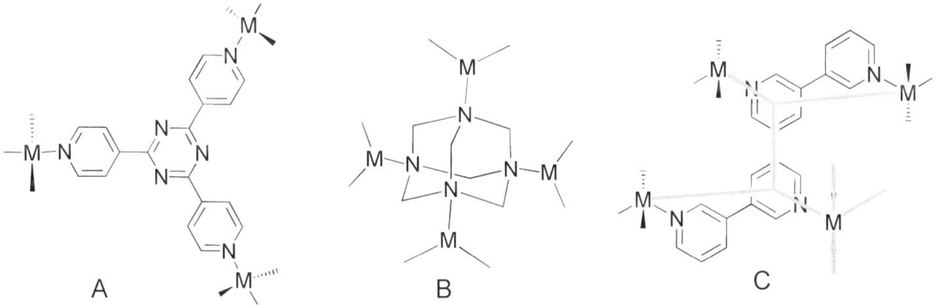

Figure 8.1 Some of the (endless) possibilities to design 3,4-connected nets. A: Triangular linking ligands with metal ions with tetrahedral coordination. B: Tetrahedral linking ligands with trigonal metal ions. C: Partly preformed trigonal nodes enable nets that do not have alternating trigonal and tetrahedral nodes.

As for the four-connected nets we can make a distinction between approximately tetrahedral four-connected nodes and square planar nodes.

We can also divide these nets into those that have all three-connected nodes binding to three four-connected nodes and all four-connected nodes binding to four three-connected nodes (alternating 3,4-nets, A and B in Figure 8.1) and all other combinations, for example as in C in Figure 8.1.

This in its turn has consequences for the stoichiometry of the nets. In the alternating nets every three-connected node binds to three four-connected nodes that it shares with three other three-connected nodes, giving the stoichiometric ratio ns_3/ns_4:

$$ns_4 = \frac{3 \cdot ns_3}{4} \Rightarrow \frac{ns_3}{ns_4} = \frac{4}{3} \tag{8.1}$$

Thus these compounds must have the formula (three-connected nodes)$_4$(four-connected nodes)$_3$ and we usually designate them with the short formula such as $(6^3)_4(6^2.8^4)_3$ for the boracite or **bor** net that we will look at next.

8.1. Alternating nets with stoichiometry (3-conn)$_4$(4-conn)$_3$ ($ns_3/ns_4 = 4/3$)

8.1.1. Boracite, $(6^3)_4(6^2.8^4)_3$-**bor** and twisted boracite $(6^3)_4(6^2.8^2.10^2)_3$-**tbo**

The **bor** net has vertex symbols 6·6·6, $6_2·6_2·8·8·8·8$ and genus 6 and with short notation this becomes $(6^3)_4(6^28^4)_3$ (taking the stoichiometry into account by the subscripts). Two possible symmetric versions of this net are shown in Figure 8.2, in one we have perfect tetrahedral nodes and in the other we have perfect trigonal nodes, it is not possible to have both at the same time.

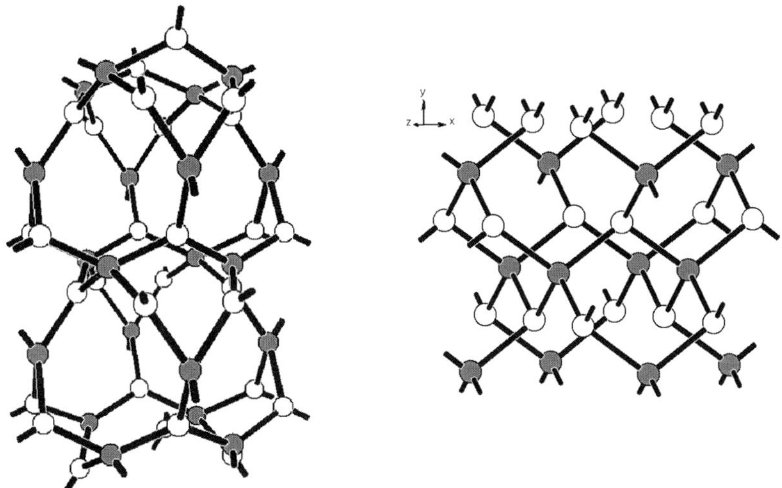

Figure 8.2 The boracite or $(6^3)_4(6^2.8^4)_3$ **bor** net in two configurations, left: perfect tetrahedral nodes and right: perfect trigonal nodes.

A hydrogen bonded version of this net was found in $[B(OCH_3)_4]_3[C(NH_2)_3]_4X \cdot nS$, X = Cl$^-$ or PF$_6^-$, S = H$_2$O or CH$_3$OH, see Figure 8.3 [2]. Note that we have already seen the guanidinium ion, $(C(NH_2)_3)^+$, play a secondary role in the sodalite **sod** net formed by Cu(II) and carbonates (Figure 8.8), here it is clearly directing the structure. For other examples of the use of the guanidinium ion in hydrogen bonded nets see for example reports by Ward and collaborators [3,4].

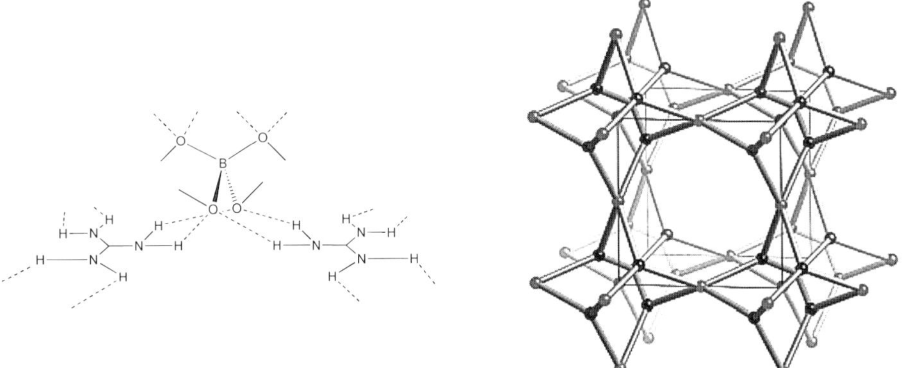

Figure 8.3A The hydrogen-bonded **bor** net in $[B(OCH_3)_4]_3[C(NH_2)_3]_4X \cdot nS$, X = Cl$^-$ or PF$_6^-$, S = H$_2$O or CH$_3$OH. Each guanidinium cation (black) binds to three borate methyl ester anions and each borate ester (grey) binds pair wise to four guanidinium ions. Reprinted with permission from [2]. Copyright 2005 American Chemical Society.

Note that the "cages" in the **bor** net is arranged in a **pcu** fashion and that this couold be an alternative assignment. There is also a **bor** net based on a copper triazine coordination polymer [5]. When the four-connected nodes are square planar this net turns into the twisted boracite net, $(6^3)_4(6^2.8^2.10^2)_3$-**tbo**, found in $[Cu_3(benzene-1,3,5-tricarboxylate)_2(H_2O)_3]$ [6].

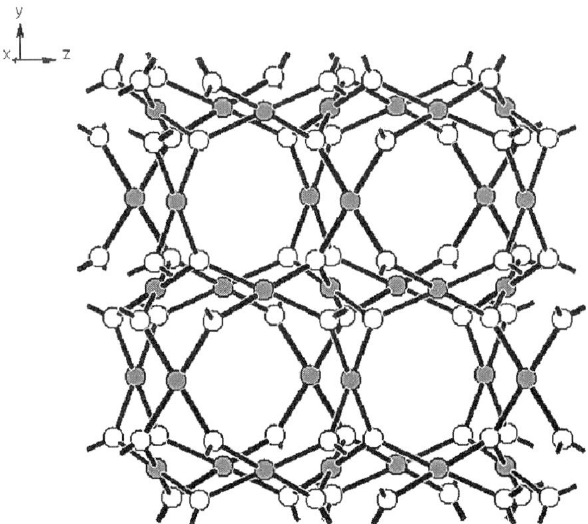

Figure 8.3B The **tbo** net has vertex symbols 6·6·6 and $6_2 \cdot 6_2 \cdot 8_2 \cdot 8_2 \cdot 12_2 \cdot 12_2$ and genus 11.

8.1.2. *The Pt₃O₄ or (8³)₄(8⁶)₃-**pto** net*

This is a net that can be constructed from perfectly symmetrical square planar and trigonal nodes and it is named after the arrangements of Pt and O atoms in $Na_xPt_3O_4$, see Figure 8.4. The **pto** net has vertex symbols $8_5 \cdot 8_5 \cdot 8_5$, $8_2 \cdot 8_2 \cdot 8_4 \cdot 8_4 \cdot 8_4 \cdot 8_4$ and genus 11.

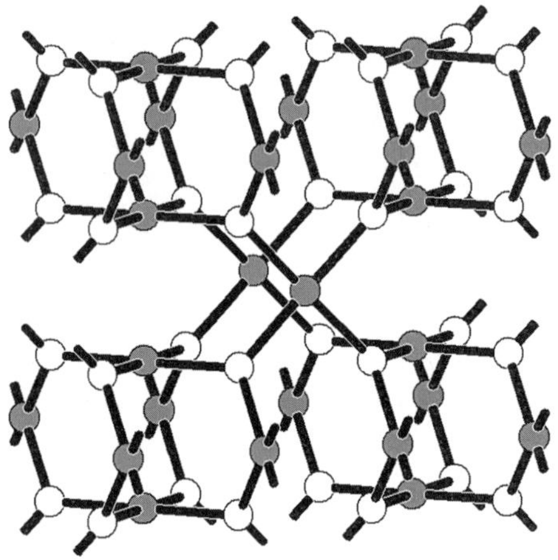

Figure 8.4 The **pto** net can be constructed from perfectly symmetric square planar and trigonal nodes.

An example is found in [Cu₃(4,4,4″-benzene-1,3,5-triyl-tribenzoate)₂(H₂O)] ·9DMF·2H₂O, [7] see Figure 8.5.

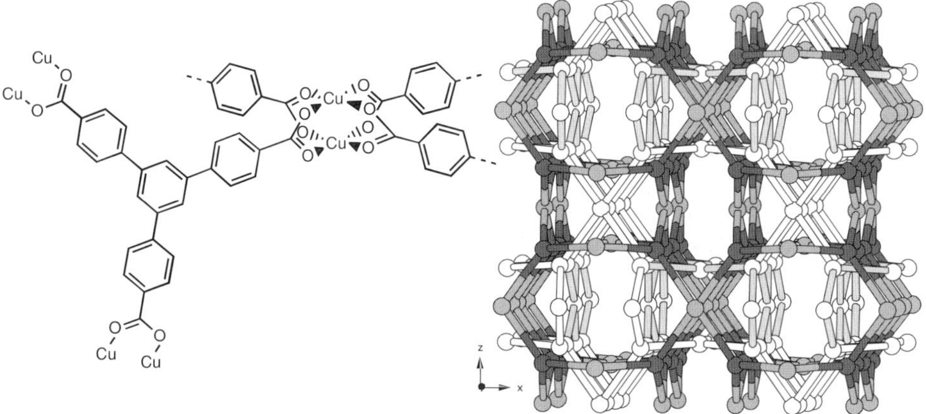

Figure 8.5 Doubly interpenetrated (second net in white) **pto** nets in [Cu₃(4,4,4″-benzene-1,3,5-triyl-tribenzoate)₂ (H₂O)]·9DMF·2H₂O. The solvent guests can be removed with the network kept intact and gases and other guest molecules can be reversibly sorbed [7].

8.1.3. The C_3N_4 or $(8^3)_4(8^6)_3$-**ctn** net

The **ctn** net is named after the C_3N_4 structure and can be constructed from near symmetric tetrahedral and trigonal nodes, see Figure 8.6. It has vertex symbols $8_5 \cdot 8_5 \cdot 8_5$, $8_3 \cdot 8_3 \cdot 8_3 \cdot 8_3 \cdot 8_4 \cdot 8_4$ and genus 11.

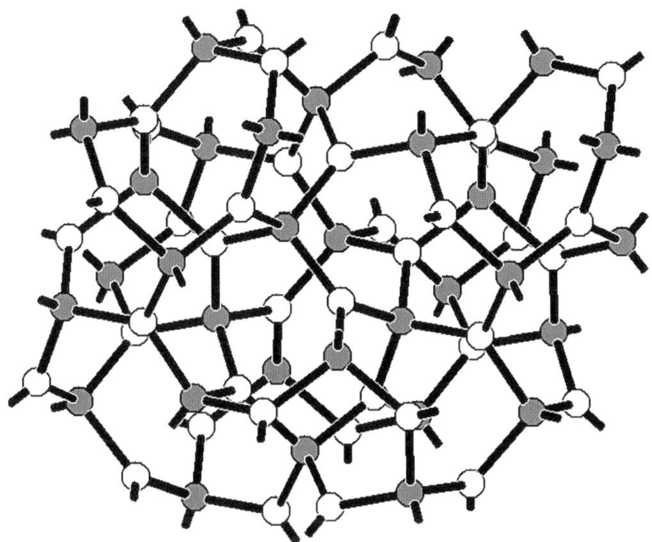

Figure 8.6 The **ctn** net is named after the C_3N_4 structure and can be constructed from close to symmetric tetrahedral and trigonal nodes.

We meet an interesting contradiction of sorts with this net; it is clearly less symmetric than the others we have seen of this sort, yet it contains the most symmetric nodes! For all the nets with only one type of connectivity, from two to six, it is possible to construct a high symmetry net (genus 3-4) but this seems impossible for the 3,4-connected nets with alternating connectivities.

The **ctn** net can be found in [Cu$_3$(2,4,6-tris(4-pyridyl)1,3,5-triazine)$_4$](BF$_4$)$_3\cdot$2/3(2,4,6-tris(4-pyridyl)1,3,5-triazine)$\cdot$5H$_2$O, see Figure 8.7 [8]. This appears to be the first reported synthesis of a 3D-net from coordination polymers using ionic liquids, and the authors note that the "With their poor coordination ability ionic liquids may also be excellent and safe media for the synthesis of metal–organic frameworks" [8]. More "normal" reaction conditions using the same reagents gave a **bor** net [5].

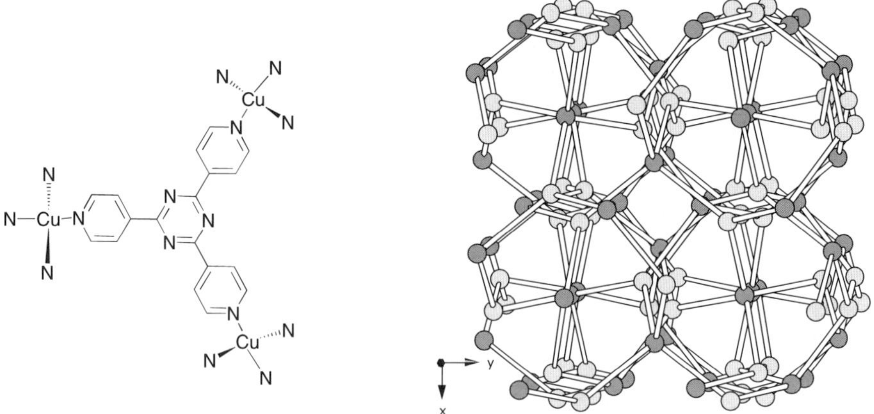

Figure 8.7. The **ctn** or C_3N_4 net can be found doubly interpenetrating in [Cu$_3$(2,4,6-tris(4-pyridyl)1,3,5-triazine)$_4$](BF$_4$)$_3$·2/3(2,4,6-tris(4-pyridyl)1,3,5-triazine)·5H$_2$O [8].

8.1.4. Other alternating nets with stoichiometry (3-conn)$_4$(4-conn)$_3$

In one of his very last scientific articles Wells goes back to the three- and four-connected nets and presents six different alternating nets [9]. Of these, "net 1" is the **bor** net, "net 6" is the **pto** net (the space group symbol is misprinted in the article, the net is also known as (8, 3;4)-a) and the **ctn** net does not appear at all in Wells' work. The "nets 3" and "5" are presented in Figure 8.8 and while "5" is equivalent to the **sln** (β-Si$_3$N$_4$) net, "nets 3" is not yet found in the RCSR database.

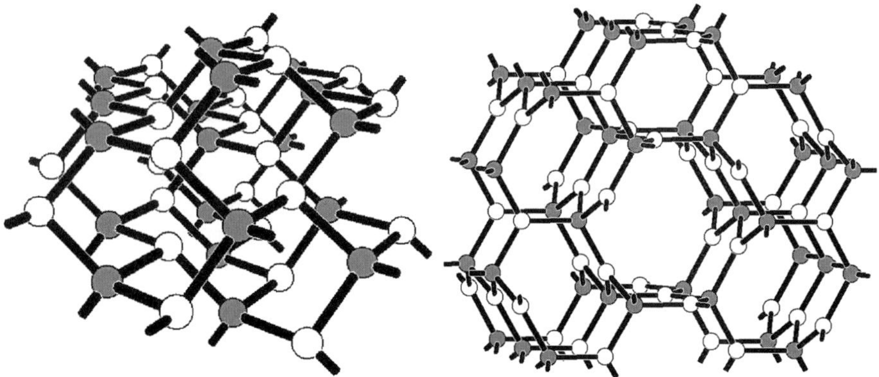

Figure 8.8. Left: Wells "net 3" from ref [9] note that this net can be derived from the **dia** (diamond) net if 1/8 of the tetrahedral nodes are deleted. Right: Wells "net 5" that is the same as the **sln** net.

8.2. Nets with 1:1 stoichiometery ($ns_3/ns_4 = 1$)

These nets contain one link between the four-connected nodes while the three-connected nodes bind only to four-connected nodes.

8.2.1. The InS or $(6^3)(6^5.8)$-**ins** net.

The **ins** net has vertex symbols $6\cdot6\cdot6_2$, $6\cdot6\cdot6\cdot6\cdot6\cdot10_6$ and genus 7 and is shown in Figure 8.9. It has in its most symmetric form trigonal pyramidal nodes and distorted plane-trigonal nodes.

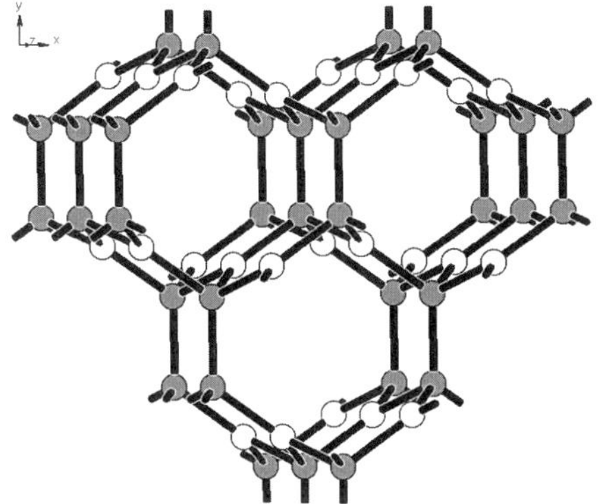

Figure 8.9 The $(6^3)(6^5.10)$ **ins** net

This net is found doubly interpenetrated in [Ag(tricyanomethanide)(phenazine)$_{0.5}$], see Figure 8.10 [10]. Another example is found in [Ag$_3$(hexamethylenetetramine)$_2$](ClO$_4$)$_3\cdot$2H$_2$O as indicated in Figure 8.1B, [11] indeed varying the stoichiometry and the anion in this system gives a range of different coordination polymers with 2D- and 3D-nets [12,13].

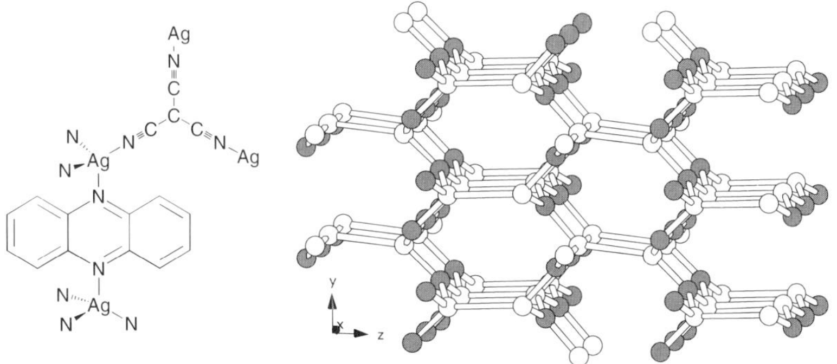

Figure 8.11 One of the two interpenetrated **ins** nets in [Ag(tricyanomethanide)(phenazine)$_{0.5}$] [10].

8.2.2. The $(4.8^2)(4.8^5)$-**dmc** net

The **dmc** net has vertex symbols $4 \cdot 8_3 \cdot 8_3$, $4 \cdot 8_2 \cdot 8_2 \cdot 8_3 \cdot 8_2 \cdot 8_3$ and genus 7 and is shown in Figure 8.12.

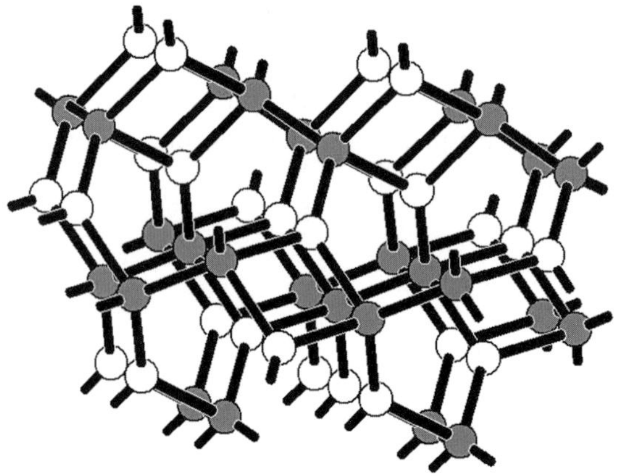

Figure 8.12 the **dmc** net.

This   net   has   been   assigned   [14]   in   the   structure   of $[Zn(HPO_3)H_2N(CH_2)_2NH_2)_{0.5}]$ [15] as seen in Figure 8.13.

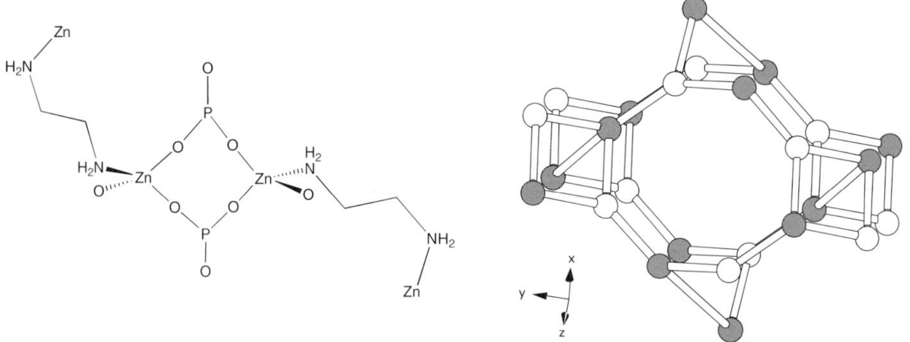

Figure 8.13 The **dmc** net has been assigned [14] in the structure of $[Zn(HPO_3)H_2N(CH_2)_2NH_2)_{0.5}]$ [15] Zn nodes are grey and phosphorous nodes are white.

However, one might argue that the four-rings formed are too small to be considered as rings in this context. The Zn-Zn and P-P separations are 4.52 Å and 4.67 Å respectively and it is thus impossible to fit but the smallest atoms into the centre of the four-ring, see Figure 8.14A. Alternatively, one could consider putting the nodes at the centroids of these rings, and thus obtaining a six-connected **pcu** (primitive cubic packing) net (section 5.2.8), see Figure 8.14B.

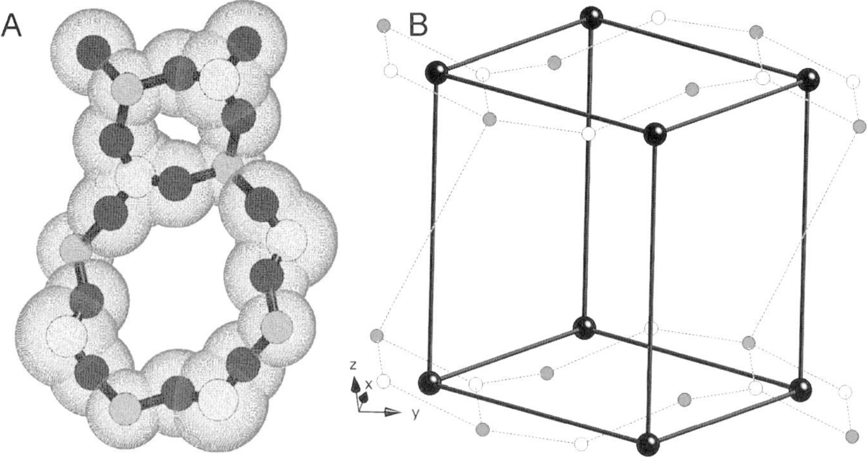

Figure 8.14 A: Shows the tiny hole (1.7 Å · 0.8 Å) formed by the -Zn-O-O-Zn-O-O- six-atom rings when atomic van der Waals radii are used to generate surfaces. B: An alternative net assignment of [Zn(HPO$_3$)H$_2$N(CH$_2$)$_2$NH$_2$)$_{0.5}$] [15] puts six-connected nodes at the centres of the -Zn-O-O-Zn-O-O- six-atom rings give a **pcu** (primitive cubic packing, section 5.2.8) net.

8.3. Nets with 1:2 stoichiometery ($ns_3/ns_4 = 1/2$)

These nets contain one link between the three-connected nodes while the four-connected nodes bind to two three-connected nodes and two four-connected nodes.

8.3.1. The $(5.8^2)(4.5^2.6.7.8)_2$-dme net

The **dme** net has vertex symbols 5·8·8, 4·5·5·6·8·10 and genus 6 and is shown in Figure 8.15. This looks like a nice and symmetric net, which it is, but it contains a multitude of different ring sizes as is evident from the vertex symbols. We have not found any examples of this net in the literature, but on the other hand, as emphasised in Chapter 1, we have not done a comprehensive search of the CSD database.

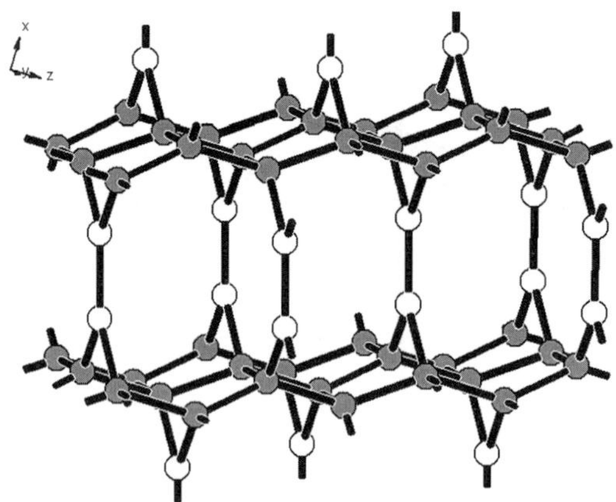

Figure 8.15 The **dme** net contains a multitude of different ring sizes as is evident from the vertex symbols 4·8₃·8₃ and 4·8₂·8₂·8₃·8₂·8₃.

8.3.2. A (6.8²)(6⁴.8.10)₂-net

What we have found however, is a net with vertex symbols $6_2 \cdot 8_2 \cdot 8_2$ and $6 \cdot 6 \cdot 6 \cdot 6 \cdot 8_2 \cdot 10_8$, in the compound [Cd₃(hexamethylenetetramine)₂(maleato)₂(H₂O)₆] SO₄·4H₂O, [16], see Figure 8.16.

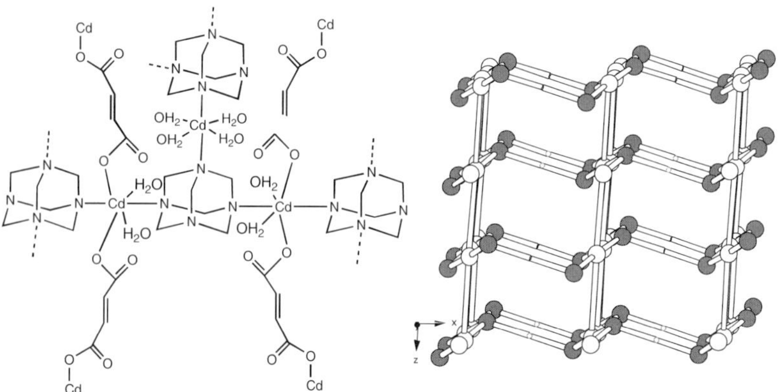

Figure 8.16 A (6 8²)(6⁴8 10)₂ net found in [Cd₃(hexamethylenetetramine)₂(maleato)₂(H₂O)₆] SO₄·4H₂O, [16]

8.4. Nets with 4:1 stoichiometery ($ns_3/ns_4 = 4$)

These nets contain two links between the three-connected nodes while the four-connected nodes bind only to three-connected nodes.

8.4.1. The $(6^2.10)_4(6^4.10^2)$-**jph** net

The **jph** net has vertex symbols $6\cdot6\cdot10_2$ and $6\cdot6\cdot6\cdot6\cdot12_2\cdot12_2$, and genus 7. Here we meet again a reoccurring theme from Chapter 6, the four-fold helices, this time forming a racemic net, see Figure 8.17. Note that square planar nodes link the helices.

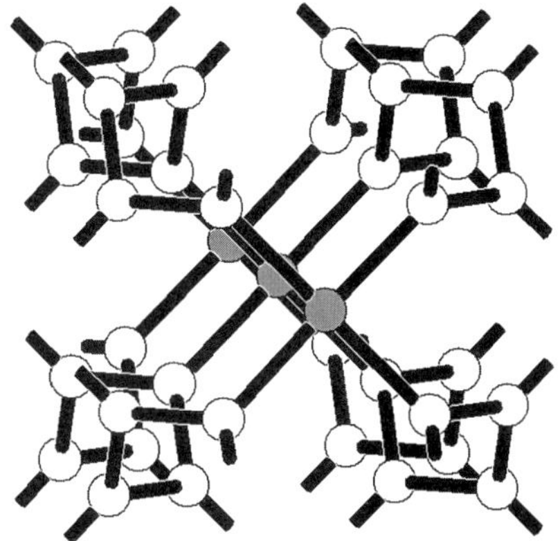

Figure 8.17 The **jph** net has four-fold helices, as so many three-connected nets we saw in Chapter 6. Note the different chirality of neighbouring helices giving a racemic net.

An interesting variation of this net with tetrahedral nodes linking the helices is found in the diaminotriazine substituted pentaerythrityltetraphenylether shown in Figure 8.18 [17]. The net assignment of this structure was discussed in section 3.3 and we favour a description based on three- and four-connected nodes. In this net tetrahedral nodes instead of square planar nodes link the helices.

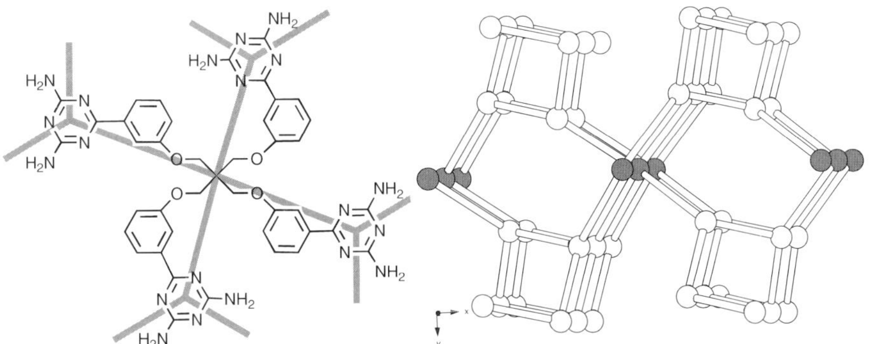

Figure 8.18 Tetrahedral nodes link the helices in this variation of the **jph** net found in the diaminotriazine substituted pentaerythrityltetraphenylether. An alternative assignment strictly based on molecular or tecton connectivity gives a diamond **dia** net [17].

8.4.2. The $(10^3)_4(10^6)$-**dmf** net

The **dmf** net contains ten-rings only, see Figure 8.19. It has vertex symbols $10_3 \cdot 10_4 \cdot 10_6$ and $10_2 \cdot 10_2 \cdot 10_3 \cdot 10_4 \cdot 10_4 \cdot 10_4$, and genus 4. Note also here the characteristic four-fold helices, this time all with the same chirality, making the **dmf** net chiral.

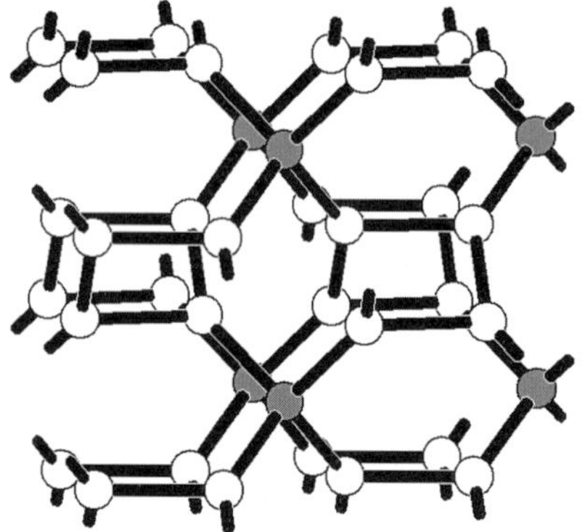

Figure 8.19 The **dmf** net also has four-fold helices although all with the same chirality, making this net chiral.

An example is found in the doubly interpenetrated structure of $K_2[CdCu_4(CN)_8]\cdot 2.5H_2O$, containing nets of different chirality, (thus racemic) see Figure 8.20 [18].

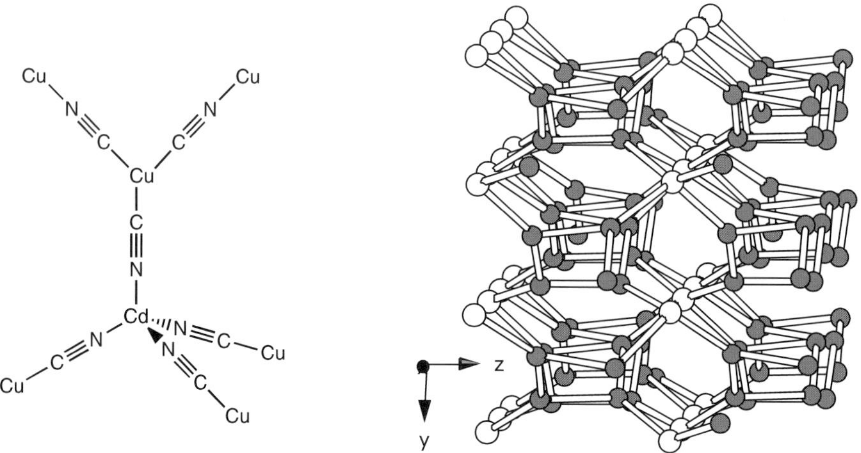

Figure 8.20 The **dmf** net is found doubly interpenetrated in the structure of $K_2[CdCu_4(CN)_8]\cdot$ $2.5H_2O$ [18].

8.5. Nets with 2:1 stoichiometery ($ns_3/ns_4 = 2$)

These nets contain one link between the three-connected nodes while the four-connected nodes bind only to three-connected nodes. Thus the stoichiometry becomes:

$$ns_4 = \frac{2 \cdot ns_3}{4} \Rightarrow \frac{ns_3}{ns_4} = \frac{2}{1} \qquad (8.2)$$

Or alternatively we have two links between the three-connected nodes while the four-connected nodes bind to two three-connected and two four connected nodes.

In this class we find two high symmetry nets with genus 3 that we nevertheless have not found any molecular examples of. One of these is based on tetrahedral, the other on square planar four-connected nodes.

8.5.1. The $(8^3)(8^6)_2$-**tfa** net based on tetrahedral nodes

The **tfa** net contains only eight-rings, and has vertex symbols $8_4 \cdot 8_4 \cdot 8_4$ and $8_2 \cdot 8_2 \cdot 8_3 \cdot 8_3 \cdot 8_3 \cdot 8_3$, and genus 3, see Figure 8.21.

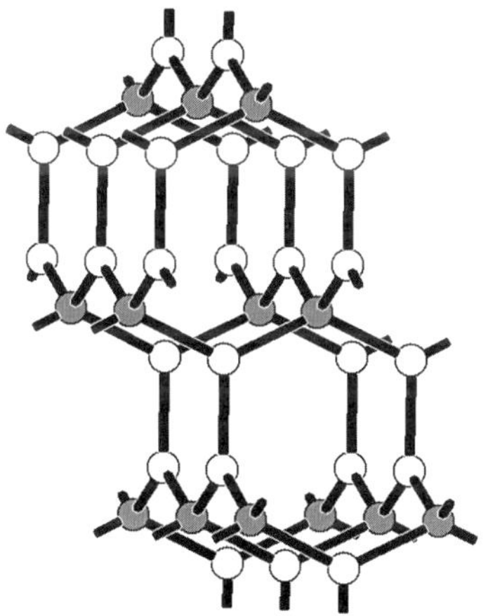

Figure 8.21 The **tfa** net.

8.5.2. The $(8^3)(8^5.12)_2$-**tfc** net based on square planar four connected nodes

The **tfc** net has vertex symbols $8 \cdot 8_3 \cdot 8_3$ and $8_2 \cdot 8_2 \cdot 8_2 \cdot 8_2 \cdot 8_2 \cdot {}^*$, genus 3 and is plotted in Figure 8.22.

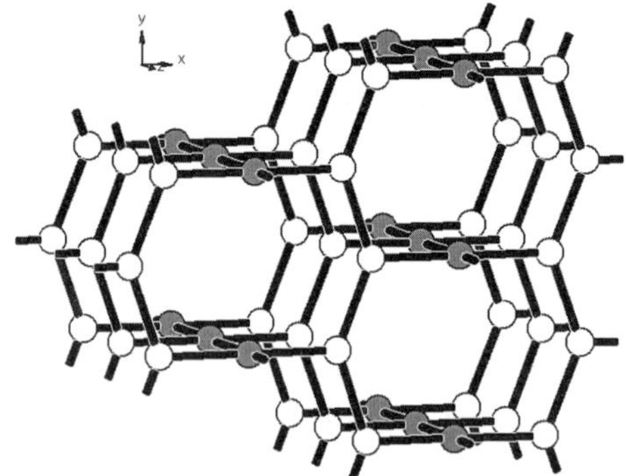

Figure 8.22 The **tfc** net has vertex symbols $8 \cdot 8_3 \cdot 8_3$ and $8_2 \cdot 8_2 \cdot 8_2 \cdot 8_2 \cdot 8_2 \cdot *$ and genus 3.

8.6. Summary of three- and four-connected nets

An important factor for these nets is the stoichiometry between the three- and four-connected nodes. To some extent it should be possible to control the formation of these types of nets by choice of node ratios, but there seems to be no study of this type in the literature. .

The reader may have found this chapter less thorough and lacking in analysis. This reflects the fact that not much theoretical analysis of these nets is available, and the deliberate synthetic efforts into creating structures based on these nets are even scarcer.

A number of additional compounds with three- and four-connected nets can be found in the recent literature, [19-22] and a number of interpenetrating structures have been listed by Proserpio et al. [14] Recently Wang et al. reported on the deliberate "self-assembly of triangles, squares, and tetrahedral" [23]. The RCSR database currently includes some 30 different three- and four-connected nets [24]. The Ge3N4 net described by Wells will probably also be made in the future.

The table on the next page concludes this chapter.

Table 8.1 Summary of the nets with three- and four-connected nodes discussed in this chapter

Net	Other names	Vertex Symb.	Short symb.	Genus	Stoichiometry[a]	
					ns_4	ns_3
bor	boracite	$6{\cdot}6{\cdot}6$ $6_2{\cdot}6_2{\cdot}8{\cdot}8{\cdot}8{\cdot}8$	6^3 $6^2.8^4$	6	3	4
tbo	twisted boracite	$6{\cdot}6{\cdot}6$ $6_2{\cdot}6_2{\cdot}8_2{\cdot}8_2{\cdot}12_2{\cdot}12_2$	6^3 $6^2.8^2.10^2$	11	3	4
pto	Pt_3O_4	$8_5{\cdot}8_5{\cdot}8_5$ $8_2{\cdot}8_2{\cdot}8_4{\cdot}8_4{\cdot}8_4{\cdot}8_4$	8^3 8^6	11	3	4
ctn	C_3N_4	$8_5{\cdot}8_5{\cdot}8_5$ $8_3{\cdot}8_3{\cdot}8_3{\cdot}8_3{\cdot}8_4{\cdot}8_4$	8^3 8^6	11	3	4
sln	(β-Si_3N_4)	$6{\cdot}6{\cdot}6$ $8_4{\cdot}8_4{\cdot}8_4$ $6{\cdot}8_2{\cdot}6{\cdot}8_3{\cdot}6{\cdot}8_3$	8^3 6^3 $6^3.8^3$	11	3	4
-	"net 3"	$6_2{\cdot}6_2{\cdot}8{\cdot}8{\cdot}8{\cdot}8$ $6_2{\cdot}6_2{\cdot}8{\cdot}8{\cdot}8{\cdot}8$ $6{\cdot}6{\cdot}6$	$6^2.8^4$ $6^4.8^2$ 6^3	?	3	4
ins	*InS*	$6{\cdot}6{\cdot}6_2$ $6{\cdot}6{\cdot}6{\cdot}6{\cdot}6{\cdot}10_6$	6^3 $6^5.8$	7	1	1
dmc		$4{\cdot}8_3{\cdot}8_3$ $4{\cdot}8_2{\cdot}8_2{\cdot}8_3{\cdot}8_2{\cdot}8_3$	4.8^2 4.8^5	7	1	1
dme		$4{\cdot}8_3{\cdot}8_3$ $4{\cdot}8_2{\cdot}8_2{\cdot}8_3{\cdot}8_2{\cdot}8_3$	5.8^2 $4.5^2\,6.7.8$	6	2	1
-	$(6\,8^2)(6^4 8\,10)_2$	$6_2{\cdot}8_2{\cdot}8_2$ $6{\cdot}6{\cdot}6{\cdot}6{\cdot}8_2{\cdot}10_8$	$6\,8^2$ $6^4 8\,10$	?	2	1
jph		$6{\cdot}6{\cdot}10_2$ $6{\cdot}6{\cdot}6{\cdot}6{\cdot}12_2{\cdot}12_2$	$6^2.10$ $6^4.10^2$	7	1	4
dmf		$10_3{\cdot}10_4{\cdot}10_6$ $10_2{\cdot}10_2{\cdot}10_3{\cdot}10_4{\cdot}10_4{\cdot}10_4$	10^3 10^6	4	1	4
tfa		$8_4{\cdot}8_4{\cdot}8_4$ $8_2{\cdot}8_2{\cdot}8_3{\cdot}8_3{\cdot}8_3{\cdot}8_3$	8^3 8^6	3	1	2
tfc		$8{\cdot}8_3{\cdot}8_3$ $8_2{\cdot}8_2{\cdot}8_2{\cdot}8_2{\cdot}8_2{\cdot}*$	8^3 $8^5.12$	3	1	2

[a] Refers to the relation between the number of four-connected and three-connected nodes.

References

[1] L. Carlucci, G. Ciani, D. M. Proserpio, A. Sironi, Inorg. Chem. 36 (1997) 1736.

[2] B. F. Abrahams, M. G. Haywood, R. Robson, J. Am. Chem. Soc. 127 (2005) 816.

[3] K. T. Holman, S. M. Martin, D. P. Parker, M. D. Ward, J. Am. Chem. Soc. 123 (2001) 4421.

[4] K. T. Holman, A. M. Pivovar, J. A. Swift, M. D. Ward, Acc. Chem. Res. 34 (2001) 107.

[5] B. F. Abrahams, S. R. Batten, H. Hamit, B. F. Hoskins, R. Robson, Angew. Chem. Int. Ed. 35 (1996) 1690.

[6] S. S. Y. Chui, S. M. F. Lo, J. P. H. Charmant, A. G. Orpen, I. D. Williams, Science 283 (1999) 1148.

[7] B. Chen, M. Eddaoudi, S. T. Hyde, M. O'Keeffe, O. M. Yaghi, Science 291 (2001) 1021.

[8] D. N. Dybtsev, H. Chun, K. Kim, Chem. Commun. (2004) 1594.

[9] A. F. Wells, Acta Cryst. A 42 (1986) 133.

[10] S. R. Batten, B. F. Hoskins, R. Robson, New J. Chem. 22 (1998) 173.

[11] L. Carlucci, G. Ciani, D. W. vonGudenberg, D. M. Proserpio, A. Sironi, Chem. Commun. (1997) 631.

[12] L. Carlucci, G. Ciani, D. M. Proserpio, S. Rizzato, J. Solid State Chem. 152 (2000) 211.

[13] M. Bertelli, L. Carlucci, G. Ciani, D. M. Proserpio, A. Sironi, J. Mat. Chem. 7 (1997) 1271.

[14] V. A. Blatov, L. Carlucci, G. Ciani, D. M. Proserpio, Crystengcomm 6 (2004) 377.

[15] J. A. Rodgers, W. T. A. Harrison, Chem. Commun. (2000) 2385.

[16] Q. Liu, Y. Z. Li, H. J. Liu, F. Wang, Z. Xu, J. Mol. Struct. 733 (2005) 25.

[17] D. Laliberte, T. Maris, J. D. Wuest, J. Org. Chem. 69 (2004) 1776.

[18] S. I. Nishikiori, J. Coord. Chem. 37 (1996) 23.

[19] M. L. Tong, J. Wang, S. Hu, S. R. Batten, Inorg. Chem. Comm. 8 (2005) 48.

[20] Y. Z. Zheng, M. L. Tong, X. M. Chen, New J. Chem. 28 (2004) 1412.

[21] M. L. Tong, X. M. Chen, S.-L. Zheng, Chem.-Eur. J. 6 (2000) 3729.

[22] J.-M. Zheng, S. R. Batten, M. Du, Inorg. Chem. 44 (2005) 3371.

[23] Z. Q. Wang, V. C. Kravtsov, M. J. Zaworotko, Angew. Chem. Int. Ed. 44 (2005) 2877.

[24] M. O'Keeffe, O. M. Yaghi, Reticular Chemistry Structure Resource, Tucson, Arizona State University, 2005, http://okeeffe-ws1.la.asu.edu/RCSR/home.htm

Chapter 9

Nets with higher connectivity than four

Nets with higher connectivity than four are not so common, possibly since there are no obvious commercially available organic building blocks that can be used. This one of the reason often cited for turning to coordination chemistry when it comes to network building. The number of coordination geometries is much extended compared to carbon, and for example six-connected nodes are readily available. The problem is, however, that the commonly used coordination polymer linkers such as 4,4'-bipyridine are sterically demanding and it is difficult to fit six of these around a small transition metal ion as Fe^{3+}.

In Chapter 5 we saw the important class of compounds based on the six-connected pcu (primitive cubic packing, or α-Polonium) net, and we also encountered the five-connected bnn or boron nitride net. In this chapter we will explore higher connectivities further, starting with the five-connected and six connected nets and then going on to the mixed 3;5, 3;6, 4;5 4:6 and 5;6 nets. We will also briefly look at some seven and eight connected nets.

Wells did not treat the higher connected nets in as much detail as the others in his books, [1-3] so our major source of data for the nets in this chapter is the RCSR database [4]. There is also a website dedicated to six-connected nets [5].

9.1. Five connected nets

9.1.1. The $4^4.6^6$-**sqp** net

This net is based on square planar pyramidal nodes and is less common than the **bnn** net (section 5.2.7) based on nodes with trigonal bipyramidal coordination. This is likely an effect of the fact that the trigonal bipyramid is slightly energetically favoured over the square planar pyramid in VSEPR (valence shell electron pair repulsion) theory [6].

The **sqp** net has vertex symbol $4·4·4·4·6·6·6_5·6_5·6_5·6_5$ and genus 4, see Figure 9.1. (For comparison the **bnn** net has vertex symbol $4·4·4·4·4·4·6·6·6·*$.)

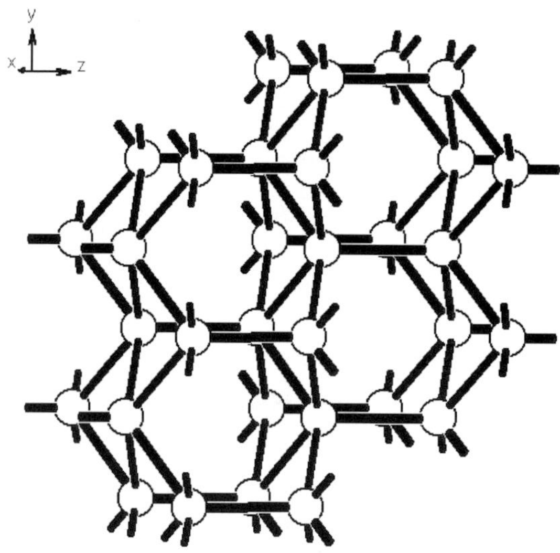

Figure 9.1 The **sqp** net is based on nodes with square planar pyramidal geometries.

This net can be found in the compounds $M_2(C_3H_2O_4)_2(H_2O)_2(\mu_2\text{-}$hexamethylenetetramine)] (M=Zn(II), Cu(II)), see Figure 9.2 [7].

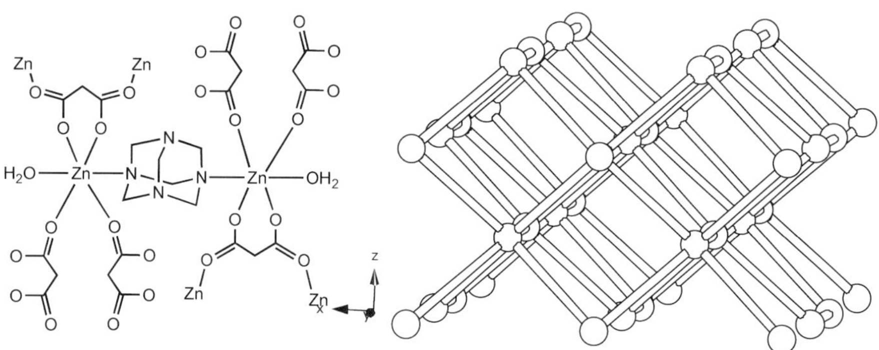

Figure 9.2 The **sqp** net can be found in the compounds $M_2(C_3H_2O_4)_2(H_2O)_2(\mu_2\text{-}$hexamethylenetetramine)] (M=Zn(II), Cu(II)) [7].

9.1.2. The $4^4.6^6$-**nov** net

The **nov** net has vertex symbol $4\cdot4\cdot4\cdot4\cdot6\cdot6\cdot6_3\cdot6_5\cdot6_5\cdot6_5$ and genus 4, see Figure 9.3.

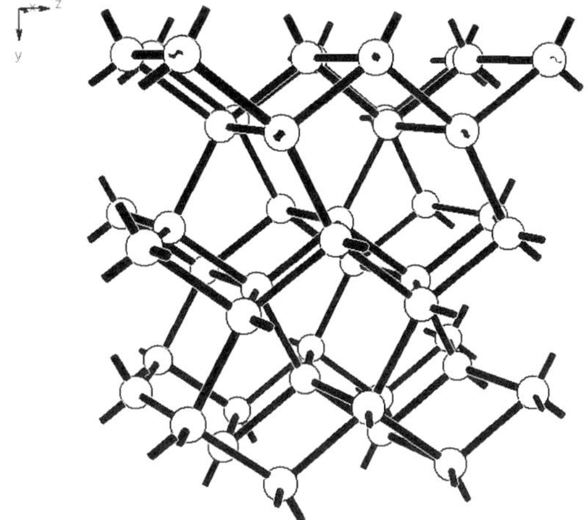

Figure 9.3 The **nov** net also has approximately square pyramidal nodes.

This net can be found in the compound [Cu$_2$(2,5-dimethylpyrazine)(N(CN)$_2$)$_4$], see Figure 9.4 [8]. This compound exhibits antiferromagnetic order below 5K.

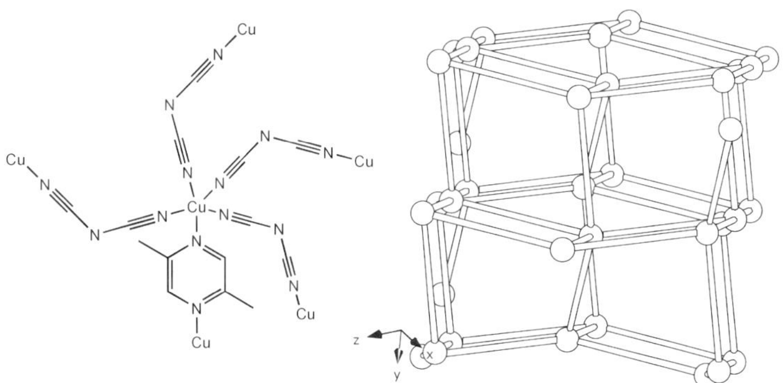

Figure 9.4 The **nov** net can be found doubly interpenetrated in the compound [Cu$_2$(2,5-dimethylpyrazine)(N(CN)$_2$)$_4$] [8].

A large number of other five-connected nets have been described (the RCSR lists 184 uninodal five-connected nets) but only a handful of other molecule based nets are known [9]. We will just pick one of these 184 and show in the next section.

9.1.3. The $4^6.6^4$-**bcu-l** net

The **bcu-l** net has vertex symbol $4 \cdot 4 \cdot 4 \cdot 4 \cdot 6 \cdot 6 \cdot 6 \cdot 6 \cdot 6_2 \cdot 6_2$ and genus 37, see Figure 9.5. You may see that it is based on the so called body centred cubic packing, that is a cube with one atom in the centre and one in each corner. It is nicely describe as an infinite polyehdron (Figure 9.5 right).

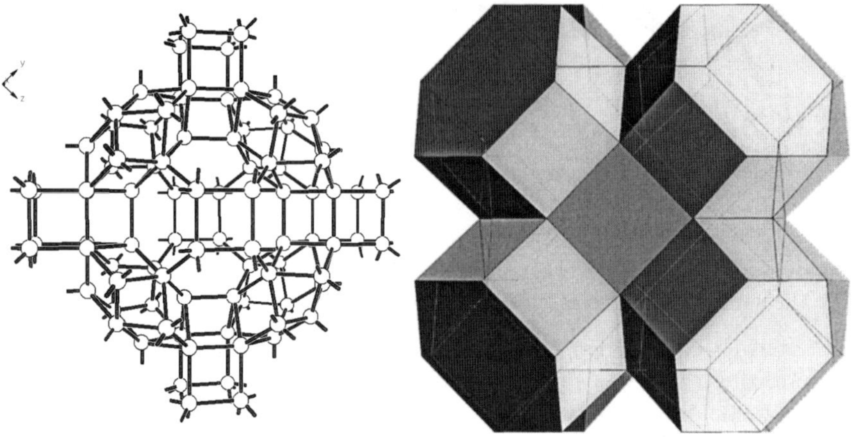

Figure 9.5 The **bcu-l** net has vertex symbol $4 \cdot 4 \cdot 4 \cdot 4 \cdot 6 \cdot 6 \cdot 6 \cdot 6 \cdot 6_2 \cdot 6_2$ and genus 37.

9.1.4. The $3^4.4^2.8^4$-**cab** net and other five-connected nets with cubic symmetry

Wells lists a number of five-connected nets with cubic symmetry with cubic symmetry, the **cab**, **pcu-i** and **ubt** nets [1]. They all contain three-rings (vertex symbols $3 \cdot 3 \cdot 3 \cdot 3 \cdot 4 \cdot 4 \cdot 8 \cdot 8 \cdot 8 \cdot 8$, $3 \cdot 4 \cdot 4 \cdot 4 \cdot 4 \cdot 4 \cdot 8 \cdot 8 \cdot 8 \cdot 8$, $3 \cdot 3 \cdot 4 \cdot 4 \cdot 6 \cdot 6 \cdot 6 \cdot 6 \cdot 6 \cdot 6$, genus 10, 37, 19 respectively), however we do not know of any molecular based examples of these nets. The **cab** net is illustrated in Figure 9.6.

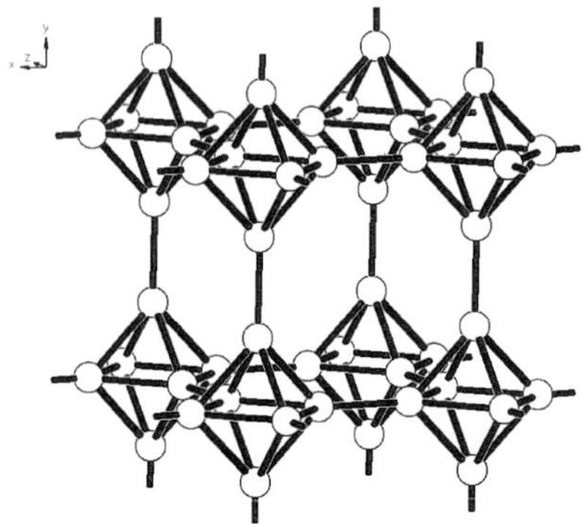

Figure 9.6 The **cab** net. A five-connected net with cubic symmetry

9.2. Six-connected nets

There are two basic geometries for six-connected nodes; octahedral and trigonal prismatic, one or both may be present in these nets. We encountered octahedral nodes in the **pcu**-net in Chapter 5, now we will look at some other nets.

9.2.1. The NiAs or the $(4^{12})(4^9.6^6)$-**nia** net

This is the net formed by the nickel and arsenic atoms in NiAs (a classical inorganic "type structure" [10]) at it is shown in Figure 9.7. The vertex symbols are 4·4·4·4·4·4·4·4·4·4·4·4·*·*·* and 4·4·4·4·4·4·4_2·4_2·4_2·6_4·6_4·6_4·6_4·6_4·6_4 and the genus is 9. Note that half of the six-connected nodes are not octahedral but have a trigonal prismatic geometry.

This means that ordinary transition metal complexes will not normally adopt this geometry since their six-coordinated geometries are almost exclusively octahedral. However, the ions of the rare-earth metals are larger and have often higher coordination numbers and for example with eight coordination it is normal that six of the ligands are arranged in a trigonal prismatic geometry.

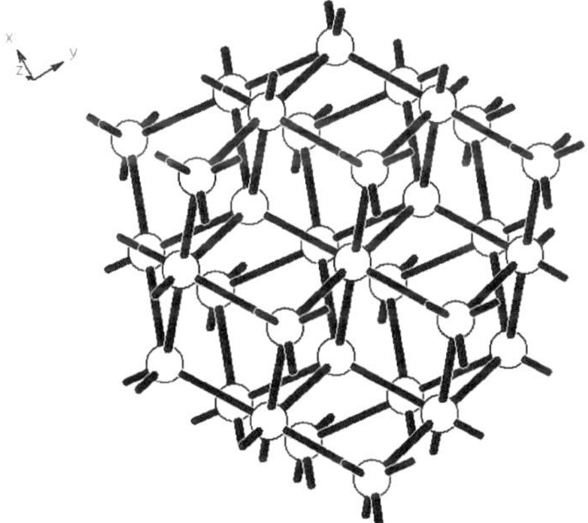

Figure 9.7. The NiAs or **nia** net. Note that half of the six-connected nodes have a trigonal prismatic geometry, and thus not octahedral.

This is exactly what was found in praseodymium hexacyanoferrate(III) tetrahydrate, see Figure 9.8 [11].

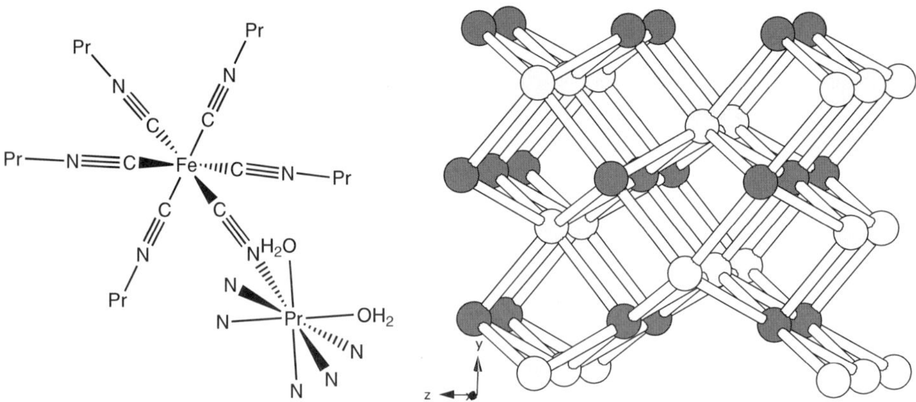

Figure 9.8 The NiAs or **nia** net in praseodymium hexacyanoferrate(III) tetrahydrate [11]. White nodes are eight coordinated praseodymium ions where the cyanide ligands adopt a trigonal prismatic geometry. Grey nodes are octahedral Fe(III) nodes.

9.2.2. The $4^8.5^4.6^3$-**bsn** net

The **bsn** net from one angle looks exactly like the **pcu** net, however, viewed at 90° we can clearly see that it is twisted (Figure 9.9). It has vertex symbol $4\cdot4\cdot4\cdot4\cdot4\cdot4\cdot4\cdot4\cdot5_3\cdot5_3\cdot5_3\cdot5_3\cdot*\cdot*\cdot*$ and genus 5.

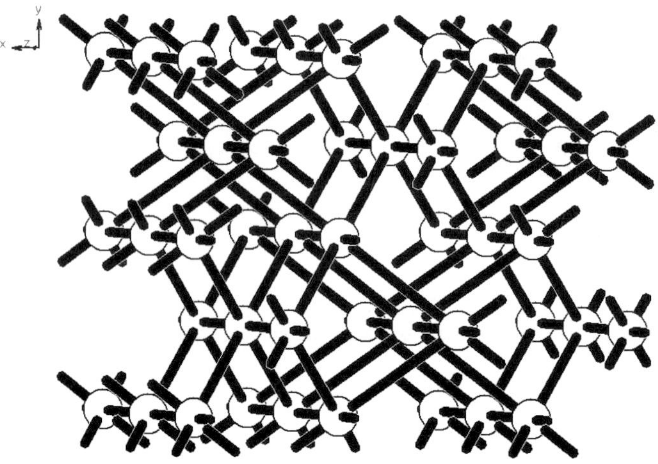

Figure 9.9 The **bsn** net from one angle look very much like the **pcu**-net, but in this view it is evident that they are different!

This net can be found doubly interpenetrated in [Cd(biphenyl-4,4'-dicarboxylate)(1,2-bis(4-pyridyl)ethane)(H$_2$O)], prepared via a hydrothermal route, [12] see Figure 9.10.

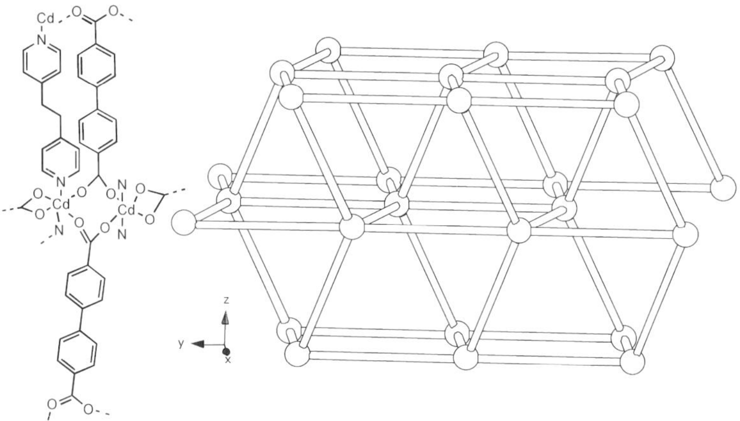

Figure 9.10 The **bsn** net can be found doubly interpenetrated in [Cd(biphenyl-4,4'-dicarboxylate) (1,2-Bis(4-pyridyl)ethane)(H_2O)] [12]. Nodes are placed at the centre of the $Cd_2O_4C_2$ units.

9.2.3. The $4^9.6^6$-**acs** net

The **acs** net is a way to connect nodes with trigonal prismatic geometry, see Figure 9.11. It has vertex symbol $4\cdot4\cdot4\cdot4\cdot4\cdot4\cdot4_2\cdot4_2\cdot4_2\cdot6_4\cdot6_4\cdot6_4\cdot6_4\cdot6_4\cdot6_4$ and genus 5.

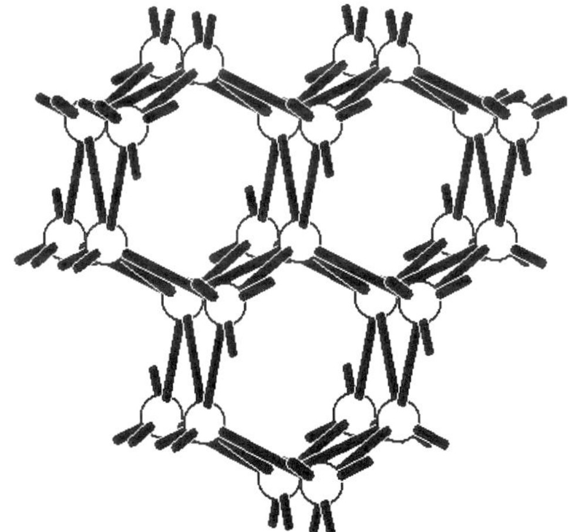

Figure 9.11. The **acs** net is a way to connect nodes with trigonal prismatic geometry.

This net was recently found in [Fe_3O(1,4-benzenedicarboxylate)$_3$(DMF)$_3$] [$FeCl_4$]·3DMF (see Figure 9.11) and also in [Fe_3O(1,3-benzenedicarboxylate)$_3$ (pyridine)$_3$]·0.5pyridine 1.5H_2O [13]. There are also other recent reports of this net [14,15]. Interestingly, the very similar building blocks in $[V_3O(H_2O))(1,3$-benzenedicarboxylate)$_3]Cl,\cdot 9H_2O$ instead give the **pcu** net, [16] see discussion in ref. 13.

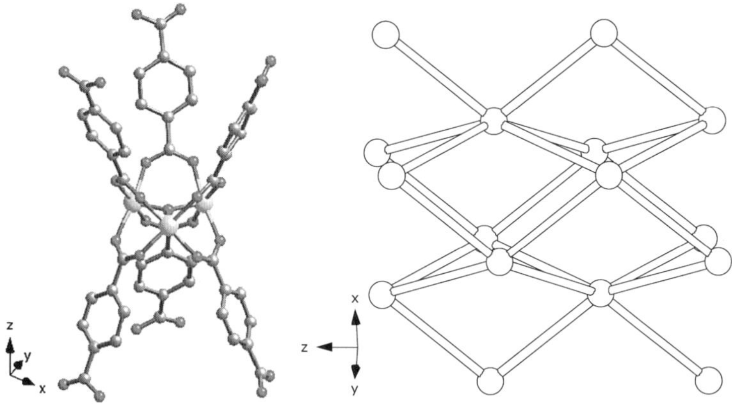

Figure 9.11 The **acs** net was recently found in $[Fe_3O(1,4$-benzenedicarboxylate)$_3(DMF)_3]$ $[FeCl_4]\cdot 3DMF$ (this picture) and also in $[Fe_3O(1,3$-benzenedicarboxylate)$_3(pyridine)_3]$ 0.5pyridine $1.5H_2O$ [13].

9.2.4. The $4^8.5^3.6^4$-**smn** net

The **smn** net has vertex symbols $4\cdot4\cdot4\cdot4\cdot4\cdot4\cdot4\cdot4\cdot5_2\cdot5_2\cdot5_5\cdot6_4\cdot6_4\cdot*\cdot*$ and genus 13, see Figure 9.12.

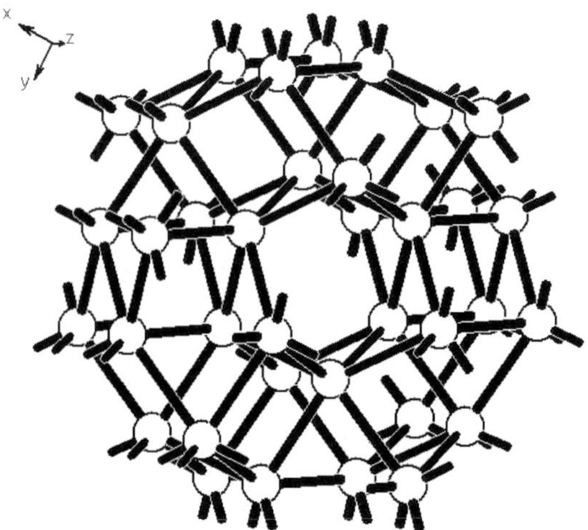

Figure 9.12. The **smn** net.

An example is found as the copper dimer net in [Cu₂(isonicotinato)₃]·(I₅·2/3I₂)·H₂O, see Figure 9.13. Note that this compound, obtained under hydrothermal conditions, also contain a polyiodide network [17].

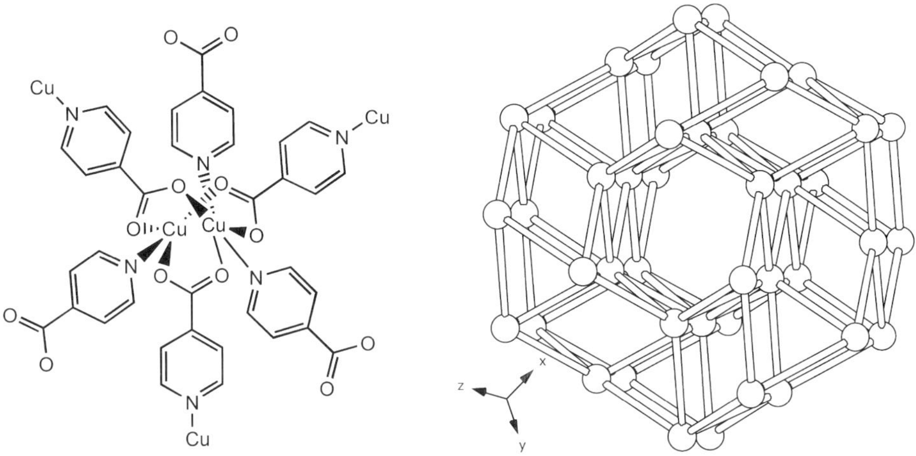

Figure 9.13 The **smn** net can be found in a copper dimer net (nodes in the middle of the mixed square-planar/square-pyramidal binuclear units) in [Cu₂(isonicotinato)₃]·(I₅·2/3I₂)·H₂O, Note that this compund contain a polyiodide network in the channels [17].

9.3. Seven-connected nets

The RCSR currently contains some 75 uninodal seven-connected nets, but only two contain four-membered rings, **svn** and **wfq**, the remaining having three-membered rings as shortest circuits. In general, we get closer and closer to close packing as we draw more and more lines, so that the links between the nodes need to be long if we should consider the structure as 3D-net.

Chemical considerations also move us away from the transition metals, at least if we consider building coordination networks where the nodes are single metal ions, since they are too small to accommodate sufficient number of ligands. However, there is clearly a possibility to have cluster compounds as nodes with higher connectivity, although at present we know of no such examples.

9.3.1. The $4^{16}.5^4.6^2$-*wfq* net

We have found only one example of molecule based seven-connected nets and it seems to contain the **wfq** net. This net is shown in its ideal form in Figure 9.14, it is based on body centred cubic packing, but with one link missing, making the nodes seven-connected. It has vertex symbol $4\cdot4\cdot4\cdot4\cdot4\cdot4\cdot4\cdot4\cdot4\cdot4\cdot4\cdot4_3\cdot4_3\cdot4_3\cdot4_3\cdot5\cdot5\cdot5_3\cdot5_3\cdot*\cdot*$ and genus 21.

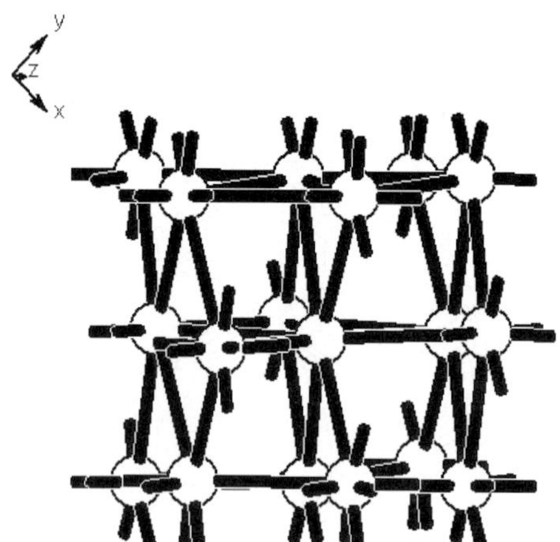

Figure 9.14. The **wfq** seven-connected net.

The example of this net is found in the eight coordinated compound [La(2,2'-bipyridine-N,N'-dioxide)$_4$](BPh$_4$)(ClO$_4$)$_2$·2.75MeOH, [18] see Figure 9.15.

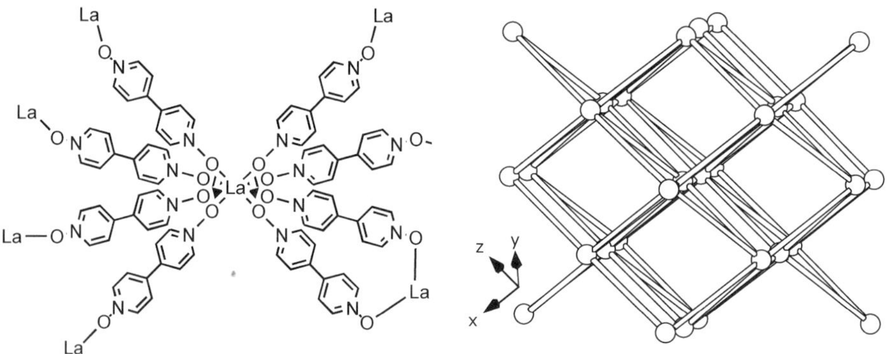

Figure 9.15 The **wfq** net is found in the eight coordinated compound [La(2,2'-bipyridine-N,N'-dioxide)$_4$](BPh$_4$)(ClO$_4$)$_2$·2.75MeOH [18]. Note that one pair of N-oxide bipy ligands binds to the same lanthanum ions making the nodes seven-connecting.

9.4. Eight-connected nets

Eight connected nets are extremely rare, but strategies to synthesis them have evolved the last years, so that we now know of a handful of examples [18-21].

9.4.1. The CsCl or $4^{24}.6^4$-**bcu** net

First we will consider the net based on the body centred cubic structure, or, as it is also known, the CsCl structure, giving the **bcu** net, see Figure 9.16.

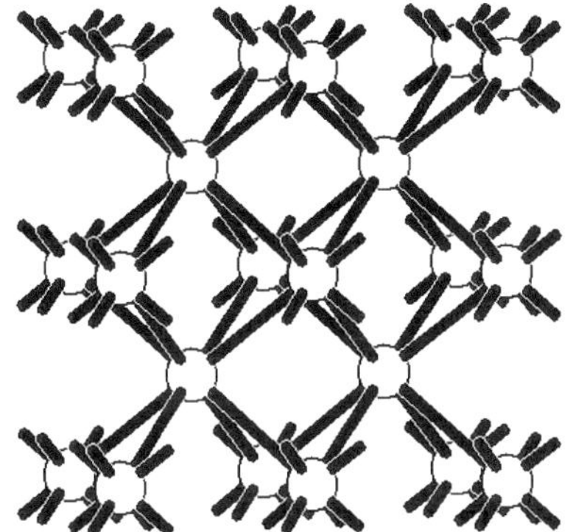

Figure 9.16 The body centred cubic structure, or, as it is also known, the CsCl structure, is the origine of the **bcu** net.

Interestingly, our example of the **bcu** net is taken from the same publications as the seven-connected **wfq** net, and it contains the same ligands but other counter ions, the formula being [La(2,2'-bipyridine-N,N'-dioxide)$_4$](CF$_3$SO$_3$)$_3$·4.2MeOH, [18] see Figure 9.17.

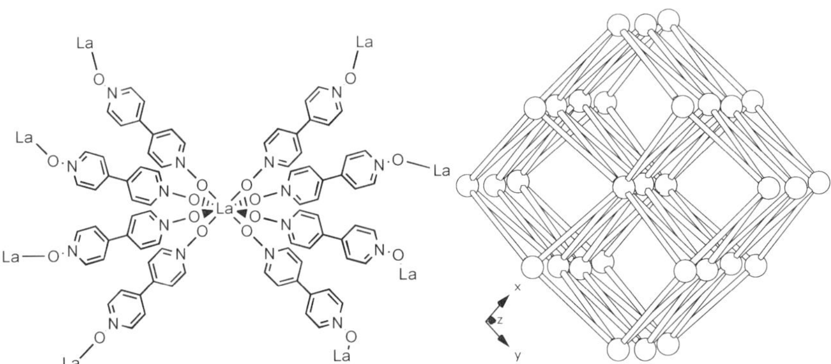

Figure 9.17 [La(2,2'-bipyridine-N,N'-dioxide)$_4$](CF$_3$SO$_3$)$_3$·4.2MeOH, [18] contains the **bcu** net (compare Figure 9.15).

9.4.2. A $3^3.4^{15}.5^8.6^2$ net

A $3^3 4^{15} 5^8 6^2$ net was recent found by the same group. Using different reaction conditions (slow diffusion crystal growth) but identical ligands the compound

[La(2,2'-bipyridine-N,N'-dioxide)$_4$](ClO$_4$)$_3$ was obtained and its 3D nets analysed, see Figure 9.18 [19]. As the nets become quite complicated with these high connectivities it may be some advantage to view them as "combinations of interconnected layered 2-D sheets or subnet tectons" [22].

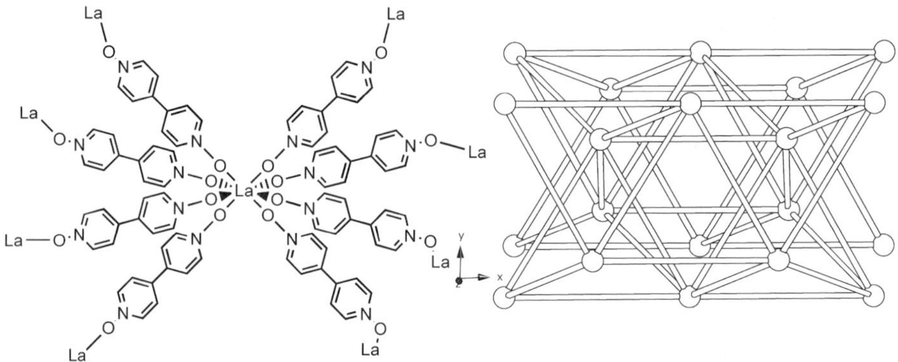

Figure 9.18 Using different reaction conditions (slow diffusion crystal growth) but identical ligands as in Figure 9.17 the compound [La(2,2'-bipyridine-N,N'-dioxide)$_4$](ClO$_4$)$_3$ containing a $3^34^{15}5^86^2$ net was recent found by the same group [19].

9.5. Nets with three- and five-connected nodes

This is another unusual way to build 3D-nets, although one of the members of this class is the very symmetric net built by stacked hexagonal layers.

9.5.1. The stacked hexagonal layer net, or $(6^3)(6^9.8)$-**hms** net

The **hms** net has vertex symbols $6\cdot6\cdot6\cdot6_2\cdot6_2\cdot6_2\cdot6_2\cdot6_2\cdot6_2\cdot*$, $6_3\cdot6_3\cdot6_3$ and genus 3, see Figure 9.19.

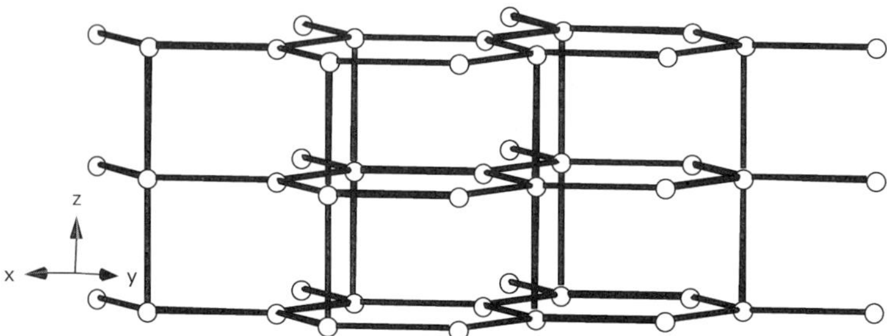

Figure 9.19 The **hms** net built by stacked hexagonal layers.

We can find this net in a few compounds, [23,24] among them Ag(tricyanomethanide)(4,4'-bipyridine) where it is doubly interpenetrated, [24] see Figure 9.20.

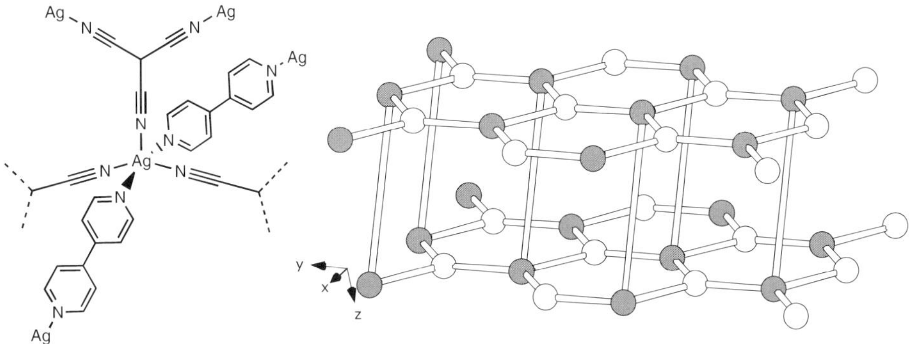

Figure 9.20 The **hms** net can be found doubly interpenetrated in Ag(tricyanomethanide)(4,4'-bipyridine) [24].

9.5.2. The graphite or $(6^3)(6^9.8)$-**gra** net

Closely related to the **hms** net is the graphite or **gra** net. In this net the hexagonal layers are no longer stacked on top of each other, but are eclipsed, see Figure 9.21. The vertex symbols are identical, but the genus is 5. We know of no examples of a molecular variation of this net, and this may be due to the different arrangement of the voids in the two structures, the nice channels that can harbor a second net within the **hms** net (all known **hms** nets are interpenetrated) are clearly blocked in the **gra** net.

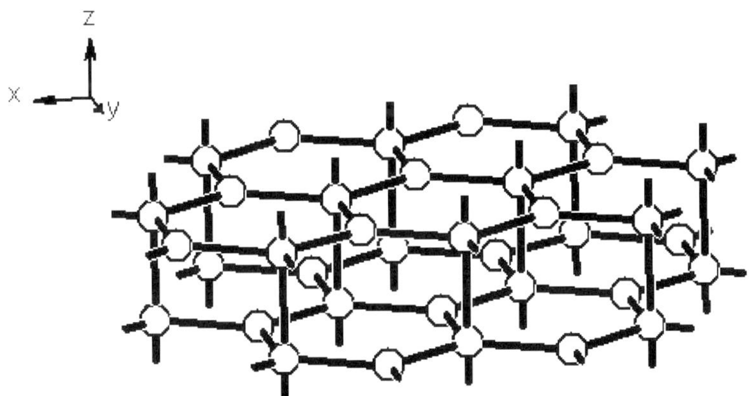

Figure 9.21 The **gra** net, compare to the **hms** net in Figure 9.19.

9.5.3. The $(4.8^2)(4.5^6.6.8^2)$-**mcf-d** net

Another net for which we lack examples, but that seem relevant, is the **mcf-d** net, see Figure 9.22. This net has vertex symbols $4 \cdot 5 \cdot 5 \cdot 5 \cdot 5 \cdot 5_2 \cdot 5_2 \cdot 8 \cdot 8 \cdot *$, $4 \cdot 8_3 \cdot 8_3$ and genus 5. Note that this net has not alternating nodes of different connectivity, thus making a coordination polymer with nodes based on metal coordination and μ^3-ligands less likely to adopt this net.

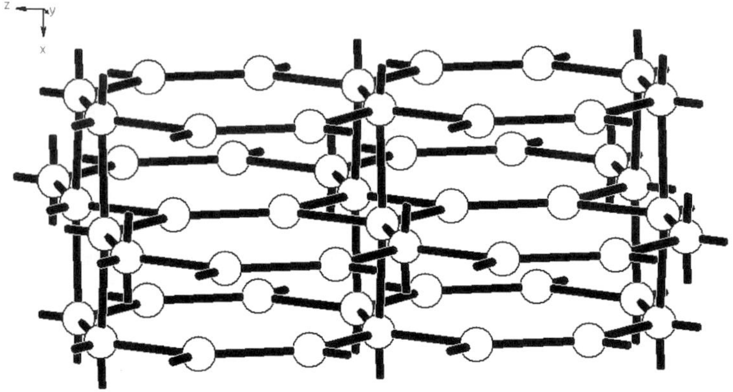

Figure 9.22 The **mcf-d** net.

9.6. Nets with three- and six-connected nodes

This is a more common class of nets, and it includes a net that may be familiar from your undergraduate studies, the rutile net.

9.6.1. The rutile or $(4.6^2)(4^2.6^{10}.8^3)$-**rtl** net

The name of this net come from the Latin word, "rutilus" meaning reddish, and that is often the colour of the TiO_2 mineral where the arrangement of Ti and O atoms form the **rtl** net, see Figure 9.23. This net has vertex symbols $4 \cdot 4 \cdot 6 \cdot 6 \cdot 6 \cdot 6 \cdot 6 \cdot 6 \cdot 6 \cdot 6_2 \cdot 6_2 \cdot * \cdot * \cdot *$, $4 \cdot 6_2 \cdot 6_2$ and genus 7. Note that the stoichiometry is $n_{\text{trigonal}}/n_{\text{octahedral}} = 2$.

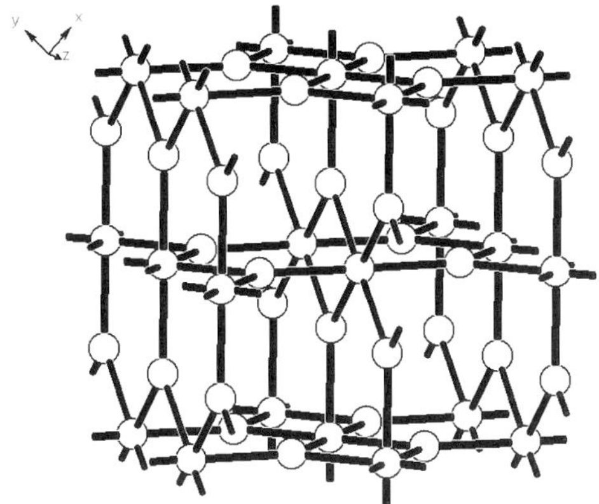

Figure 9.23 The rutile or **rtl** net.

The **rtl** net can be found in [Zn(1,3,5-benzenetricarboxylate)]NH$_2$(CH$_3$)$_2$·DMF prepared by solvothermal methods, see Figure 9.24. The DMF molecules can be removed by heating and vacuum without breaking the framework [25]. Note that the net stoichiometry is maintained since each six-connected node needs two Zn(II) ions, thus giving the correct node ratio 2/1.

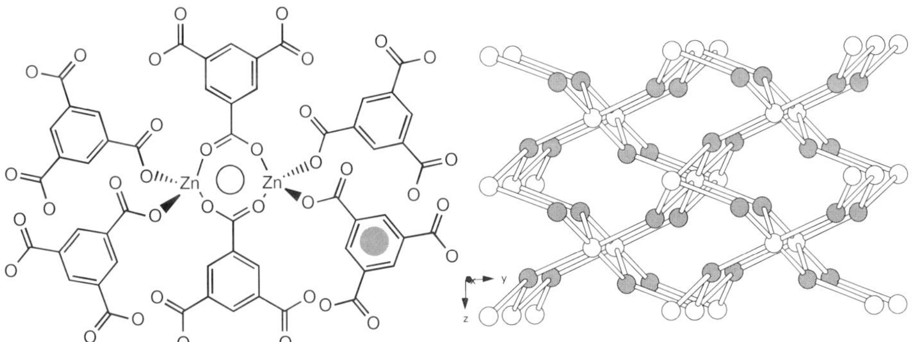

Figure 9.24 [Zn(1,3,5-benzenetricarboxylate)]NH$_2$(CH$_3$)$_2$·DMF contain the rutile or **rtl** net. The DMF molecules can be removed by heating and vacuum without breaking the framework [25]. Grey and white circles in the left picture indicate the nodes used in the **rtl** net to the right.

9.6.2. The pyrite or $(6^3)(6^{12}.8^3)$-**pyr** net

Another of the "inorganic type structures" has given name to the pyrite (FeS$_2$) or **pyr** net, see Figure 9.25. This net has vertex symbols $6\cdot6\cdot6\cdot6\cdot6\cdot6\cdot6_2\cdot6_2\cdot6_2\cdot6_2\cdot6_2\cdot6_2\cdot*\cdot*\cdot*$, $6_3\cdot6_3\cdot6_3$ and genus 13.

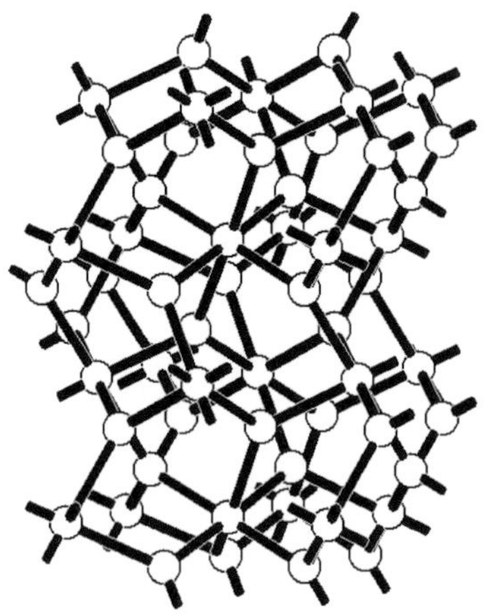

Figure 9.25 The pyrite or **pyr** net.

An example of the **pyr** net can be found as the two nets in [Zn$_4$O(4,4,'4'-nitrilotrisbenzoate)$_2$]·*N,N'*-diethylformamide·EtOH, see Figure 9.26 [26,27]. This doubly interpenetrated structure absorbs hydrogen (1.9 wt % at 77 K and 1 atm) remains crystalline even at 400 °C and 10^{-5} Torr, and the ligands undergo reversible dynamics, mainly rotational motion, in response to removal and rebinding of the guest molecules. Figure 9.27 shows all atoms (except hydrogens) for the two nets and the channel structure is readily seen.

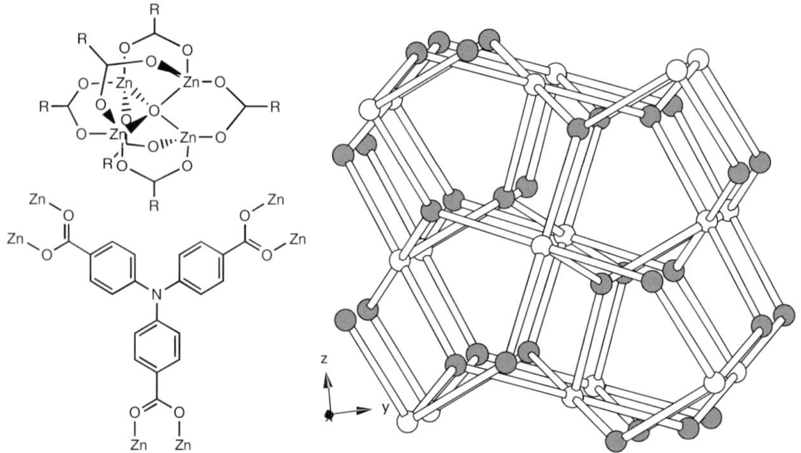

Figure 9.26 The **pyr** net can be found doubly interpenetrated in [Zn$_4$O(4,4,'4'-nitrilotrisbenzoate)$_2$]· *N,N'*-diethylformamide·EtOH, see also Figure 9.27 [26,27].

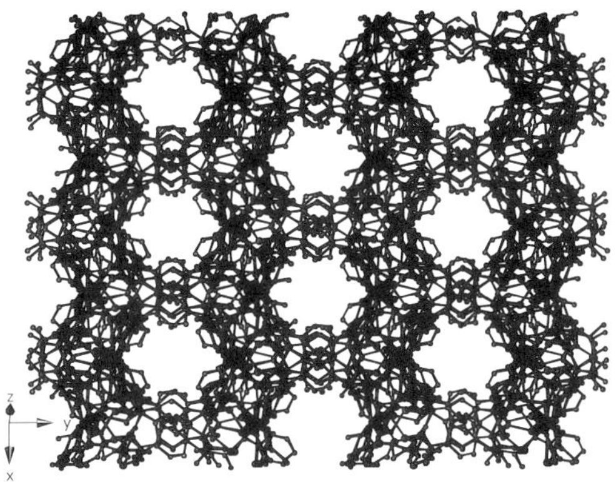

Figure 9.27 All atoms (except hydrogens) for the two interpenetrated **pyr** nets (see Figure 9.26) in [Zn$_4$O(4,4,′4′-nitrilotrisbenzoate)$_2$]·*N,N*′-diethylformamide·EtOH [26,27]. The channels are approximately 4 Å wide.

6.2.1. The five nodal **qom** net containing only six-rings

The **qom** net was recently demonstrated in [Zn$_4$O(1,3,5-benzenetribenzoate)$_2$]· 15*N,N*-diethylformamide·3H$_2$O containing the familiar Zn$_4$O-core coordinating six carboxylates. This compound has a record surface area estimated at 4,500 m^{2}g^{-1} [28].

The **qom** net has genus 25 but as the net has five nodes the vertex symbols become quite cumbersome and we will not cite them here. The net is shown in Figure 9.28 and a comparison between the **pyr** and the **qom** net is shown in Figure 9.29.

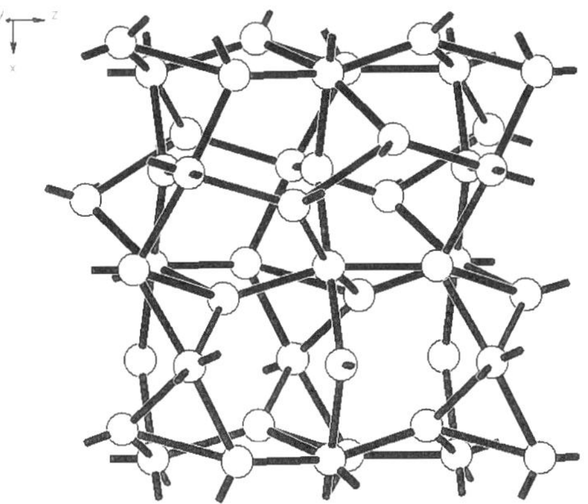

Figure 9.28 The five-nodal **qom** net found in [Zn$_4$O(1,3,5-benzenetribenzoate)$_2$]·15(*N,N*-diethylformamide)·3(H$_2$O) [28].

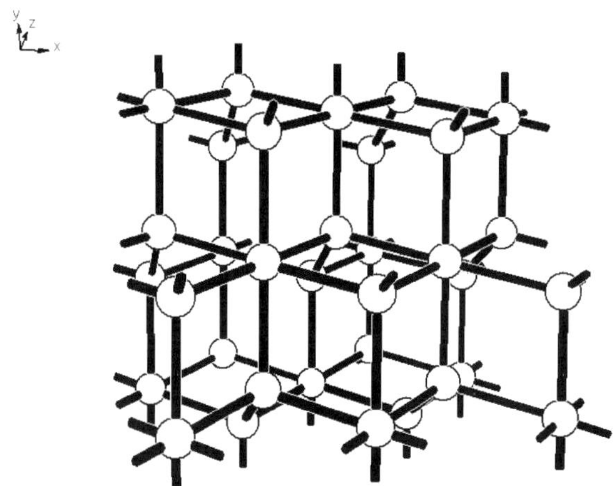

Figure 9.35 The **fsh** net.

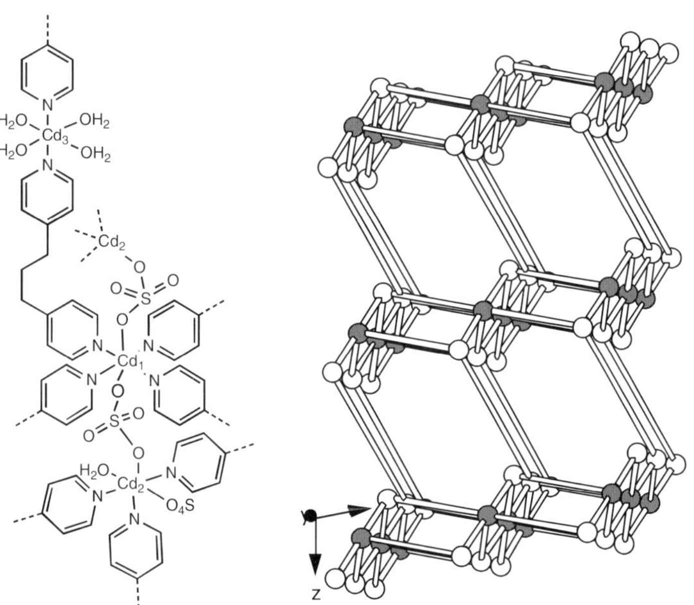

Figure 9.36 doubly interpenetrated in $[Cd_2(SO_4)_2(1,2\text{-bis(4-pyridyl)ethane,})_3(H_2O)_{2.7}]\cdot 4.5H_2O$. The structure contains three different Cd(II) ions, one is a six-connected node, another the four-connected node and the third is just bridging [31].

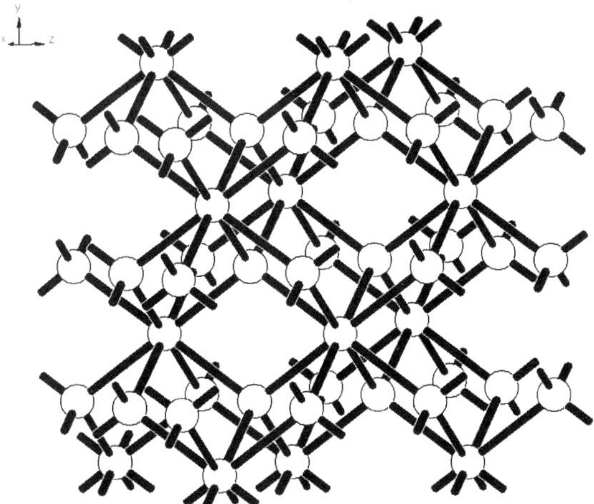

Figure 9.37 The four- and eight-connected fluorite (CaF$_2$) or **flu** net is built from perfect tetrahedra and cubes.

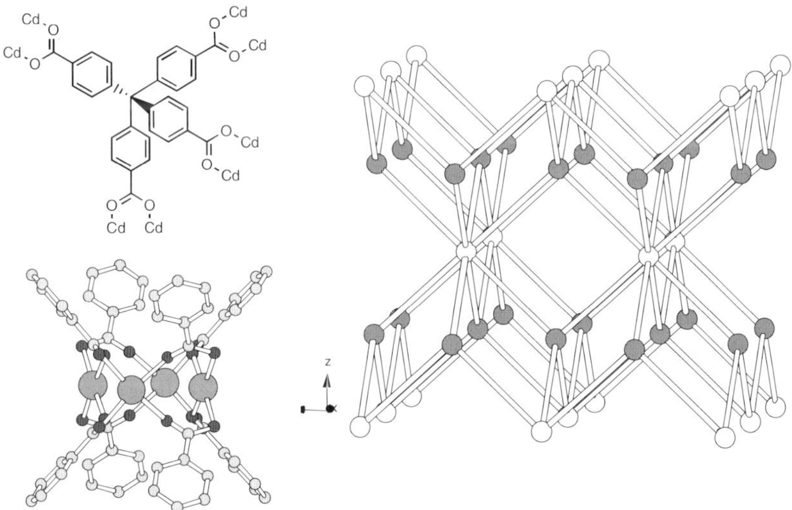

Figure 9.38 [Cd$_4$(tetrakis(4-carboxyphenyl)methane)$_2$(DMF)$_4$]·4DMF·4H$_2$O contains the **flu** net [32]. Upper left shows the ligand structure, bottom left show the Cd$_4$-cluster giving the eight-connected nodes.

9.9. Summary of higher connected nets

The table on the next page concludes this chapter.

                                                                 L. Öhrström & K. Larsson

Table 9.1 Summary of the four-connected nets discussed in this chapter and in chapter 5

Net	Vertex Symbol	Short Symbol	Connect	Nodes[a]
sqp	$4{\cdot}4{\cdot}4{\cdot}4{\cdot}6{\cdot}6{\cdot}6_5{\cdot}6_5{\cdot}6_5{\cdot}6_5$	$4^4,6^6$	5	Sqpy
nov	$4{\cdot}4{\cdot}4{\cdot}4{\cdot}6{\cdot}6{\cdot}6_3{\cdot}6_5{\cdot}6_5{\cdot}6_5$	$4^4.6^6$	5	Trbp
bcu-l	$4{\cdot}4{\cdot}4{\cdot}4{\cdot}6{\cdot}6{\cdot}6{\cdot}6{\cdot}6_2{\cdot}6_2$	$4^6.6^4$	5	Trbp
cab	$3{\cdot}3{\cdot}3{\cdot}3{\cdot}4{\cdot}4{\cdot}8{\cdot}8{\cdot}8{\cdot}8$	$3^4.4^2.8^4$	5	Sqpy
nia	$4{\cdot}4{\cdot}4{\cdot}4{\cdot}4{\cdot}4{\cdot}4{\cdot}4{\cdot}4{\cdot}4{\cdot}4{\cdot}4{\cdot}*{\cdot}*{\cdot}*$ $4{\cdot}4{\cdot}4{\cdot}4{\cdot}4{\cdot}4{\cdot}4_2{\cdot}4_2{\cdot}4_2{\cdot}6_4{\cdot}6_4{\cdot}6_4{\cdot}6_4{\cdot}6_4{\cdot}6_4$	$4^{12}.6^3$ $4^9 6^6$	6	Oct+Trpr
bsn	$4{\cdot}4{\cdot}4{\cdot}4{\cdot}4{\cdot}4{\cdot}4{\cdot}4{\cdot}5_3{\cdot}5_3{\cdot}5_3{\cdot}5_3{\cdot}*{\cdot}*{\cdot}*$	$4^8.5^4.6^3$	6	Oct
acs	$4{\cdot}4{\cdot}4{\cdot}4{\cdot}4{\cdot}4{\cdot}4_2{\cdot}4_2{\cdot}4_2{\cdot}6_4{\cdot}6_4{\cdot}6_4{\cdot}6_4{\cdot}6_4{\cdot}6_4$	$4^9.6^6$	6	Trpr
smn	$4{\cdot}4{\cdot}4{\cdot}4{\cdot}4{\cdot}4{\cdot}4{\cdot}4{\cdot}5_2{\cdot}5_2{\cdot}5_5{\cdot}6_4{\cdot}6_4{\cdot}*{\cdot}*$	$4^8.5^3.6^4$	6	
wfq	$4_2{\cdot}4{\cdot}4_2{\cdot}4_2{\cdot}4{\cdot}4_2{\cdot}4_2{\cdot}4_2{\cdot}4{\cdot}4_2{\cdot}4{\cdot}4{\cdot}4_2{\cdot}4_2{\cdot}4{\cdot}$ $6_4{\cdot}6_4{\cdot}6_4{\cdot}6_6{\cdot}6_6{\cdot}6_6$	$4^{15}.5^5.6^2$	7	
bcu	$4{\cdot}4{\cdot}4{\cdot}4{\cdot}4{\cdot}4{\cdot}4{\cdot}4{\cdot}4{\cdot}4{\cdot}4{\cdot}4{\cdot}4{\cdot}4{\cdot}4{\cdot}4_3{\cdot}4_3{\cdot}4_3{\cdot}4_3{\cdot}$ $4_3{\cdot}4_3{\cdot}4_3{\cdot}4_3{\cdot}4_3{\cdot}4_3{\cdot}4_3{\cdot}4_3{\cdot}*{\cdot}*{\cdot}*{\cdot}*$	$4^{24}.6^4$	8	Cub
-	$3{\cdot}3{\cdot}3{\cdot}4{\cdot}4{\cdot}4{\cdot}4{\cdot}4{\cdot}4{\cdot}4_2{\cdot}4_2{\cdot}4_2{\cdot}4_2{\cdot}4_2{\cdot}4_2{\cdot}4_2{\cdot}$ $4_2{\cdot}4_2{\cdot}5{\cdot}5{\cdot}5{\cdot}5{\cdot}5_2{\cdot}5_2{\cdot}5_2{\cdot}5_2{\cdot}*{\cdot}*$	$3^3 4^{15} 5^8 6^2$	8	Cub
hms	$6{\cdot}6{\cdot}6{\cdot}6_2{\cdot}6_2{\cdot}6_2{\cdot}6_2{\cdot}6_2{\cdot}6_2{\cdot}*, 6_3{\cdot}6_3{\cdot}6_3$	$(6^3)(6^9.8)$	3;5	Trig+Trbp
gra	$6{\cdot}6{\cdot}6{\cdot}6_2{\cdot}6_2{\cdot}6_2{\cdot}6_2{\cdot}6_2{\cdot}6_2{\cdot}*, 6_3{\cdot}6_3{\cdot}6_3$	$(6^3)(6^9.8)$	3:5	Trig+Trbp
mcf-d	$4{\cdot}5{\cdot}5{\cdot}5{\cdot}5{\cdot}5_2{\cdot}5_2{\cdot}8{\cdot}8{\cdot}*, 4{\cdot}8_3{\cdot}8_3$	$(4.8^2)(4.5^6.6.8^2)$	3;5	Trig+Trbp
rtl	$4{\cdot}4{\cdot}6{\cdot}6{\cdot}6{\cdot}6{\cdot}6{\cdot}6{\cdot}6{\cdot}6{\cdot}6_2{\cdot}6_2{\cdot}*{\cdot}*{\cdot}*, 4{\cdot}6_2{\cdot}6_2$	$(4.6^2)(4^2.6^{10}.8^3)$	3;6	Oct+Trig
pyr	$6{\cdot}6{\cdot}6{\cdot}6{\cdot}6{\cdot}6{\cdot}6_2{\cdot}6_2{\cdot}6_2{\cdot}6_2{\cdot}6_2{\cdot}6_2{\cdot}*{\cdot}*{\cdot}*,$ $6_3{\cdot}6_3{\cdot}6_3$	$(6^3)(6^{12}.8^3)$	3;6	Oct+Trig
qom	five-nodal, see appendix	see appendix	3;6	
sit	$4{\cdot}4{\cdot}6{\cdot}6{\cdot}6{\cdot}6{\cdot}6{\cdot}6{\cdot}6{\cdot}6{\cdot}6_2{\cdot}6_2{\cdot}8{\cdot}8{\cdot}8_6, 4{\cdot}6_2{\cdot}6_2$	$(4.6^2)(4^2.6^{10}.8^3)$	3;6	Trig+Trpr
cor	$4{\cdot}4{\cdot}4{\cdot}4{\cdot}4{\cdot}4{\cdot}6_2{\cdot}6_2{\cdot}6_2{\cdot}6_3{\cdot}6_3{\cdot}6_3{\cdot}*{\cdot}*{\cdot}*, 4{\cdot}4{\cdot}4_2{\cdot}$ $6_4{\cdot}6_2{\cdot}6_2$	$(4^3.6^3)(4^6.6^9)$	4;6	Tetr+Oct
fsg	$4{\cdot}4{\cdot}4{\cdot}4{\cdot}4{\cdot}4{\cdot}4{\cdot}4{\cdot}6{\cdot}6{\cdot}6{\cdot}6{\cdot}*{\cdot}*{\cdot}*, 4{\cdot}4{\cdot}4{\cdot}4{\cdot}6_2{\cdot}*$	$(4^2.6^2)(4^8.6^7)$	4;6	Tetr+Oct
-	$4{\cdot}4{\cdot}4{\cdot}4{\cdot}4{\cdot}4{\cdot}4{\cdot}4{\cdot}6_2{\cdot}6_2{\cdot}6_2{\cdot}6_2{\cdot}*{\cdot}*{\cdot}*, 4{\cdot}4{\cdot}4{\cdot}4{\cdot}$ $6_2{\cdot}6_2$	$(4^4.6^2)(4^8.6^6.8)$	4;6	Tetr+Oct
fsh	$4{\cdot}4{\cdot}4{\cdot}4{\cdot}4{\cdot}4{\cdot}6_2{\cdot}6_2{\cdot}6_2{\cdot}6_2{\cdot}6_2{\cdot}6_2{\cdot}*{\cdot}*{\cdot}*,$ $4{\cdot}6_2{\cdot}4{\cdot}6_2{\cdot}4{\cdot}6_2$	$(4^3.6^3)(4^6.6^6.8)$	4;6	Tetr+Oct
flu	$4{\cdot}4{\cdot}4{\cdot}4{\cdot}4{\cdot}4,$ $4{\cdot}4{\cdot}4{\cdot}4{\cdot}4{\cdot}4{\cdot}4{\cdot}4{\cdot}4{\cdot}4{\cdot}4{\cdot}4{\cdot}6_2{\cdot}6_2{\cdot}6_2{\cdot}6_2{\cdot}6_2{\cdot}$ $6_2{\cdot}6_2{\cdot}6_2{\cdot}6_2{\cdot}6_2{\cdot}6_2{\cdot}6_2{\cdot}*{\cdot}*{\cdot}*{\cdot}*$	$(4^6)(4^{12}.6^{12}.8^4)$	4;8	Tetr+Cub

[a] Tetrahedral, octahedral, square planar, trigonal, trigonal prismatic, trigonal bipyramidal, square pyramidal, and cubic (eight-connected) respectively.

References

[1] A. F. Wells, Three-dimensional nets and polyhedra, John Wiley & Sons, New York, 1977.

[2] A. F. Wells, Further Studies of Three-Dimensional Nets, Polycrystal book service, Pittsburgh, 1979.

[3] A. F. Wells, Structural Inorganic Chemistry, 5th ed. Clarendon Press, Oxford, 1984.

[4] M. O'Keeffe, O. M. Yaghi, Reticular Chemistry Structure Resource, Tucson, Arizona State University, 2005, http://okeeffe-ws1.la.asu.edu/RCSR/home.htm

[5] A. Le Bail, Journées de la Division Chimie du Solide (SFC), Paris, 1996. http://sdpd.univ-lemans.fr/vrml/6c3d/6c3dnets.html

[6] P. W. Atkins, L. Jones, Chemical Principles, the Quest for Insight, 3rd ed. W.H. Freeman, 2004.

[7] Q. Liu, Y. Z. Li, Y. Song, H. J. Liu, Z. Xu, J. Solid State Chem. 177 (2004) 4701.

[8] W. F. Yeung, S. Gao, W. T. Wong, T. C. Lau, New J. Chem. 26 (2002) 523.

[9] N. W. Ockwig, O. Delgado-Friedrichs, M. O'Keeffe, O. M. Yaghi, Acc. Chem. Res. 38 (2005) 176.

[10] D. F. Shriver, P. W. Atkins, Inorganic Chemistry, 2001.

[11] V. Langer, L. Smrcok, Y. Masuda, Acta Cryst. C. 60 (2004) I104.

[12] F. A. A. Paz, Y. Z. Khimyak, A. D. Bond, J. Rocha, J. Klinowski, Eur. J. Inorg. Chem. (2002) 2823.

[13] A. C. Sudik, A. P. Cote, O. M. Yaghi, Inorg. Chem. 44 (2005) 2998.

[14] C. Serre, F. Millange, S. Surble, G. Ferey, Angew. Chem. Int. Ed. 43 (2004) 6286.

[15] G. Yang, R. G. Raptis, Chem. Commun. (2004) 2058.

[16] K. Barthelet, D. Riou, G. Ferey, Chem. Commun. (2002) 1492

[17] J. Y. Lu, V. Schauss, Eur. J. Inorg. Chem. (2002) 1945.

[18] D. L. Long, A. J. Blake, N. R. Champness, C. Wilson, M. Schroder, Angew. Chem. Int. Ed. 40 (2001) 2444.

[19] D. L. Long, R. J. Hill, A. J. Blake, N. R. Champness, P. Hubberstey, D. M. Proserpio, C. Wilson, M. Schroder, Angew. Chem. Int. Ed. 43 (2004) 1851.

[20] F. A. A. Paz, J. Klinowski, Inorg. Chem. 43 (2004) 3882.

[21] H. L. Sun, S. Gao, B. Q. Ma, F. Chang, W. F. Fu, Microporous Mesoporous Mater. 73 (2004) 89.

[22] R. J. Hill, D. L. Long, N. R. Champness, P. Hubberstey, M. Schroder, Acc. Chem. Res. 38 (2005) 335.

[23] S. R. Batten, B. F. Hoskins, R. Robson, New J. Chem. 22 (1998) 173.

[24] B. F. Abrahams, S. R. Batten, B. F. Hoskins, R. Robson, Inorg. Chem. 42 (2003) 2654.

[25] L. H. Xie, S. X. Liu, B. Gao, C. D. Zhang, C. Y. Sun, D. H. Li, Z. M. Su, Chem. Commun. (2005) 2402.

[26] E. Y. Lee, S. Y. Jang, M. P. Suh, J. Am. Chem. Soc. 127 (2005) 6374.

[27] H. K. Chae, J. Kim, O. D. Friedrichs, M. O'Keefe, O. M. Yaghi, Angew. Chem. Int. Ed. 42 (2003) 3907.

[28] H. K. Chae, D. Y. Siberio-Perez, J. Kim, Y. B. Go, M. Eddaoudi, A. J. Matzger, M. O'Keeffe, O. M. Yaghi, Nature 427 (2004) 523.

[29] R. Natarajan, G. Savitha, P. Dominiak, K. Wozniak, J. N. Moorthy, Angew. Chem. Int. Ed. 44 (2005) 2115.

[30] A. M. Kutasi, A. R. Harris, S. R. Batten, B. Moubaraki, K. S. Murray, Cryst. Growth Des. 4 (2004) 605.

[31] M. J. Plater, M. R. S. Foreman, T. Gelbrich, S. J. Coles, M. B. Hursthouse, J. Chem. Soc., Dalton Trans. (2000) 3065.
[32] H. Chun, D. Kim, D. N. Dybtsev, K. Kim, Angew. Chem. Int. Ed. 43 (2004) 971.

Chapter 10

Some mathematics related to 3D-nets

The main reasons to develop a mathematical formalism of 3D-nets are:

- To derive different types of 3D-nets.
- To classify these nets in terms of symmetry, topology or other criteria.

This will be useful in order:

- To find new nets as possible synthetic targets.
- To differentiate nets from each other. As the same net may occur with different symmetries (or configurations), this is not always easy to identify by just looking at them.
- To equate nets with each other in order to find similarities that might otherwise go unnoticed.
- To analyse phenomena such as interpenetration, porosity, and surface area.

We will not attempt to make a rigorous derivation of the mathematical concepts and equations used in the description of 3D-nets. A number of texts are already devoted to this [1-3]. What we want to do is to provide a more popular description of the mathematics behind nets and perhaps this will be a useful introduction to such more advanced works for the practicing chemist. For those without these ambitions we will simply try to show some of the beauty and usefulness of this approach.

There are also non-trivial computational aspects of nets, especially when it comes to the calculation of vertex symbols which quickly gets very heavy and need efficient algorithms in order not to become too time consuming [4]. The reader should be aware of this, but this subject is so much outside our competences that we will not treat it further in this text.

For convenience we recollect here some of the definitions already given in earlier chapters.

10.1. Nets, Polyhedra and Topology

10.1.1. Definitions

For some definitions we start with the 2D-nets that have nodes (or vertices) with the same connectivity and where all the smallest polygons are the same. The connectivity at the nodes we call p and the number of sides of the smallest polygon we call n (n-gons). The resulting net is called an (n,p)-net, and it is easy to convince oneself that the examples in Figure 10.1 are the only ones containing one type of vertex (node) and one type of polygon.

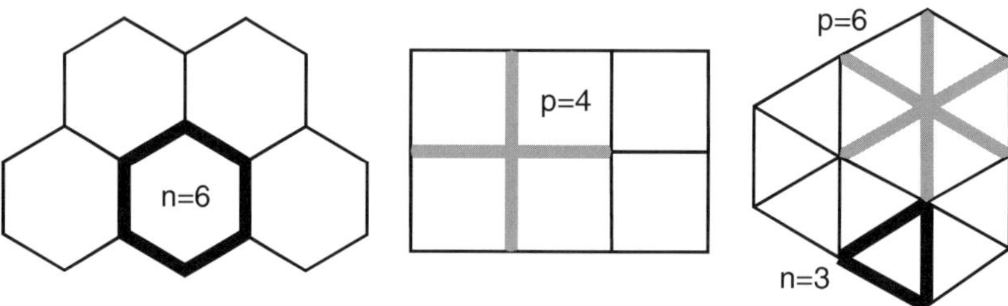

Figure 10.1 The (n,p) nets (6,3), (4,4) and (3,6), where n designates the smallest polygon found in the net and p is the connectivity at each node. These are the only 2D-nets with one value of n and one value of p.

This nomenclature can also be used in three dimensions and those nets are also called (n,p)-nets. There is thus nothing in this description that will tell us if a particular (n,p) combination is a 3D or 2D net. Indeed, it could also be a 0D-net, that is, a polyhedron.

This is the simplest kind of systematic naming of nets, but as we have already seen in preceding chapters, although only a rough description of the type of net we are dealing with, it provides a reasonable starting point.

10.1.2. The Platonic bodies and the (n,p) relation to dimensionality

Even though we are interested in 3D-nets, the collection of symmetric polyhedra called the Platonic bodies, shown in Figure 10.2, is a convenient starting point for our discussion [5][1].

---

[1] The reference here is the famous book by English-Canadian mathematician H.S.M. Coexter, "Regular polytopes", but we actually admit not getting much further than the first chapter (especially not into the fourth dimension…).

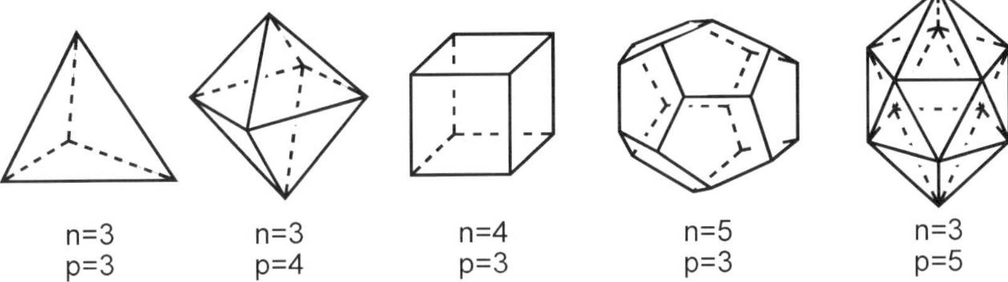

Figure 10.2 The Platonic bodies with their corresponding (n,p) designation. The icosahedron was once thought to be impossible to find in inorganic structures, but, as pointed out by Coexter in the second edition of his book, is in fact the structure of the B_{12} molecule [5,6].

It was shown more than 2000-years ago by Euklides in Alexandria (Euklides' "Elementa") that these are the only possible polyhedra that have integer values of n and p. Of course, many other polyhedra exist, but these all have either more than two nodes with different connectivity, or two or more n-gons (or both). One chemically interesting examples is the fullerene C_{60} polyhedron shown in Figure 10.3.

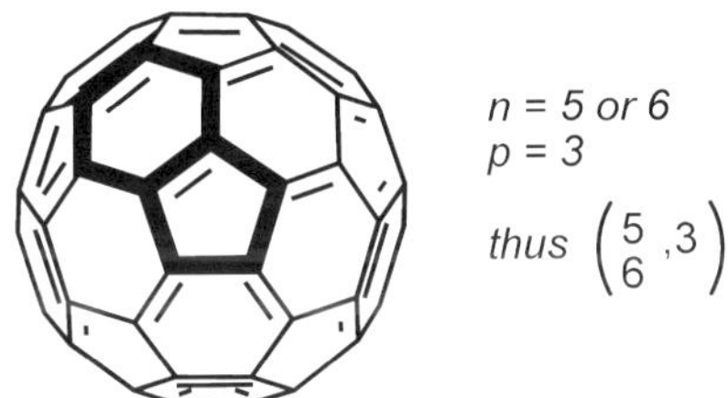

Figure 10.3 The fullerene C_{60} polyhedron has an average n value of $(12 \cdot 5 + 20 \cdot 6)/32 = 45/8 = 5.625$[2]

While the (n,p) designation does not immediately tell us the dimensionality of our system, it is nevertheless completely determined by these numbers. The way to work this out is to consider the sum of the p angles meeting at each node. For a polyhedron that completely encapsulates a volume of space, the sum of these angles has to be less than 360°. For a plane net, propagating in two dimensions only, the sum has to be exactly 360°. With a connectivity of n there has to be n such angles at the node.

The value of this angle will be determined by the n-gon it is part of, since the sum of the angles in an n-gon is $(n-2) \cdot 180°$.[3] Assuming a regular polygon each

---

[2] The number of five and six rings in C_{60} can be calculated by dividing the contribution in atoms for each ring with the connectivity (the number of rings sharing each atom) and summing up to a total of 60. The equation $60 = x_5 \cdot 5/3 + x_6 \cdot 6/3$ has integer solutions for $x_5 = 12$ and $x_6 = 20$.

angle will be $(n-2)\cdot180°/n$ and for the angles meeting at each node we can thus write:

$$p\,\frac{(n-2)180°}{n}=360°\ \text{ for planes and }\ p\,\frac{(n-2)180°}{n}<360°\ \text{ for polyhedra}\ \textit{(10.1-2)}$$

$$\text{This can be rearranged to }p\,\frac{(n-2)}{n}=2\ \text{ for planes}\qquad\qquad\textit{(10.3)}$$

This expression is somewhat inconvenient since it contains n in both denominator and numerator. It can be rearranged to

$$\begin{aligned}&p(n-2)-2n=0\ \Rightarrow\ pn-2p-2n=0\Rightarrow(p-2)(n-2)-4=0\Rightarrow\\&(p-2)(n-2)=4\ \text{(planes) and }(p-2)(n-2)<4\ \text{(polyhedra)}\end{aligned}\qquad\textit{(10.4-5)}$$

We can now see that for polyhedra the possible combinations of p and n according to equation 10.4 are: (3,3), (3,4), (3,5), (4,3) and (5,3) which correspond to the Platonic solids in Figure 10.2. For the plane nets the possible integer solutions to 10.5 are (3,6), (4,4) and (6,3) giving the nets in Figure 10.1.

It appears that these equations also hold for mean values of n and p. Thus, for C_{60} we get: (3-2)(5.625-2)=3.625, which clearly is less than 4.

We have now dealt with the plane nets and the polyhedra. The condition for a 3D-net is consequently:

$$\text{For 3D-nets}:\ p\,\frac{(n-2)}{n}>2\ \text{ and thus }(p-2)(n-2)>4\qquad\qquad\textit{(10.6)}$$

From earlier examples, notably the **srs** or (10,3)-a net, we know that 3D-nets can be constructed from planar nodes. The condition must be that the corresponding n-gons are no longer flat. Restricting ourselves to completely symmetric (all sides and angles equal) n-gons, it is still time consuming to work out the angle sum for such non-flat polygons. We thus only note that for a three-connected net with 120° angles, then n has to be larger than 6 to satisfy equation 10.6. We will come back to this equation shortly, but note in passing that Wells states that the solutions for

$$(p-2)(n-2)=8\qquad\qquad\textit{(10.7)}$$

"…include some simple and important 3D-nets." [3] i.e. (10,3), (4,6) and (6,4). One wonders if there is underlying message in this. Geometry will,

---

[3] An n-gon can always be divided into n-2 triangles angle each with a sum of are 180°. This is done by starting at point 1, drawing a straight line to point 3, and then from point 3 to point 4, an so forth. When the last line is to be drawn we will have to leave out one node again to make a final triangle. Thus we have drawn lines to n-3 points in the n-gon, giving a total of n-3 lines and n-2 triangles. The sum of the angles of these triangles is equal to the angle sum of the polygon.

however, take us only a part of our journey, and we now turn to the field of *Topology*.

10.1.3. A few words on topology

Our everyday notion of the word *topology* is somewhat synonymous with shape, *i.e.* the notion that a golf ball and the moon share some characteristics, namely their approximately spherical shape, never mind the size difference. Topology has also been called "rubber sheet geometry" and is described in Encyclopaedia Britannica as " (topology) ...studies those properties an object retains under deformation—specifically, bending, stretching and squeezing, but not breaking or tearing" [7].

This fits our purposes since we want to recognise our nets as identical even if they are deformed, but not if any chemical bonds are broken. Now, the problem becomes how to describe these topological differences between various nets.

We first look at how the mathematical discipline of *algebraic topology* differentiates between the shapes of finite 3D-objects. For example, which objects have the same topology; a sphere, and ellipsoid or a torus (doughnut shape)?

One way of dealing with this problem is to find what is called *topological invariants*. These are sets of numbers or groups (as groups in the classification of molecules and crystals through *group theory*) that differentiate various topologies. One such invariant comes from Euler's theorem on polyhedra [5] where he established that the number of faces, F, the number of edges, E, and the number of vertices (or nodes), V for any polyhedron are connected by the equation:[4]

$$F - E + V = 2 \qquad\qquad (10.8)$$

We will not prove this, but it is important to note that the proof is purely topological, and does not rely on the summing of angles etc.

If we now express E and F in terms of E, p, and n we get:

$$F = \frac{Vp}{n} \qquad\qquad (10.9)$$

since every vertex is connected to p faces that are shared with n other vertices (one for every corner of the polygon). And:

---

[4] This equation may seem familiar to chemists remembering their first course in chemical thermodynamics since it is similar to "Gibbs phase rule", f + C - P = 2. (f= degrees of freedom, C=number of components and P number of phases [8-10]. This equation is also related to the famous "The bridges of Köningsberg" problem and the foundations of topology and graph theory. In chemistry it may also be applied to find the number of faces of a fullerene, compare footnote 2.

$$E = \frac{Vp}{2} \tag{10.10}$$

since there are p edges emerging from every vertex and every edge is shared by two vertices. We insert these expressions into (10.8) and rearrange to obtain V as a function of n and p.

$$\frac{Vp}{n} - \frac{Vp}{2} + V = 2 \Rightarrow V\left(\frac{p}{n} - \frac{p}{2} + 1\right) = 2 \Rightarrow V\left(\frac{2p}{2n} - \frac{np}{2n} + \frac{2n}{2n}\right) = 2 \Rightarrow$$

$$V = \frac{2 \cdot 2n}{2p - np + 2n} \Rightarrow V = \frac{4n}{4 - (2-n)(2-p)} \tag{10.11}$$

For a polyhedron we have a finite number of vertices, thus the denominator cannot be negative or zero, thus:

$$4 - (2-n)(2-p) > 0 \Rightarrow -(2-n)(2-p) < 4 \Rightarrow (n-2)(p-2) < 4 \tag{10.12}$$

This is the same equation as 10.5, giving the five Platonic bodies as the solutions for integer values of n and p. For an infinite number of vertices the denominator should be zero, thus yielding equation 10.4 again. In this formulation it is not evident that these solutions should give 2D-nets. However, an enclosed volume covered by an infinite number of faces also has infinite values of n and p. Since 10.4 gives finite values of n and p these solutions have to correspond to a 2D-net.

The solutions of 10.6 that correspond to 3D-nets, $(n-2)(p-2)>4$, give finite, negative values of the number of vertices according to 10.11. This is because the derivation of the numbers of faces and edges according to equations 10.9-10.10 does not hold for infinite (3D) polyhedra.

However, for some extended structures like zeolites, it may be profitable to expand the 3D-nets and give them volume, thus forming an infinite enclosure. The surface so formed can be divided into polygons (the technical term is *tessellation*) and the resulting object is called a 3D-polyhedra, see Figure 10.4.

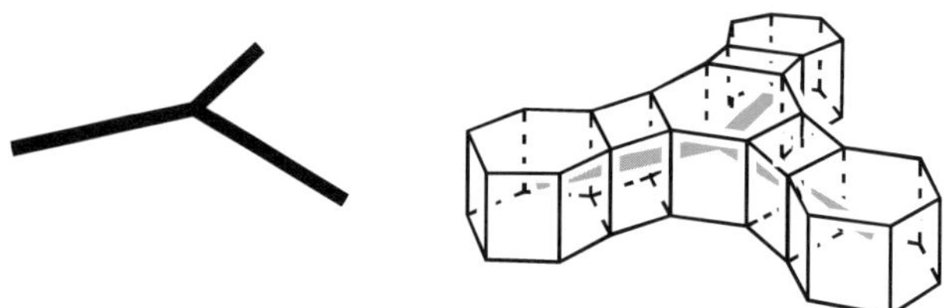

Figure 10.4 A three-connected vertex with its three neighbours (left) and a corresponding part of a 3D-polyhedron with a tessellation of hexagons and squares.

To obtain a measurement of the complexity of a geometrical object the term *genus* has been introduced. This topological genus, g, can then be calculated by an extension of the Euler formula 10.8.

$$F - E + V = 2 - 2g \qquad (10.13)$$

This can then be used to differentiate between different 3D-polyhedra, but the extension of this to infinite polyhedra is not trivial. However, for the molecule-based nets discussed in this book a related, alternative description will be adopted, see section 10.2.

However, as an example we will show how the concept of genus can be used to answer the question about the sphere, the ellipsoid and the torus posed at beginning of this section.

Since equation 10.8 is true for any polyhedron, it will also be true for any tessellation of both the sphere and the ellipsoid. Thus both of these must have genus $g=0$. In contrast, the genus for a tessellation of the torus will be $g=1$ which can be seen from Figure 10.5.

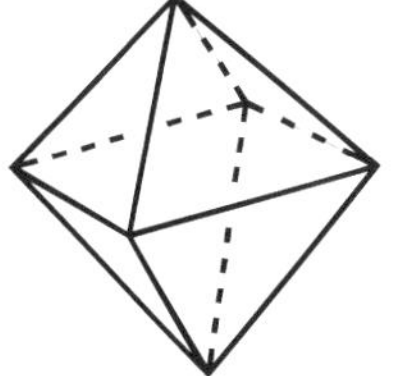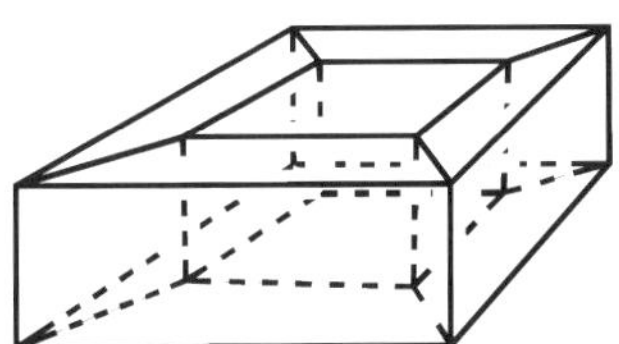

Figure 10.5 Illustrating genus and equation 10.13. The genus for the octahedron will be g=(8-12+6-2)/2=0. For the figure to the right representing a tessellation of the torus it will be g=(20-32+16-2)/2=1

Thus the torus is topologically different from the sphere and the ellipsoid and it may seem that determining the genus is only a question of counting the number of holes in the 3D-body. However, this is probably an oversimplification, as can perhaps be envisaged by considering objects such as the Möbius strip and its equivalent in 3D, or "holes" in higher dimensions.

10.2. Genus, Tilings and Nets

10.2.1. A classification of nets in terms of their genus

As alluded to in the preceding section, inflating the 3D-net and replacing it with connected polyhedra in principle makes it possible to use the concept of genus. However, Bonneau et al. have suggested a conceptually much easier definition as we will shortly see [11].

The smallest piece of the net needed to form the whole 3D-net is the unit that is repeated only by translation throughout the structure. It can be seen as the unit cell contents of the net. For a six-connected 3D net the number of vertices in such a unit, called Z_t, may be as small as one, but for a four-connected 3D net it is at least 2 since we need a minimum of six "loose ends" to build from.

Moreover, it is clear from Figure 10.6 that the smallest possible genus for a 3D-net is 3 (all have six "loose ends"), and it is possible to make a comprehensive listing of such nets, see Table 10.1 [11].

Table 10.1 The 3D-nets with genus 3 (the minimal nets) (after Bonneau et et al. [11]).

Net	Wells	Connect.	Vertex symbols	Genus
pcu	-	6	$4 \cdot 4 \cdot 4 \cdot 4 \cdot 4 \cdot 4 \cdot 4 \cdot 4 \cdot 4 \cdot 4 \cdot 4 \cdot 4 \cdot * \cdot * \cdot *$	3
hms	-	5; 3	$6 \cdot 6 \cdot 6 \cdot 6_2 \cdot 6_2 \cdot 6_2 \cdot 6_2 \cdot 6_2 \cdot 6_2 \cdot *$; $6_3 \cdot 6_3 \cdot 6_3 \cdot$	3
dia	6^6-(a)	4	$6_2 \cdot 6_2 \cdot 6_2 \cdot 6_2 \cdot 6_2 \cdot 6_2$	3
cds	$4^2 8^4$	4	$6 \cdot 6 \cdot 6 \cdot 6 \cdot 6_2 \cdot *$	3
tfa	(8, 3;4)-b	4; 3	$8_2 \cdot 8_2 \cdot 8_3 \cdot 8_3 \cdot 8_3 \cdot 8_3$; $8_4 \cdot 8_4 \cdot 8_4$	3
tfc		4; 3	$8_2 \cdot 8_2 \cdot 8_2 \cdot 8_2 \cdot 8_2 \cdot *$; $8 \cdot 8_3 \cdot 8_3$	3
srs	(10,3)-a	3	$10_5 \cdot 10_5 \cdot 10_5$	3
ths	(10,3)-b	3	$10_2 \cdot 10_2 \cdot 10_4$	3

There is of course a relation of the genus to the Z_t variable as this has its minimal value for genus 3 nets and then increases as Z_t increases. The point is, however, that that g starts at 3 for all nets whereas the minimal value for Z_t can be 1, 2, 3 or 4 depending on the connectivity. The quotient graphs for **pcu** and **dia** are given in Figure 10.8 and the nets in Table 10.1, called the minimal nets, are illustrated in Figure 10.9.

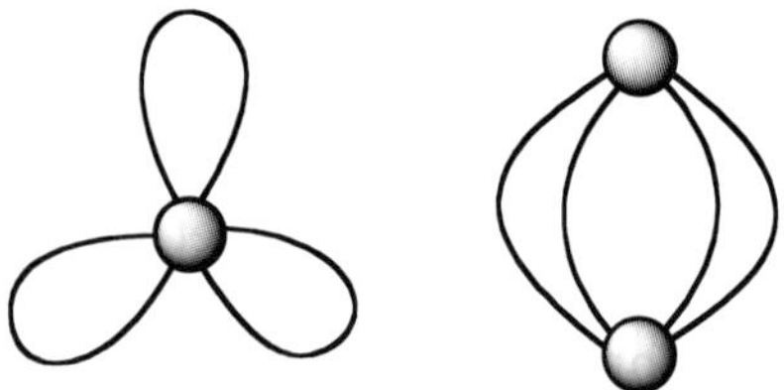

Figure 10.8 The quotient graphs for **pcu** and **dia**. Compare Figure 10.6.

For the three-connected ($p = 3$) nets the genus and the Z_t variable are related (since each edge shares two vertices) by the reformulation of equation 10.14:

$$g = \frac{pZ_t}{2} - (Z_t - 1) = \frac{3Z_t}{2} - (Z_t - 1) = \frac{Z_t}{2} + 1 \qquad (10.15)$$

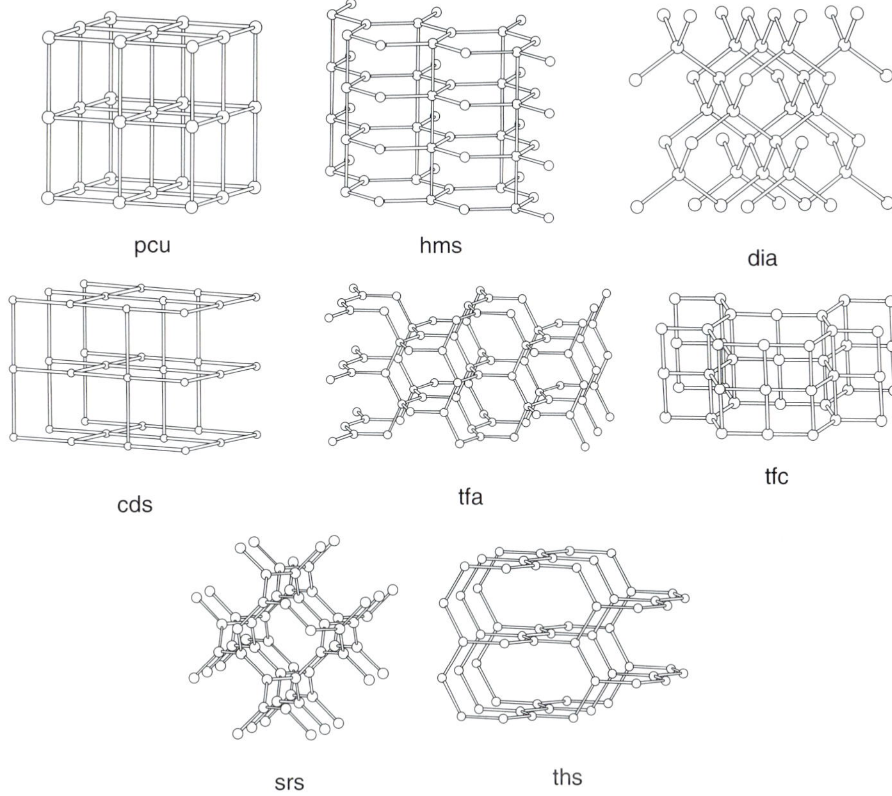

Figure 10.9 The 3D-nets with genus 3 (the minimal nets)

As an example of a net with higher genus we have drawn the four-connected **pts** net and its quotient graph in Figure 10.10

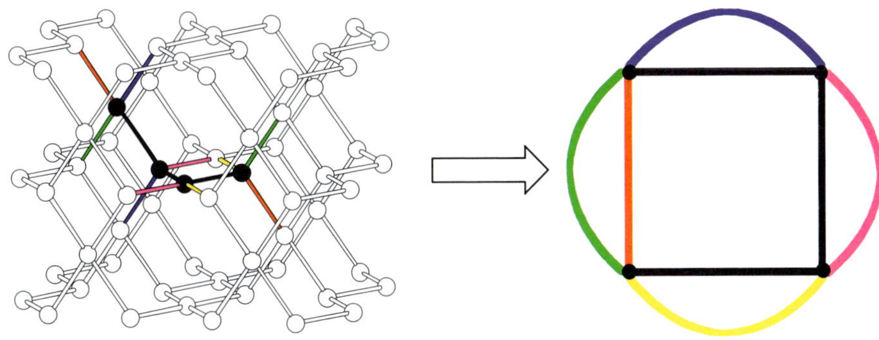

Figure 10.10 The **pts** net its quotient graph. The genus of this net is 5 and Z_t is 4.

With increasing complexity and lower symmetry we get higher values of the genus. For the three-connected nets in the RCSR database, [13] we find all values of *g* up to 13 with the exception of 6, 8 and 12. The highest known genus for such listed nets is 25, of which there are 7 examples, one shown in Figure 10.11, the 3;4 connected **rhr-a** net.

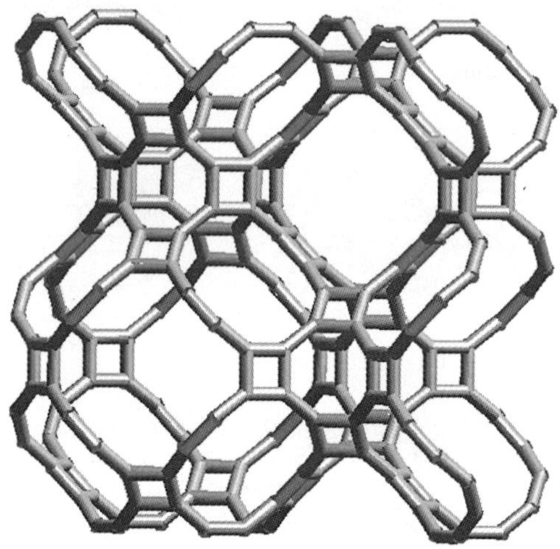

Figure 10.11 The genus of the three-connected **rhr-a** net is 25 and its vertex symbol is 4·8·12.

In conclusion, the concept of genus for a net gives a unified way of classifying nets with varying connectivity, and the quotient graphs is a neat shorthand representation of a net.

10.2.3. Nets as tilings and the concept of transitivity

It would be nice with a nomenclature with a complexity in between the convolution of the vertex symbols and the simplicity of the genus. A possible approach based on *nets as tilings* was suggested by O. Delgado-Friedrich et al. in 1999 and elaborated by O. Delgado-Friedrich, M. O'Keeffe and co-workers in subsequent years [11,14-16].

In short, it is based on constructing polyhedra (*the tiles*), using the rings formed by the net, that completely fills space. There are many ways to do this, so it was proposed that the shortest rings in the structure should be used as the faces of the tiles, so that for each net there would be only one *natural tiling*.

As can be seen in Figure 10.12 we are now talking about very different polyhedra from the five Platonic bodies. Each type of polyhedra (there can be several types of polyhedra used in the tiling) will have a certain number of faces defined by the rings (*r*), a certain number of edges (*q*) and a certain number of vertices, (*p*). The number of different types of polyhedra is *s*.

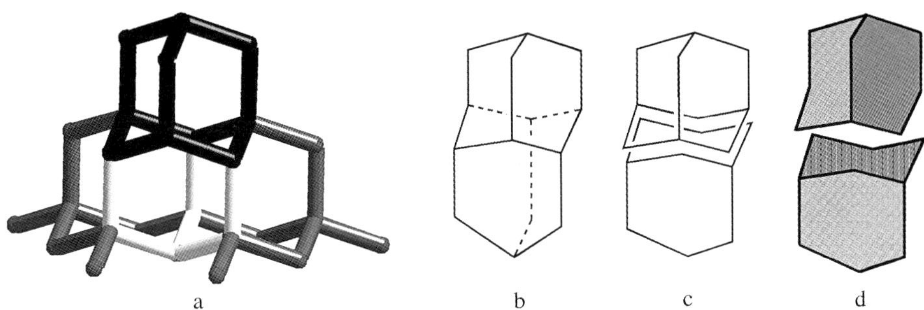

Figure 10.12 (a) Fragment of the diamond net (**dia**). Different tiles marked in black, white and grey. (b) Framework of two tiles (c) Framework of two tiles separated (d) Two tiles separated. This net has only one type of vertex (p), one type of edge (q) one type of ring (r) and one type of tile (s), so it has the *transitivity* 1111.

We can now describe the net by this set of numbers, *pqrs*, called the *transitivity* of the net. The point of this is that we will have a set of numbers, basically describing the complexity of the net, that again are independent of the type of connectivity at each node, and also of the size of the rings formed.

It has been shown that there are only five nets with transitivity 1111, called the *regular nets* [15]. These are given below in Table 10.2, and illustrated in Figure 10.13.

Table 10.2 The regular nets, After O. D. Friedrichs, M. O'Keeffe and O. M. Yaghi, *Acta Cryst. A* **2003**, *59*, 22-27.

Net	Coordination	Vertex symbol	Genus	Transitivity
srs	3	$10_5 \cdot 10_5 \cdot 10_5$	3	1111
nbo	4	$6_2 \cdot 6_2 \cdot 6_2 \cdot 6_2 \cdot 8_2 \cdot 8_2$	4	1111
dia	4	$6_2 \cdot 6_2 \cdot 6_2 \cdot 6_2 \cdot 6_2 \cdot 6_2$	3	1111
pcu	6	$4 \cdot 4 \cdot 4 \cdot 4 \cdot 4 \cdot 4 \cdot 4 \cdot 4 \cdot 4 \cdot 4 \cdot 4 \cdot 4 \cdot {*} \cdot {*} \cdot {*}$	3	1111
bcu	8	$4 \cdot 4 \cdot 4 \cdot 4 \cdot 4 \cdot 4 \cdot 4 \cdot 4 \cdot 4 \cdot 4 \cdot 4 \cdot 4 \cdot 4_3 \cdot 4_3 \cdot$ $4_3 \cdot 4_3 \cdot 4_3 \cdot 4_3 \cdot 4_3 \cdot 4_3 \cdot 4_3 \cdot 4_3 \cdot 4_3 \cdot 4_3 \cdot {*} \cdot {*} \cdot {*} \cdot {*}$	4	1111

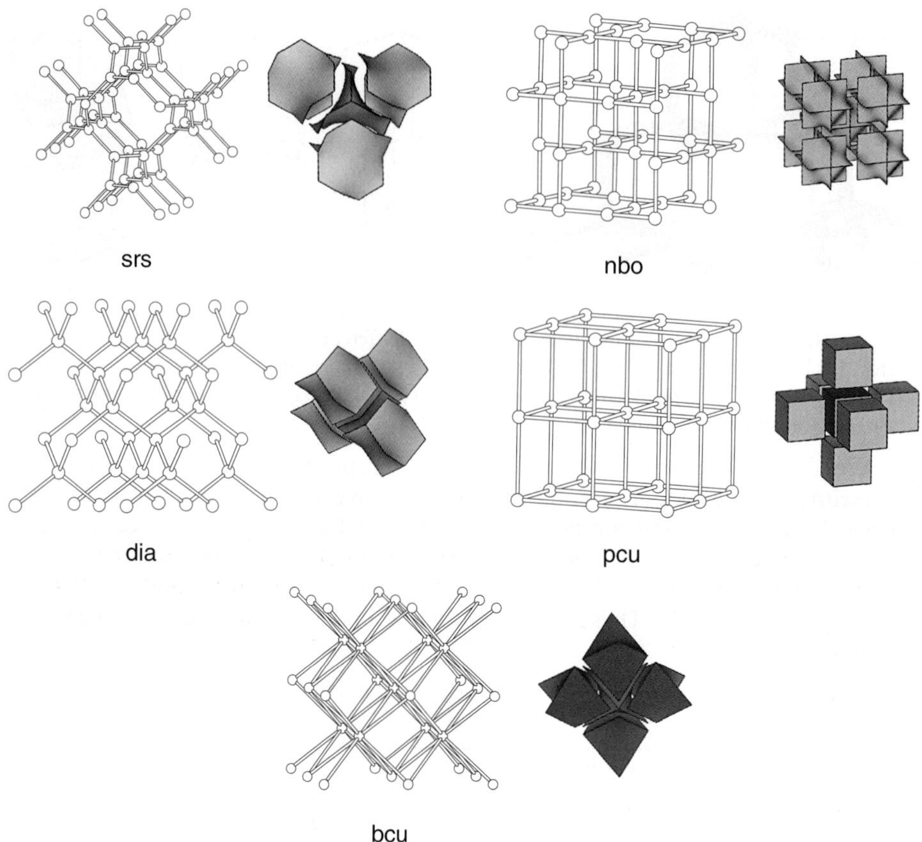

Figure 10.13 The regular nets are the nets with transitivity 1111, that is they are formed by one kind of tile, having only one kind of vertex, edge and ring. This figure shows the regular nets with their natural tilings. Adapt from ref. [15]

It has been proposed to name different classes of 3D-nets in descending order of symmetry: [15,16]

- **Regular nets** have transitivity 1111
- **Quasiregular nets** have transitivity 1112 (there is only one these, **fcu**[6].)
- **Semiregular nets** have transitivity 11rs (one kind of vertx, one kind of edge)

10.2.4. The analysis of the voids as the dual of a net

The definition of the tiles of a net means that they enclose a principal hole, or an empty space, in the net. As we can make many tilings apart from the natural

---

[6] See appendix A.

titling we can define many kinds of holes, and indeed, many of these nets have empty spaces that are best described as channels.

Two things make these empty spaces important. First, they are the *raison d'être* for many of these compounds; the porosity is one of the desired properties. Secondly, the fact that many network structures are interpenetrated makes the analysis of the empty space of a net important.

Interpenetration occurs in many degrees, two, three or four fold interpenetration being the most common, and is notoriously hard to predict, see Chapter 11 [17-20]. A way of finding out if a second net would fit inside the first would be to insert new vertices in the centre of the natural tilings and connect these new vertices through new edges passing through the face shared by the two vertices. If the new net so formed is identical to the first (have the same tiling) then the chances for double interpenetration should be good (see Figure 10.14). We say that this net is self-dual [15,16,21].

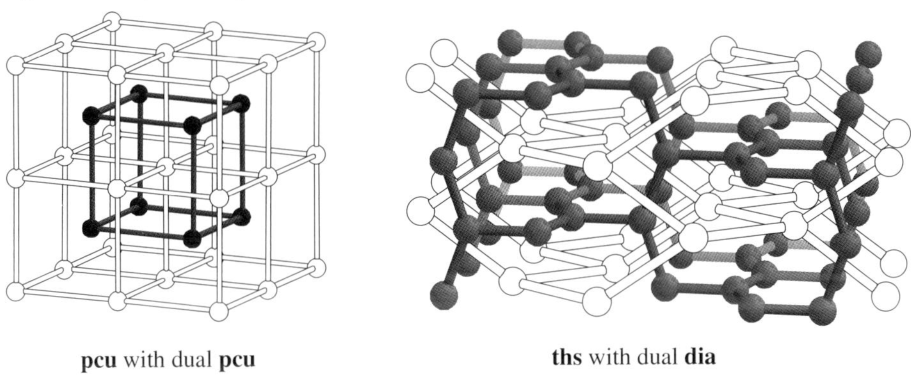

pcu with dual pcu   ths with dual dia

Figure 10.14 Left: The **pcu** net with its dual net that is an identical **pcu** net (self-dual). Right: The **the** net with its dual the **dia** net. Thus, the **pcu** net can easily be interpenetrated by an identical net, while this possibility is not obvious for **ths**.

We will examine four cases: the self dual nets **pcu** and **srs**, and the nets **pts** and **ths**, whose dual nets are unknown and **dia** respectively, through data from the CSD search for interpenetration by Proserpio et al. [17] and data from the CSD search for coordination polymers by O'Keeffe et al. [22], see Table 10.3

Table 10.3 Data for coordination polymers (metal organic frameworks) from O'Keeffe et al. [22] and for interpenetrated networks from Proserpio et al. [17][7]

Net	*Self dual*	*Total found [21]*	*Interpenetrated [17]*		
			Doubly	*Other*	*Total %*
pcu	yes	145	42	5	32
pts	no	21	4	0	16
srs	yes	72	17	6	32
ths	no	24	6	6	50

While the statistics of Table 10.3 is not completely reliable, much fewer **pts** nets than **pcu** nets where found, but the reverse trend was found for the **srs** vs. **ths**.

However, one can wonder if self duality may have something to do with the total frequency of nets found. Interpenetration is a way to obtain efficient packing, and if interpenetration is difficult to achieve, then the structure may not form at all, thus accounting for the lower number of **pts** and **ths** nets found. Also, as we shall see in the next chapter, not all interpenetrated nets achieve maximum separation of the nets, intertwining by double helix formation is one example.

We anticipate further studies in this area, both theoretical and by statistical analysis of data from the CSD.

10.2.5. The pqrs to srqp relation of dual nets

It may not be obvious to find the dual of a net, but it turns out that the transitivity of the dual net can be obtained directly from the transitivity of the original net.

First, we have one new vertex (p) inside each tile (s) thus directly giving us:

$$p_{\text{dual net}} = s_{\text{original net}} \qquad (10.16)$$

Then we consider each link (edge) that penetrates the face between two vertices. The number of different links will be equal to the number of different faces present, and each type of face consists of one type of ring. Thus we get:

$$q_{\text{dual net}} = r_{\text{original net}} \qquad (10.17)$$

This finally gives:

---

[7] Note that these two groups performed their CSD searches in different ways, and thus the interpenetrated set of nets is not necessarily a subset of the total number. The comparison, thus, is only approximate.

$$pqrs_{\text{dual net}} = srqp_{\text{original net}} \qquad (10.18)$$

This give also give a condition for self-duality; To have a self-dual net, the original net has to have a palindromic (reads the same forwards and backwards) transitivity sequence. In table 10.5 we present some nets that are self-duals (.

Table 10.4 The nets that are known to have dual nets that are identical to the original nets. (self-dual). Extracted from the Reticular Chemistry Structural Resource database [13].

Net	Transitivity	Vertex symbol	Genus
cds	1221	$6{\cdot}6{\cdot}6{\cdot}6{\cdot}6_2{\cdot}*$	3
dia	1111	$6_2{\cdot}6_2{\cdot}6_2{\cdot}6_2{\cdot}6_2{\cdot}6_2$	3
ftw	2112	See appendix[8]	9
hms	2222	$6{\cdot}6{\cdot}6{\cdot}6_2{\cdot}6_2{\cdot}6_2{\cdot}6_2{\cdot}6_2{\cdot}6_2{\cdot}*; \; 6_3{\cdot}6_3{\cdot}6_3{\cdot}$	3
pcu	1111	$4{\cdot}4{\cdot}4{\cdot}4{\cdot}4{\cdot}4{\cdot}4{\cdot}4{\cdot}4{\cdot}4{\cdot}4{\cdot}4{\cdot}*{\cdot}*{\cdot}*$	3
pyr	2112	$6{\cdot}6{\cdot}6{\cdot}6{\cdot}6{\cdot}6{\cdot}6_2{\cdot}6_2{\cdot}6_2{\cdot}6_2{\cdot}6_2{\cdot}6_2{\cdot}*{\cdot}*{\cdot}*; \; 6_3{\cdot}6_3{\cdot}6_3$	13
rtw	2332	$5{\cdot}5{\cdot}5{\cdot}5{\cdot}6_2{\cdot}6_2; \; 4{\cdot}4{\cdot}5{\cdot}5{\cdot}5 \; 6_3{\cdot}6_3{\cdot}6_3{\cdot}6_3{\cdot}*$	9
mcf	2442	$6{\cdot}6{\cdot}6{\cdot}6{\cdot}6_2{\cdot}6_2{\cdot}6_2{\cdot}6_2{\cdot}6_3{\cdot}6_3; \; 6_3{\cdot}6_3{\cdot}6_3$	5
srs	1111	$10_5{\cdot}10_5{\cdot}10_5$	3

Much more can be said about the dual nets, and much research no doubt remains to be done. We refer our interested readers to the original literature.

10.3. Glossary

Texts devoted to the more mathematical side of 3D-nets often contain numerous words and concepts unfamiliar to the chemist, and maybe even to the crystallographer. Here we try to explain some of them.

---

[8] This net is twelve- and four-connected and thus has 6+66 terms in the vertex symbol! (equation 4.1)

archimedean	An archimedean net is uninodal but contains different sizes of rings.
borromean	Rings linked together in such a fashion that there is no interpenetration of any ring into any other ring, but still the rings are inseparable.[9]

catalan	A catalan net has all rings of the same size but more than one kind of node.
Connectivity & coordination	We use the word connectivity to describe the number of nodes connected to each node. As "connectivity" is used in a different sense in graph theory "coordination" is sometimes used instead. However, since this word is used in again another way in coordination chemistry (a metal ion may be a three connected node but still six-coordinated) we are left with a situation where someone is bound to be unhappy.
dual	An object described by certain parameters *ab* has a dual described by the parameters *ba*. The cube is the dual of the octahedron.
dual net	A net fitted inside a net with transitivity *pqrs* respecting certain symmetry requirements of the first net so that the new transitivity is *srqp*.
edges	Connection (links) between nodes.
Euler's theorem	The relation between faces, edges, and vertices in a polyhedron: F-E+V = 2
genus	A way of dividing objects into classes, describing them with integer numbers. Roughly represents the number of holes in a 3D body.
n-periodic net	n-dimensional net
Platonic bodies	The five polyhedra that can be constructed from one kind of face and one kind of vertex.
quotient graph	A finite graph retaining the properties of a (*n*-dimensional) net.
spanning tree	The graph connecting the Z_t vertices (nodes) of the smallest translational repeating unit of a net.
Stable sphere packings	Packing of uniform spheres with at least four contacts between spheres and not all the contacts on the same hemisphere.
tessellation	Dividing a surface into polygons.
topology	Studies of properties an object retains under deformation.
transitivity	Describes the tiling of a net by *s* different polyhedra having a certain

---

[9] The Borromean ring was named after the crest of the Borromeo family in 15th-century Italy.

	number of faces defined by the rings (r), a certain number of edges (q) and a certain number of vertices, (p)..
uniform	Composed of the same type of rings (or polygons)
vertex	node

                                    L. Öhrström & K. Larsson

References

[1] M. O'Keeffe, B. G. Hyde, Crystal Structures I: Patterns and Symmetry, Mineral Soc. Am., Washington, 1996.

[2] A. F. Wells, Further Studies of Three-Dimensional Nets, Polycrystal book service, Pittsburgh, 1979.

[3] A. F. Wells, Three-dimensional nets and polyhedra, John Wiley & Sons, New York, 1977.

[4] X. Yuan, A. N. Cormack, Comp. Mat. Sci. 24 (2002) 343.

[5] H. S. M. Coexter, Regular Polytopes, 2nd ed Dover, New York, 1973.

[6] N. N. Greenwood, A. Earnshaw, Chemistry of the Elements, 2nd Pergamon Press, Oxford, 1997.

[7] Encyclopædia Britannica Online, Encyclopædia Britannica Inc., 2005,

[8] J. Turulski, J. Niedzielski, J. Mathem. Chem. 36 (2004) 29.

[9] J. Turulski, J. Niedzielski, J. Chem. Inf. Comp. Sci. 42 (2002) 534.

[10] T. P. Radhakrishnan, J. Mathem. Chem. 5 (1990) 381.

[11] C. Bonneau, O. Delgado-Friedrichs, M. O'Keeffe, O. M. Yaghi, Acta Cryst. A 60 (2004) 517.

[12] W. E. Klee, Cryst. Res. Technol. 39 (2004) 959.

[13] M. O'Keeffe, O. M. Yaghi, Reticular Chemistry Structure Resource, Tucson, Arizona State University, 2005, http://okeeffe-ws1.la.asu.edu/RCSR/home.htm

[14] O. Delgado-Friedrichs, A. W. M. Dress, D. H. Huson, J. Klinowski, A. L. Mackay, Nature 400 (1999) 644.

[15] O. Delgado-Friedrichs, M. O'Keeffe, O. M. Yaghi, Acta Cryst. A 59 (2003) 22.

[16] O. Delgado-Friedrichs, M. O'Keeffe, O. M. Yaghi, Acta Cryst. A 59 (2003) 515.

[17] V. A. Blatov, L. Carlucci, G. Ciani, D. M. Proserpio, Crystengcomm 6 (2004) 377.

[18] S. R. Batten, Crystengcomm (2001) 1.

[19] S. R. Batten, R. Robson, Angew. Chem. Int. Ed. 37 (1998) 1461.

[20] L. Carlucci, G. Ciani, D. M. Proserpio, Coord. Chem. Rev. 246 (2003) 247.

[21] O. Delgado-Friedrichs, M. O'Keeffe, O. M. Yaghi, Solid State Sciences 5 (2003) 73.

[22] N. W. Ockwig, O. Delgado-Friedrichs, M. O'Keeffe, O. M. Yaghi, Acc. Chem. Res. 38 (2005) 176.

Chapter 11

Interpenetration - strategies and nomenclature

We first encountered interpenetration (also known as catenation, see however discussion in Carlucci et al. [1]) in Chapter 3 and through Chapters 5-9 we have noted that may structures contain interpenetrating nets, and occasionally we have shown the interpenetrating patterns. We have avoided detailed discussion of this since the emphasis has been on the nets themselves, but in this chapter we will venture a bit further, and we partly base this discussion on some recent reviews [1-6] and the web pages by Batten [7]. Note, though, that also Wells has a rather extensive discussion of interpenetration in his book, although at that time only a couple of examples were known [8].

11.1. What is interpenetration

The large voids formed by some single nets are seldom empty. Besides solvent molecules and counter ions, these voids can be filled with one or more additional nets. To be classified as interpenetrating, these nets should not be able to separate without breaking any bonds. Self-penetration on the other hand is a topological property of a single net (see for example section 6.2.1) and not dealt with in this chapter.

Depending on the constituting molecules the interpenetrated nets can be entwined together, located in the complimentary voids of the other net or a combination of both. Depending on the number of networks in the complete structure, they are classified as N-fold networks where N is two or more. Thus for a structure with four interpenetrating networks it will be a four-fold network. We also talk synonymously about degree of interpenetration.

Currently, we note that the highest degree of interpenetration is the eleven-fold hydrogen bonded diamond net of the 1:2 adduct of tetrakis(4-(3-hydroxyphenyl)phenyl)methane and benzoquinone, [9] and the ten fold diamond net in $[Ag(1,12\text{-}dodecanedinitrile)_2]NO_3$ [10].

We may also distinguish between cases where the nets are maximally displaced from each other, and cases where there are attractive forces between the nets to give for example double helices or other kinds of *interweaving* [6].

We will discuss nomenclature of interpenetration at the end of this chapter because in order to understand the "topology of interpenetration" and thus the complete crystal structure we need a language to describe it with. However, we will first deal with strategies to avoid (or to promote) interpenetration.

11.2. Strategies to control interpenetration

It may be desirable to avoid interpenetration, for example if we want to create a porous structure. Note, however, that porosity will not be automatically destroyed by interpenetration, and that sometimes interpenetration may actually be advantageous. It may provide additional stability to the net and it may also give the structure a larger surface area.

In any case, if the synthetic chemist wants to have control over the material produced, both the net and the degree of interpenetration have to be mastered. There may be two slightly different ways to (eventually) get there, we can either analyse the nets themselves, or we can analyse the voids within them.

11.2.1. Density of nets

It seems perhaps reasonable that nets built from the same kinds of nodes and having the same lengths of the links should also have the same density (counted as nodes per unit volume[1]), but this is not the case. In Table 11.1 we show some data from the RCSR database [11] for four-connected nets.

Table 11.1 Nodal density of some four-connected nets [11].

Net	*Vertex Symbol*	*Nodes/ unit cell*	*Density[nodes/$Å^3$]* [a]	*Volume[Å]* [b]
sod	$4{\cdot}4{\cdot}6{\cdot}6{\cdot}6{\cdot}6$	12	0.5303	22.6268
sra	$4{\cdot}6{\cdot}4{\cdot}6{\cdot}6{\cdot}8_2$	8	0.5535	14.4535
lon	$6_2{\cdot}6_2{\cdot}6_2{\cdot}6_2{\cdot}6_2{\cdot}6_2$	4	0.6494	6.1593
dia	$6_2{\cdot}6_2{\cdot}6_2{\cdot}6_2{\cdot}6_2{\cdot}6_2$	8	0.6495	12.3168
qtz	$6{\cdot}6{\cdot}6_2{\cdot}6_2{\cdot}8_7{\cdot}8_7$	3	0.7500	4.0001

[a] Number of nodes in the unit cell divided by the volume [b] Unit cell volume for the ideal net calculated for a node-node distance of 1.00 Å.

There are some possible conclusions from this table. For example, the higher density of the quartz (**qtz**) net compared to the diamond (**dia**) net means that there is less space available for interpenetration in **qtz** structures. The table also suggest that for a given chemical system the degree of interpenetration may vary with the net adapted.

We may also make some rough estimates from the volumes given. Take the commonly used bidentate ligand 4,4′-bipyridine and let it coordinate to a metal

---

[1] The density can also be expressed as the fraction volume occupied by spheres that are placed in the nodal positions with radii equal to half the link length, so called sphere packings. The difference between the sphere packing ratio and the nodal density is a factor $4{\cdot}\pi{\cdot}0.5^3/3$. Thus the density of the **fcu** net (face centred or cubic close packing) is 1.4143 which gives the sphere packing ration as 74% which is the closest possible packing of uniform spheres, well known from any general chemistry textbook.

(M) ion in tetrahedral fashion. Suppose this will make a **dia** net. Then we can calculate the volume per node using approximate values for the atomic radii ($C_{aromatic}$H: 1.76 Å, N: 1.64 M: 2.20) [12] and scale the volume with the M-M distance (approximately 11.4 Å). We then get an atoms-in-net to volume-of-net ratio of 4600/18250 = 0.25 suggesting a maximum degree of interpenetration of four. If we take some kind of counter ion into account will get a projected degree of interpenetration at around 3.5. In real structures four-fold interpenetration was found in [Cu(4,4′-bipyridine)$_2$]PF$_6$ [13] and [Ag(4,4′-bipyridine)$_2$]CF$_3$SO$_3$ [14], but it should be stressed that calculations such as these are very approximate and very dependent on the choice of atom radii.

As we are talking about volumes, we should add that Wells also noted that a cubic close packing (face centred close packing) can be broken down into several interpenetrating nets for example four (10,3)-a or **srs** nets or three (10,3)-b or **ths** nets [8].

11.2.2. Lengths and thickness of links, size of counter ion

Another factor to take into account is of course the length of the connectors between the nodes. One would naively expect an increased degree of interpenetration with increased linker length or decreased counter ion size. This has also been observed for a series of compound made from AgBF$_4$, AgClO$_4$, AgPF$_6$, AgAsF$_6$, AgSbF$_6$ or AgSO$_3$CF$_3$ reacted with NC(CH$_2$)$_n$CN (n = 2 to 7), although other effects (some unknown) were clearly also as important. Sometimes diamond nets were formed, sometimes 2D nets, and the authors surprisingly find that "the diamondoid networks are formed only with the dinitriles having an even number of carbon atoms" [15].

Another series of structures with diamond nets were reported by Evans and Lin containing Cd(II) or Zn(II) and bridging *p*-pyridinecarboxylate ligands, and this resulted in the convincing graph in Figure 11.1 [16].

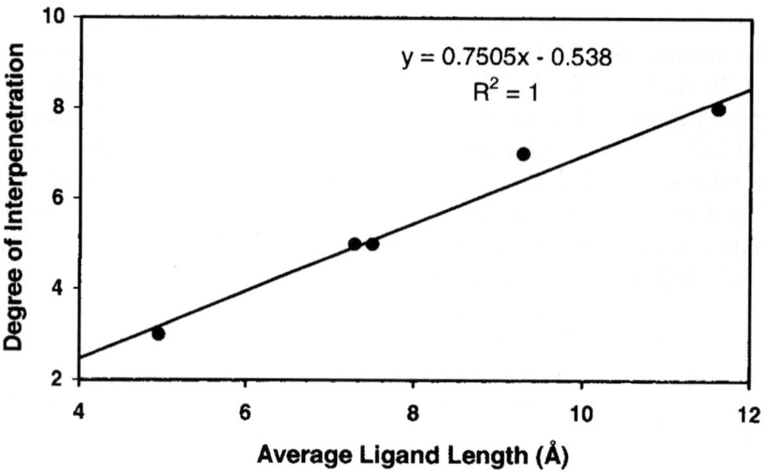

Figure 11.1 Cd(II) or Zn(II) and bridging *p*-pyridinecarboxylate ligands give diamond nets where the degree of interpenetration is dependent on the ligand length. Reprinted with permission from [16]. Copyright 2002 American Chemical Society.

They further noted that in order to get acentric solids, odd-fold interpenetration is advantageous since this prevents the crystallisation into centrosymmetric structures [17].

Yaghi, O'Keeffe and co-workers developed a simple mathematical model for interpenetration of **pcu** (primitive cubic packing) nets using the parameters in Figure 11.2 [18].

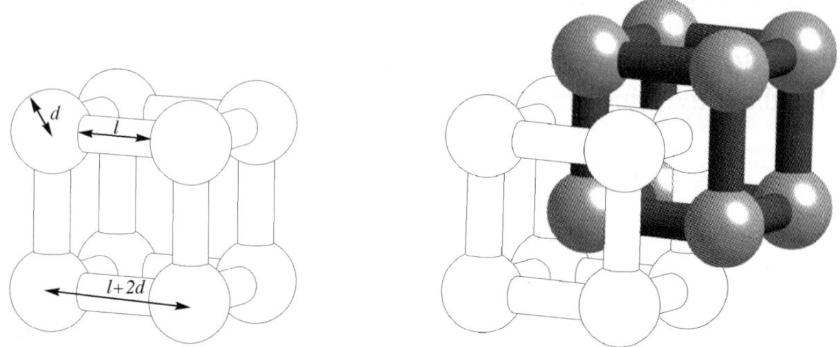

Figure 11.2 Left: Parameters used in Yaghi and O'Keeffe's mathematical model for interpenetration of **pcu** (primitive cubic packing) nets. The connectors are assumed to be infinitely thin. Right: Two-fold interpenetrated net.

By evenly spacing the interpenetrated nets along the body diagonal they could calculate the maximal degree of interpenetration for given values of *d* and *l*, and also the free volume in the structure, see Figure 11.3 The model assumes negligible thickness of the connectors and is therefore only strictly valid for systems with larger clusters as nodes.

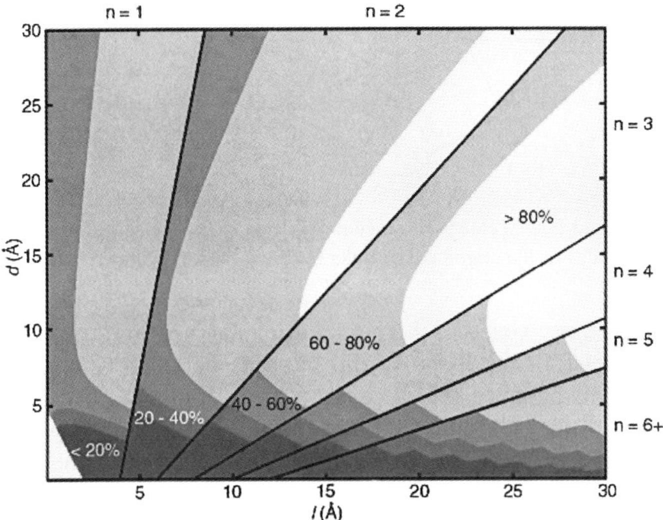

Figure 11.3 By evenly spacing the interpenetrated nets along the body diagonal the maximal degree of interpenetration was calculate for given values of d and l, (solid lines) and also the free volume in the structure was calculated (shaded areas). Reprinted with permission from [18]. Copyright 2000 American Chemical Society.

But nature does not always follow a rule of "maximal interpenetration". The well known (by now) compounds with general formula $Zn_4O(L)_3$ where L is a di-carboxylate ligand with lengths from 7 Å to 16 Å, were often found to contain interpenetrated nets, however by carrying out the syntheses under more dilute conditions non-interpenetrated nets could be obtained for all linkers [19].

11.2.3. Analysis of the voids within a net

It is obvious that also the symmetry of the net and of the empty space within it has a large influence on the interpenetration. Consider the simplified 2D examples of Figure 11.4: We will make a new net from the starting net and then displace it Δx and Δy. In the (4,4) square grid it is easy to find displacement parameters that avoid collisions between nodes or the alignment of links but much more difficult for the (3;4,5) net. We might ask if there is some specific geometric property that describes this phenomenon?

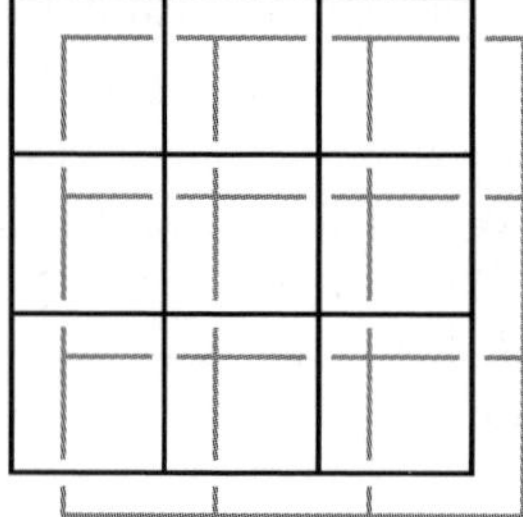 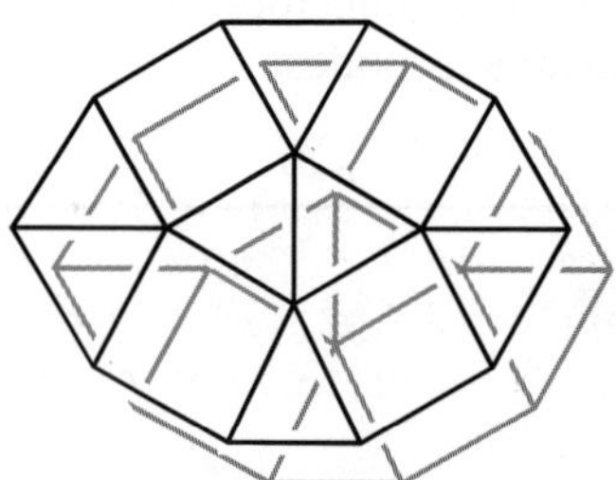

Figure 11.4 Left: It is easy to find displacement parameters Δx and Δy for the second net (grey) that avoid collisions between nodes or the alignment of links for the (4,4)-net. Right: This is much more difficult for the (3;4,5) net.

The answer is that the empty space in the net can be analysed in terms of the *dual net* of the original net [20-22]. This we did in some detail in Chapter 10, but we will demonstrate this again, using 2D examples. In Figure 11.5 we have created two new nets from the (4,4) and (3;4,5) nets in Figure 11.4 by placing new nodes (vertices) in the centre of each polygon and then connecting these nodes by links intersecting the old links at their midpoints.

These nets are called the *dual nets* and we see that for (4,4) we obtain an identical net, but for (3;4,5) we get a different (5, 3;4) net. We say that the (4,4) net is *self dual* and the procedure to obtain the dual nets from 3D nets is exactly the same (for a 3D net the links will go through the centre of the faces of the tiles defining the net) [20,22,23].

The conclusion we can draw from this is that if a net is self dual, then it should be easy for one or more identical nets to interpenetrate. The concept of using dual nets in the analysis of interpenetration is still rather recent, but seems promising, see for example the discussion of a number of four-connected nets in a recent article by Delgado Friedrichs, O'Keeffe and Yaghi [21].

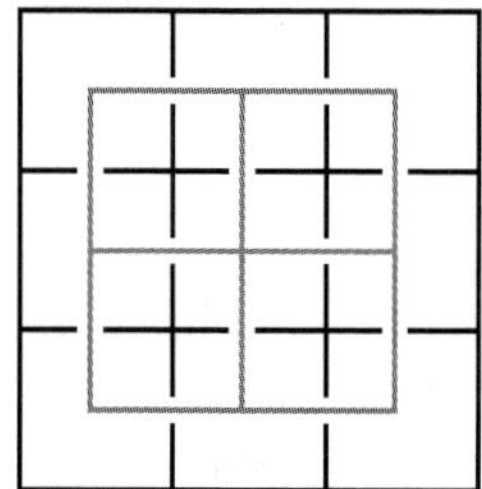 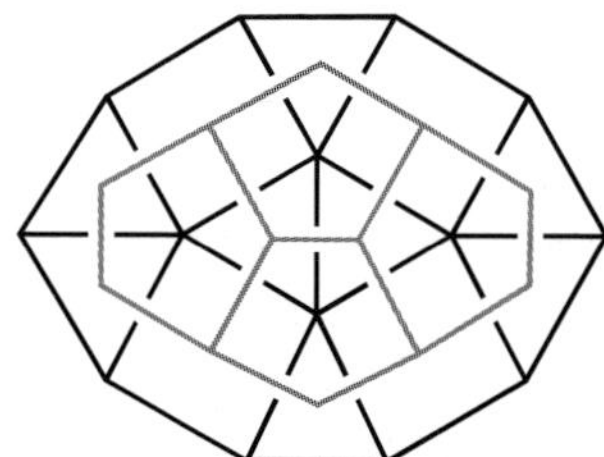

Figure 11.5 We have created two new nets (grey) from the (4,4) and (3;4,5) nets by placing new nodes in the centre of each polygon and then connecting these nodes by links intersecting the old links at their midpoints. These are the *dual nets* and we see that in the first place we obtain the same net as we started with, but in the second case we get a different net.

A strategy to avoid interpenetration would then be to deliberately chose a net that has a dual net very different from the original net. This was successfully implemented in the synthesis of $[Zn_4O(1,3,5\text{-benzenetribenzoate})_2]$ [24]

We will not discuss this further, but we note that although the number of self dual nets is limited, currently there are nine such nets known, **cds**, **dia**, **ftw**, **hms**, **pcu**, **pyr**, **rtw**, **mcf**, and **srs**, see Table 5.5, some of them they are among the most common nets. However, interpenetration is not restricted to these nets!

11.2.4. Specially designed ligands

It has been argued that ligands could be specially designed to prevent interpenetration [25]. Possibly we could think of the ligands in Figure 11.6 as containing preformed parts of the net (several nodes) and thus for interpenetration to take place much larger units with "awkward" shapes need to assemble. See also [26]

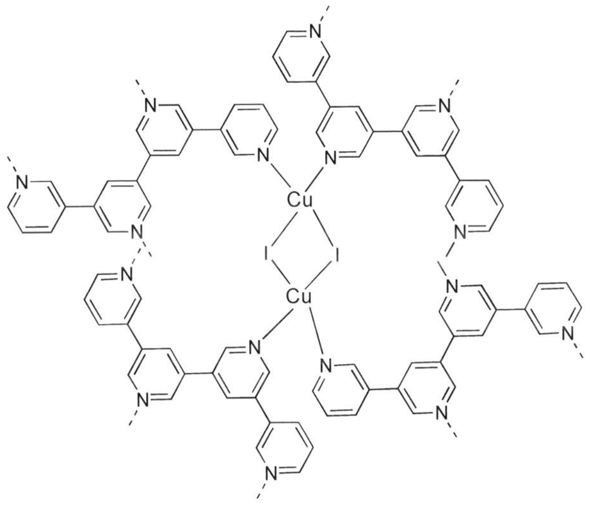

Figure 11.6 3,3':5',3":5",3"'-Quaterpyridine was used as a ligand to prevent interpenetration giving a **mof** net with CuI (see also section 6.3.1) [25].

11.3. Nomenclature

A descriptive notation has been developed by Robson and Batten covering all types of interpenetration (here we only treat 3D-3D, but there are 2D-2D, 2D-3D etc) dividing them into either parallel interpenetration, when there is no change of direction of the nets, only a parallel translation, and inclined interpenetration when for example two 2D-nets are no longer parallel [5,6].

Recently, Proserpio and co-workers formalised this and proposed a more stringent, but also more complicated, classification of interpenetrating nets.

                                                        L. Öhrström & K. Larsson

They investigated the Cambridge Structural Database for covalent or coordination nets that interpenetrate using the TOPOS software package and divided their results into three main classes and several subclasses as shown in Table 4.1 [2].

Table 11.1 Classes of interpenetration according to Proserpio et al. [2]

Class	Net relations	Z^a	Z symbol	Vector
Ia (one translation vector)	Only translations	Z_t	Z_t	FIV[b]
Ib (more than one translation vector)	Only translations	Z_t	$Z_t(\Sigma Z_{it}*\Sigma Z_{ct})$	PIV[c]
IIa (one FISE)	Symmetry operations	Z_n	Z_n	None
IIb	Symmetry operations	Z_n		None
IIIa (one translation vector)	Translations and symmetry operations	$Z_t \cdot Z_n$	$Z(Z_t*Z_n)$	TIV[d]
IIIb (more than one translation vector)	Translations and symmetry operations	$Z_t \cdot Z_n$	$Z(\Sigma(Z_{it}*\Sigma Z_{ct})*Z_n)$	PIV[c]

[a] Total degree of interpenetration [b] full interpenetration vector, one vector is enough to describe the entire interpenetration [c] partial interpenetration vectors, several vectors needed to describe the interpenetration [d] total interpenetration vector.

As can be seen in the table, the interpenetration is first divided into groups depending on if there are translational or non-translational relations between the nets. That is, if we can take the net from its first origin and then just move it, without rotating or tilting, to its next location, then there is only a translation involved, and we have class I. The direction and distance we have to move it we call the interpenetration vector.

Each class is then divided into subclasses, **a** and **b** depending on if we can build up the structure by using the original net and only one displacement vector, (subclass **a**) or as in subclass **b** when we need two or more vectors to create the entire structure.

The number Z is the total degree of interpenetration and is the product of Z_t (translational interpenetrating nets) and Z_n (non-translational interpenetrating nets).

Class **I** thus contains nets which are related only by translation and in **Ia** one vector relates all nets, called a full interpenetration vector (**FIV**).[2] There are of course an infinite number of vectors which relates the individual nets, but, only the shortest should be listed. Note that this does not imply doubly interpenetrated nets, the same vector can be employed several times to create any degree of interpenetration.

---

[2] This vector can be calculated by taking the crystallographic (not Cartesian) coordinates (x_i, y_i, z_i) of equivalent points in two adjacent nets and constructing the vector $(x_2-x_1, y_2-y_1, z_2-z_1)$.

Type **Ib** needs more than one vector to connect all nets, and these are called partial interpenetration vectors (**PIV**). These are included in the Z-symbol as $Z_t(\sum Z_{it}*\sum Z_{ct})$, where Z_{it} is the number of integer translations such as the [1,0,0], [0,1,0] and [0,0,1] vectors, and Z_{ct} the number of centering translations, that is where x,y and z in the vector [x,y,z] has a values of 0.5 or 0. In this case, the degree of interpenetration will be;

$$Z = \Sigma Z_{it} \cdot \Sigma Z_{ct} \qquad\qquad 4.2$$

where ΣZ_{it} and ΣZ_{ct} are equal to the number of translations of each type plus one (so that we count also the original net). To illustrate the **Ia** and **Ib** class, Figure 11.7 and Figure 11.8 show two **srs** (10,3)-a nets, one of each class.

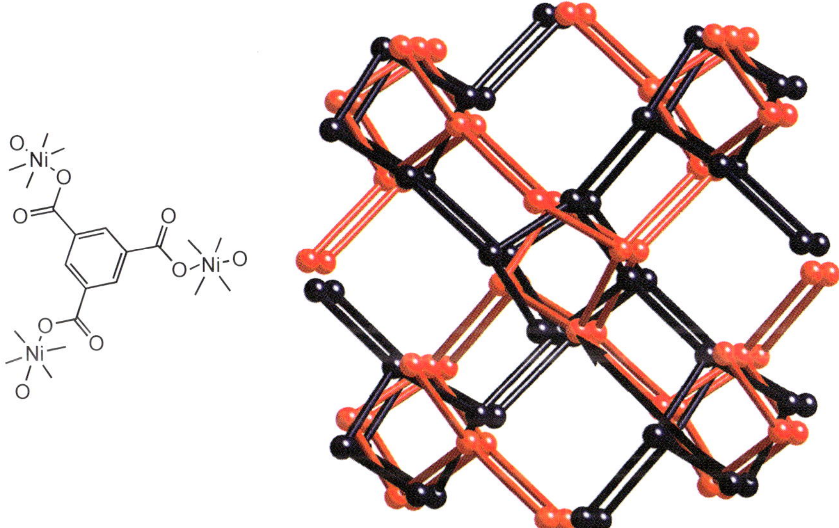

Figure 11.7 Two **srs** nets built of benzene-1,3,5-tricarboxylato anions coordinated to Ni(II) with additional (R)-phenylethane-1,2-diol and pyridine ligands completing the coordination sphere [27]. The double interpenetration is of class Ia and shows how interpenetration can reinforce a porous structure rather than just clog the channels. The nodes are benzene centroids.

symmetry element (**NISE**), similar to FISE in Class **IIa**, and partial symmetry element (**PISE**). The example in Figure 11.11 contains three centering translations ($\sum Z_{ct}=4$) and one inversion centre ($\sum Z_n=2$). This results in an 8-folded interpenetrating net and the Z symbol is written as 8[(1*4)*2].

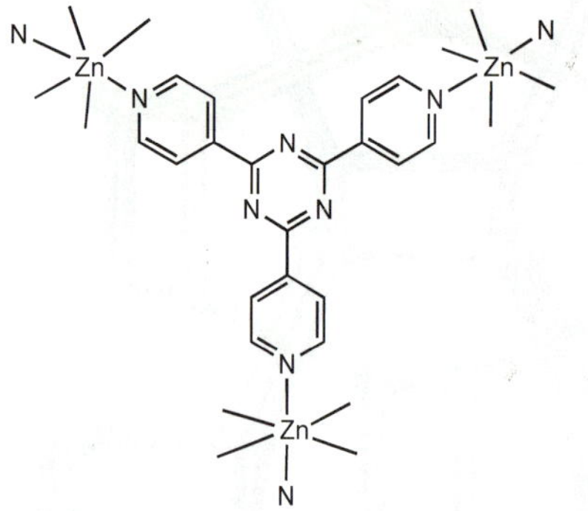

Figure 11.11A A 2,4,6-tris(4-pyridyl)-1,3,5-triazine Zn(II) network that forms an eight-fold interpenetrating **srs** net [30].

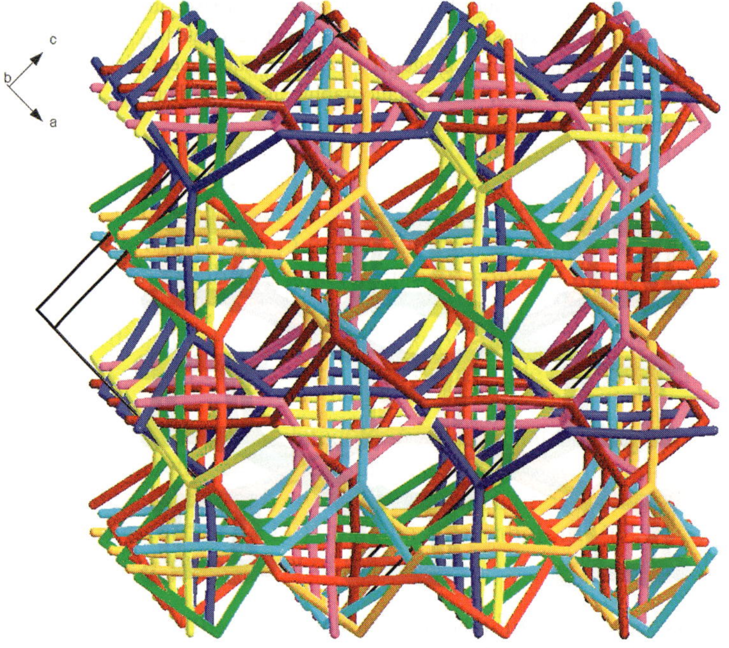

Figure 11.11B A 2,4,6-tris(4-pyridyl)-1,3,5-triazine Zn(II) network that forms an eight-fold interpenetrating **srs** net [30]. The structure also contains SiF_6^{2-} anions, methanol and water. The relationships between the different nets are shows in the following figures.

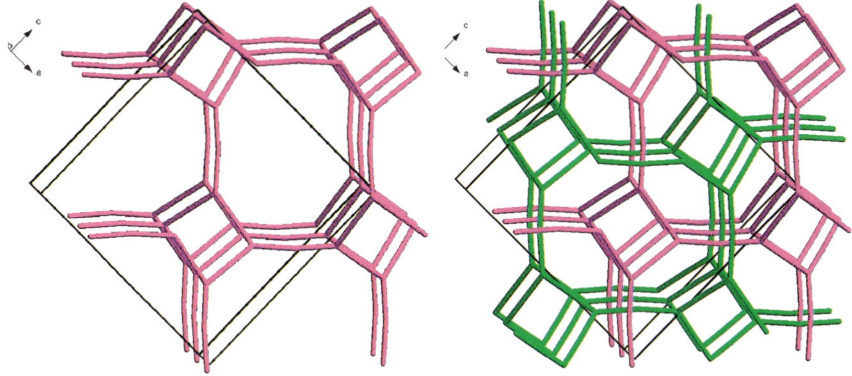

Figure 11.12 Left: One net from Figure 11.11B. Right: The green net is related to the magenta net by a [1/2,1/2,0] centring vector.

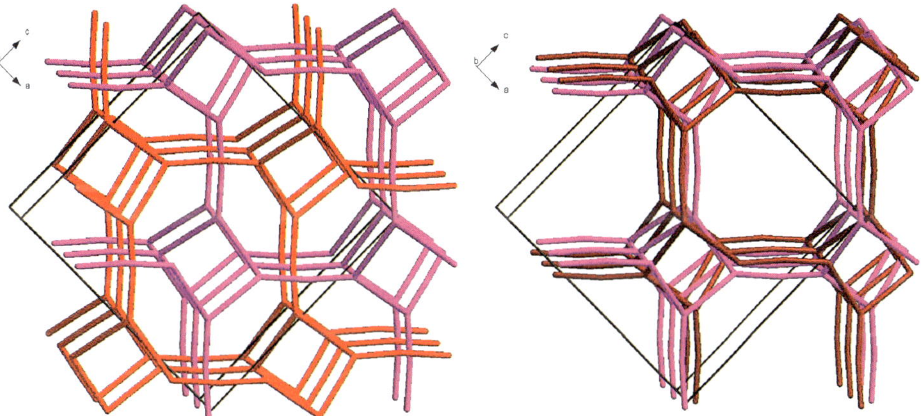

Figure 11.13 Left: The red net is related to the magenta net by a [0,1/2,1/2] centring vector. Right: The brown net is related to the magenta by a [1/2,0,1/2] centring vector.

                                          L. Öhrström & K. Larsson

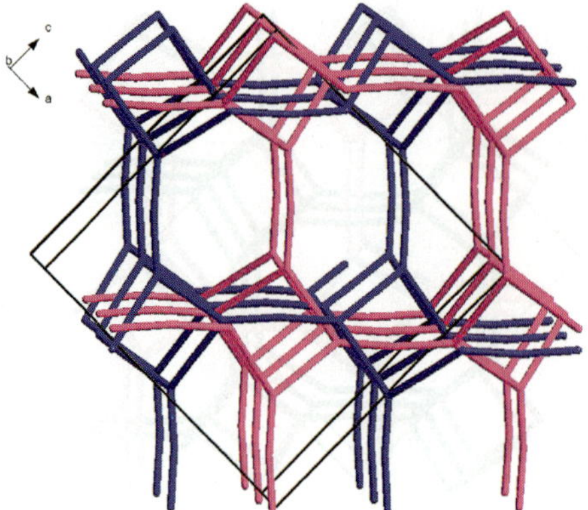

Figure 11.14 The blue net is related to the magenta by an inversion axis.

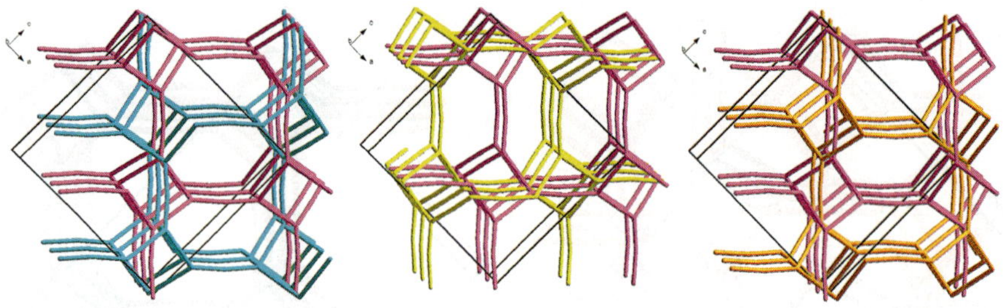

Figure 11.15 The last three pictures shows the 3 remaining nets created by the inversion axis and three centering vectors.

References

[1] L. Carlucci, G. Ciani, D. M. Proserpio, Coord. Chem. Rev. 246 (2003) 247.

[2] V. A. Blatov, L. Carlucci, G. Ciani, D. M. Proserpio, Crystengcomm 6 (2004) 377.

[3] L. Carlucci, G. Ciani, D. M. Proserpio, Crystengcomm 5 (2003) 269.

[4] S. R. Batten, Curr. Opin. Solid State Mat. Sci. 5 (2001) 107.

[5] S. R. Batten, Crystengcomm (2001) 1.

[6] S. R. Batten, R. Robson, Angew. Chem. Int. Ed. 37 (1998) 1461.

[7] S. R. Batten, Monash University, Australia, 2005,
http://web.chem.monash.edu.au/Department/Staff/Batten/Intptn.htm

[8] A. F. Wells, Three-dimensional nets and polyhedra, John Wiley & Sons, New York, 1977.

[9] D. S. Reddy, T. Dewa, K. Endo, Y. Aoyama, Angew. Chem. Int. Ed. 39 (2000) 4266.

[10] L. Carlucci, G. Ciani, D. M. Proserpio, S. Rizzato, Chem. Eur. J. 8 (2002) 1520.

[11] M. O'Keeffe, O. M. Yaghi, Reticular Chemistry Structure Resource, Tucson, Arizona State
University, 2005, http://okeeffe-ws1.la.asu.edu/RCSR/home.htm

[12] M. Gerstein, F. M. Richards, in: The International Tables for Crystallography: Vol 7.
Crystallography of biological macromolecules, M.G. Rossmann, E. Arnold (Eds.) IUCr, 2005.

[13] L. R. MacGillivray, S. Subramanian, M. J. Zaworotko, J. Chem. Soc., Chem. Commun.
(1994) 1325.

[14] L. Carlucci, G. Ciani, D. M. Proserpio, A. Sironi, J. Chem. Soc., Chem. Commun. (1994)
2755.

[15] L. Carlucci, G. Ciani, D. M. Proserpio, S. Rizzato, Crystengcomm (2002) 413.

[16] O. R. Evans, W. B. Lin, Acc. Chem. Res. 35 (2002) 511.

[17] M. J. Zaworotko, Chem. Soc. Rev. 23 (1994) 283.

[18] T. M. Reineke, M. Eddaoudi, D. Moler, M. O'Keeffe, O. M. Yaghi, J. Am. Chem. Soc. 122
(2000) 4843.

[19] M. Eddaoudi, J. Kim, N. Rosi, D. Vodak, J. Wachter, M. O'Keefe, O. M. Yaghi, Science 295
(2002) 469.

[20] O. Delgado-Friedrichs, M. O'Keeffe, O. M. Yaghi, Acta Cryst. A 59 (2003) 515.

[21] O. Delgado-Friedrichs, M. O'Keeffe, O. M. Yaghi, Solid State Sc. 5 (2003) 73.

[22] O. Delgado-Friedrichs, M. O'Keeffe, O. M. Yaghi, Acta Cryst. A 59 (2003) 22.

[23] C. Bonneau, O. Delgado-Friedrichs, M. O'Keeffe, O. M. Yaghi, Acta Cryst. A 60 (2004)
517.

[24] H. K. Chae, D. Y. Siberio-Perez, J. Kim, Y. B. Go, M. Eddaoudi, A. J. Matzger, M.
O'Keeffe, O. M. Yaghi, Nature 427 (2004) 523.

[25] K. Biradha, M. Aoyagi, M. Fujita, J. Am. Chem. Soc. 122 (2000) 2397.

[26] N. L. Rosi, M. Eddaoudi, J. Kim, M. O'Keeffe, O. M. Yaghi, Angew. Chem. Int. Ed. 41
(2002) 284.

[27] T. J. Prior, M. J. Rosseinsky, Inorg. Chem. 42 (2003) 1564.

[28] A. Johansson, M. Håkansson, S. Jagner, Chem. Eur. J. (2005).

[29] A. Mosset, M. Abboudi, J. Galy, Z. Kristallogr 164 (1983) 171.

[30] B. F. Abrahams, S. R. Batten, H. Hamit, B. F. Hoskins, R. Robson, J. Chem. Soc., Chem.
Commun. (1996) 1313.

Chapter 12

3D-nets as specific synthetic targets – Crystal Engineering

In Chapter 2 we outlined a number of properties that made molecule-based 3D nets interesting targets for new materials. In subsequent chapters we have occasionally made reference to the properties of the specific compounds discussed, but almost never touched on the synthetic strategy behind the preparations. This is something we will try to do very briefly in this chapter. Some useful general comments about porous coordination polymers can be found in Rowsell & Yaghi, [1], otherwise we refer the reader back to the original literature.

12.1. Choice of interaction (hydrogen bonds, coordinative bonds etc)

The possible interactions that we can use to build our net were discussed in Chapter 3, now there is the time to make a choice between them.

12.1.1. Covalent bonds

The reactions forming the covalent bonds that we normally associate with organic and polymer synthesis are seldom found in "net making", possibly because of their lack of reversibility that prevents "bricks" inserted in the wrong way to be removed and the error corrected. (Remember how hard it is to make diamonds!) On the other hand, zeolites and related materials are clearly covalent materials, although formed under very different conditions, and as calcinogens and other main group elements are further studied in this respect, it does not seem unlikely that we shall see 3D-nets based on at least organic-main group chemistry, for example using sulphur-sulphur bonds.

12.1.2. Metal-ligand bonds (coordination bonds)

Metal-ligand bonds, or coordination bonds, on the other hand, have been successfully used, as we have seen in the past chapters. They often combine strong, partly covalent interactions with reversibility. Single metal-ligand bonds easily build up 3D-structures but may suffer from some inherent flexibility in the coordination sphere (Barry pseudorotation etc), smaller clusters (for example [$Zn_4O(OOCR)_6$] are more inert in this respect.

12.1.3. Hydrogen bonds

If several hydrogen bonds work together to form a synthon, the interaction may be as strong as a metal-ligand bond, but usually we are talking about weaker forces connecting the net. The principal difficulty in working with hydrogen bonds is the competition of the desired hydrogen bond pattern (synthon) with other possible hydrogen bonds in the system (other parts of the building blocks, counter ions, solvents, especially water). As a guide, a couple of studies of the Cambridge Structural Database may be useful, [2,3] and you might want to perform similar searches yourself.

One of their main attractions is the possibility of (biomimetic) molecular recognition. Very stable and selective hydrogen bonded structures form in nature with DNA (helices) or polypeptides, and well developed and automated preparation methods exists for making specific small oligomers of either polypeptides, or DNA, PNA (peptide nucleic acids) or other nucleobase mimicking systems. It may be possible to assemble different types of molecular building blocks using DNA bases (or mimics) that will program them to attach in certain predefined ways, [4-8] see Figure 12.1 for a schematic illustration.

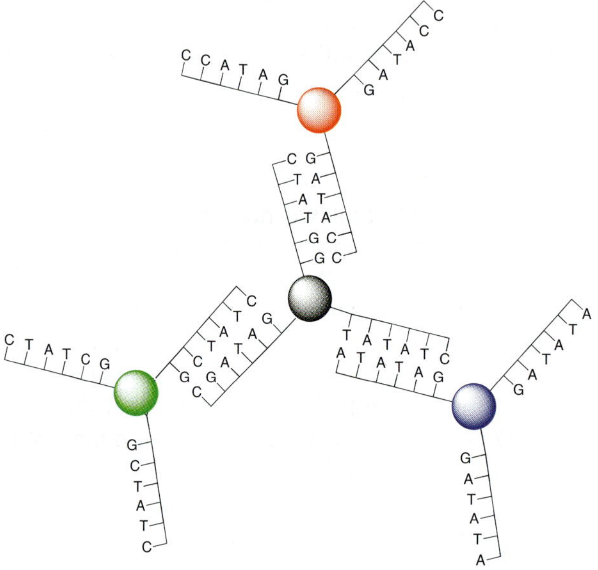

Figure 12.1 Fragment of hypothetical hydrogen bonded net based on DNA-base recognition (C=cytosine, G=guanine, T=thymine, A=adenine). The middle spheres represent for example metal clusters or organic molecules. Note that the order of attachment is "programmed" by the DNA sequence, blue cannot bind to green, red not to blue etc.

Some related recent advances have been made using DNA or RNA, see Figure 12.2 [9,10]. For some recent work on peptide assembly, see articles by Kaplan et al, [11] and Karle [12].

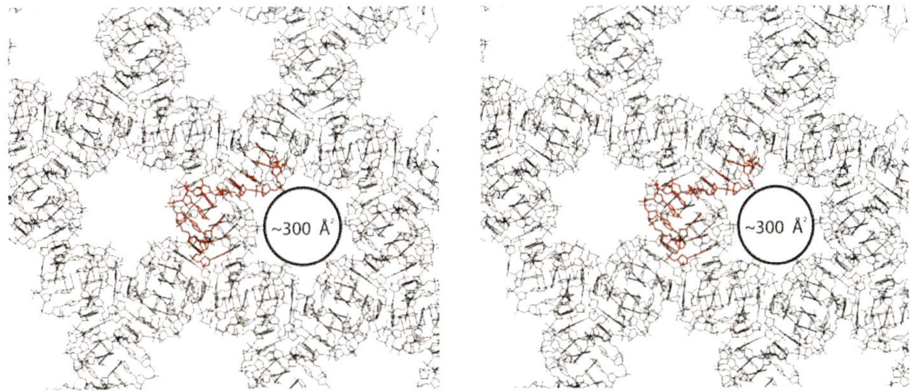

Figure 12.2 Crystal Packing of d(GGACAGTGGGAG) A stereoview down the 6-fold axis showing the hexagonal channels running the length of the crystal. Reproduced with permission [9].

Figure 12.1 also exemplifies a way of making principally different connections (different synthons), instead of head-to-head we here have a side-to-side link, see Figure 12.3. With the side-to-side approach we can increase the thermodynamic driving force for net formation while making the links in the net longer, this is not possible in the head-to-head strategy.

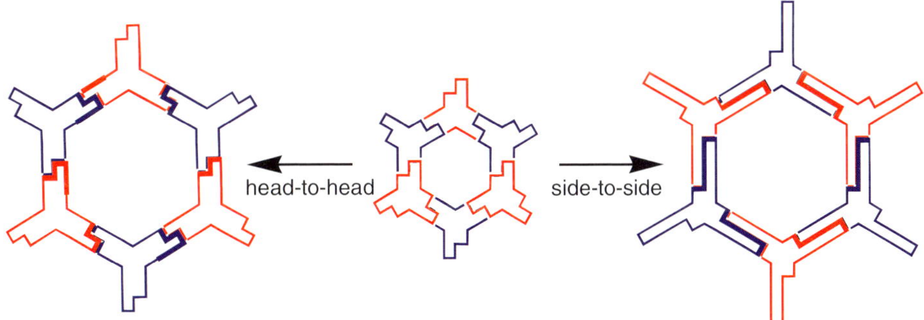

Figure 12.3 Most links in hydrogen bonded nets are head-to-head (left and centre) compared to the side-to-side link (right). With the side-to-side approach we can increase the thermodynamic driving force (the hydrogen bonds, bold lines) for net formation while making the links in the net longer, this is not possible with the head-to-head strategy.

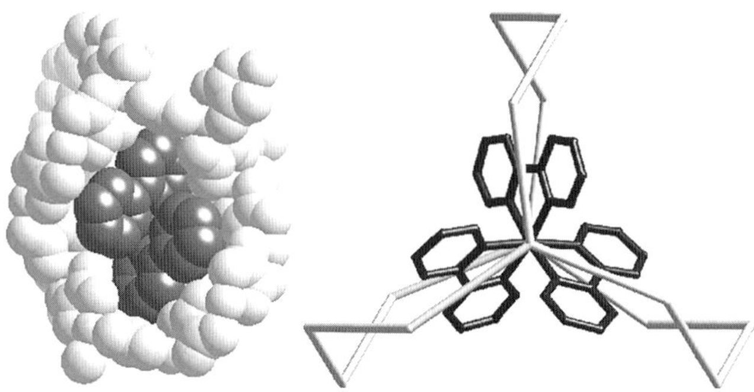

Figure 12.5. Left: Fraction of the Cu-oxalate polymer (white) forming a (10,3)-a or **srs** net in [Ru(2,2'-bipyridine)$_3$][Cu$_2$(oxalate)$_3$]. The template effect of the counter ion [Ru(2,2'-bipyridine)$_3$]$^{2+}$ (black) is clearly hinted by the close fit between the net and the counter ion [22]. Right: [Ru(2,2'-bipyridine)$_3$]$^{2+}$ shown inside a "tile" of the **srs** net demonstrating that both moieties have D_3 symmetry (a three-fold rotation axes with three two-fold rotation axes perpendicular to the three-fold axis, and no mirror planes).

A suitable counter ion for a corresponding cationic net could be the tris(tetrachlorobenzenediolato)phosphate(V) ion, [26] or related phosphate(V) catechol type derivatives, but as far as we know this has not been achieved yet.

It is of course not enough to note an inclusion molecule or ion to claim a template effect, it has to be demonstrated that the synthesis under similar conditions but without the templating agent yields a very different structure, as for example in the case of [Pr(adipic acid)$_3$(H$_2$O)$_2$]·4,4'-dipyridyl [27]. This unfortunately makes it difficult to prove templating effects of ligands and other components also contained in the net [28].

For reference we here also cite a number of recent reports on the template aided synthesis of molecule based 3D nets [29-34].

12.2.3. Methods of crystallization

To an organic chemist used to elaborate measures to exclude oxygen and water from the reactions, multi-step synthetic schemes and tedious work-up procedures, some of the preparations reported for molecule based 3D nets may seem a bit effortless. However, our game does not finish with the synthesis, we need crystals, and crystal of a quality and size suitable for X-ray diffraction.

And this, anyone who has tried can tell you, is not as a rule effortless, quite the contrary.[1]

We do not intend to give a short course in crystal growth, we just want to point out some techniques that may be useful, and give some references.

First of all, crystals should grow slowly. High concentrations of reactants that give immediate precipitation, or evaporation to dryness, give only very rarely anything useful for the crystallographer. Avoiding abrupt temperature changes is a good idea. Crystals should also, if possible, be stored in the solution from which they have grown (mother liquid).

Many more good "rules of thumb" are found on a couple of websites, [35,36] and an abundant specialist literature exist. However, be aware that many methods suppose that you have a pure product that you can re-crystallise, whereas 3D nets tend to be insoluble (and if not, the recrystallisation is likely to give a different product to the initial reaction) and you have to apply the methods so that they fit a reaction, not a re-crystallisation.

One excellent general method to do this is crystal growth in gels of different types, the general principle being that the gel slows down the process and also minimises the number of nucleation sites (sites where the crystals start to grow). One of the reactants can be dissolved in the gel, or the reagents can be separated by a gel in for example a U-tube. Both aqueous and non-aqueous solvents can be used, and the technique has been successfully applied to 3D nets [37,38].

12.2.4. Hydrothermal or solvothermal methods

Another method somewhat related to crystal growth is hydrothermal or solvothermal (if another solvent than water is used) reactions [1]. It involves heating the reactants and a suitable solvent in an sealed autoclave ("bomb") for a long time (hours to days), The pressure will built up and the solvent thus can be heated well above its normal boiling point, The advantages are several:

- Many components in network synthesis have low solubility, and the increase in temperature can help the dissolution.
- Activation energies are easily overcome with increased temperature.
- This can effectively prevent the products from oxidizing.
- Good crystals tend to form.
- 3D nets seem to be favoured by hydrothermal methods as well as higher temperatures [39,40].

---

[1] Some readers may also be under the impression that once the crystal is securely fastened on the goniometer head in the diffractometer, data collection, and structure solving and refinement is merely a matter of initiating a number of "black box" programs. Although this *might just happen* it is our impression, from our own work, and from examining a large number of CIFs' (Crystallographic Information Files) for this book, that an immense amount effort has to be invested also in the crystallography work in this field.

We might also add that for some systems *in situ* generated ligands are possible, [41] and may give products different from "normal" synthesis.

Currently, it may be hard to rationalise all differences between "normal" synthesis and solvothermal methods, but it has for example been argued that when there is competition between ligand bridge formation and water metal binding the T·ΔS term favours free water compared to coordinated water at higher temperatures [40].

Solvothermal methods have successfully been applied to the synthesis of molecule-based 3D nets for some 10 years, one of the first examples being the diamond nets of $K_2[M(2,3\text{-pyridinedicarboxyolate})_2]$ (M=Mn, Zn) [42], and the number of reports are rising and we here cite some recent work [27,43-53].

This method has some resemblance to the microwave methods currently of much interest in organic synthesis where sealed vessels are heated by microwaves [54,55]. It would probably be interesting to use this technique also for hydrogen bonded or coordination polymer 3D-nets.

12.2.5. Grinding and kneading, mechanochemical methods

As with the hydrothermal method, solid-solid grinding reactions, or mechano-chemical reactions as they are also called, is a method with long tradition in solid state chemistry[2] that has recently been applied also to molecular chemistry [56,57]. This method has several advantages:

- Solvent-free reactions are attractive from an environmental and sustainability point of view.
- Solid–solid reactions often lead to very pure products.
- The formation of solvate species can (normally) be avoided.
- Solid-state reactions can give products otherwise difficult or impossible to obtain from reactions in solution.

The main disadvantage is that single crystals are not produced in these type of reactions. However, powder X-ray diffraction (also known as XRD) can be used to solve this problem, either by direct solution of the structure from the XRD-data, (sometimes possible although not quite a standard method) or by comparison to the XRD diffractogram of a known structure. If the product is a new compound it may be possible to grow single crystals from seeds using the powder material and in that way get the X-ray structure [58].

The examples of 1D, 2D or 3D nets obtained in this way are still scarce, but recently the preparation of the coordination polymers $[Ag(N(CH_2CH_2)_3N)_2(CH_3COO)]\cdot5H_2O$ and $[Zn(N(CH_2CH_2)_3N)Cl_2]$ were reported [59].

---

[2] Superconductors are for example conveniently made by grinding the starting materials, pressing them to a table, and finally prolonged heating at high temperatures.

A small ("catalytic") amount of solvent may also be added to the grinding, in which case we call the method *kneading*.

12.3. Synthesis of chiral, porous, 3D nets

The synthesis of porous, enantiomerically pure, 3D-nets has been dubbed "the Holy Grail of Crystal Engineering". While this seems to be something of an overstatement, unless one attaches some metaphysical significance to chirality, it is still important enough to merit a special section. In effect, the literature on porous materials having some kind of chirality (attached chiral ligands, templated chiral voids from enantiomerically pure compounds, chiral porosity) is extensive and the importance of chiral synthesis, catalysis and separation is growing in the pharmaceutical industry and elsewhere.

As many people have noted, molecule-based 3D-nets has a certain advantage over zeolite materials in that cheap, enantiomerically pure, compounds from the chiral pool can be directly incorporated into the building blocks of a net, and some nets are also chiral by themselves [60-62]. As for the practical use of such compounds, enantioselective processes using coordination polymers has recently been reviewed by Lin et al. [63,64]

12.3.1. Intrinsic chiral nets and induced chiral nets

We may differentiate between nets that are chiral in themselves such as the **srs** or (10,3)-a net, and achiral nets that become chiral because of enantiomerically pure substituents in the network. While there are a number of examples of the latter type, incorporating for example polycarboxylic amino acids, there appears to be no examples of materials with permanent porosity obtained this way [65-71].

The other type, where the nets as such are chiral, have attracted interest, especially the (10,3)-a or **srs** net since this is the three-connected net with highest symmetry and the one that seem to have the highest likelihood to form from potentially three-connecting building blocks.

When trying to prepare such nets we should distinguish between these three cases:

1. An enantiomerically pure component attached (or indeed identical) to one of the molecular building blocks directs the net formation to one enantiomer because of the possible formation of two diastereomers between the net and the chiral component. If one of these diastereomers is more stable than the other we may obtain an enantiomerically pure product with only one chiral form of the net.
2. Without any additional chirality both enantiomers of the 3D-net will form and the product will be a racemic mixture. However, in the absence of any

crystal twinning of enantiomorphs each crystal picked up for analysis will be one single enantiomer.

3. Without any additional chirality *we might still get only one enantiomer* in our product. However, subsequent batches may yield the other enantiomer, and an average over several batches should tend towards a racemic mixture.[3]

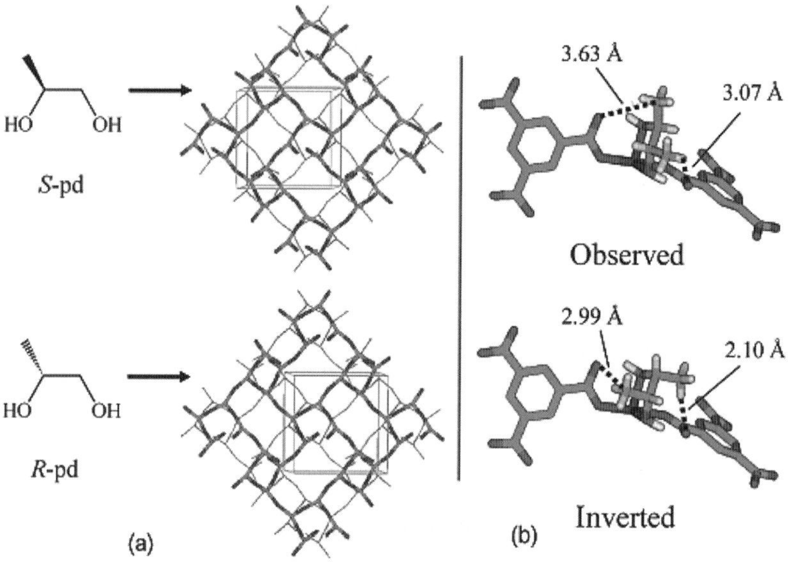

Figure 12.6 The chirality of the diol coordinated to the metal controls the helicity of the **srs** structure (a), anticlockwise with (S) and clockwise with (R) diol. In panel b, the incorrect combination of helix sense and diol produces less favourable contacts between the auxiliary and network forming ligands. Reproduced with permission from ref. [61]

Case one has been elegantly demonstrated by Rosseinsky and his group using [Ni$_3$(1,3,5-benzenetricarboxylate)$_2$(μ^2-1,2-propanediol)$_3$(pyridine)$_6$] giving **srs** or (10,3)-a nets (see Figure 11.7) with chirality controlled by the *R* or *S* form of the diol [74]. The diastereomeric interactions responsible to this effect is shown in Figure 12.6 [61].

An approach that has hitherto not been successful is the use of enantiomerically pure "chiral-at-metal" bischelating self-complimentary hydrogen bonding octahedral complexes as building blocks to form the **srs** net. In theory, complexes such as Δ-[Co(III)(2,2'-biimidazolato)$_3$] should easily form the **srs** net because of the perfect match between the torsions angles of the

---

[3] There are some notable exceptions where racemic or achiral starting materials consequently give the same enantiomer [72]. The origin of this is not clear, but sometimes the concept of *cryptochirality* is used as an explanation, that is, due to the presence for a very long time on earth of basically enantiopure living organisms our environment is chiral, but at concentrations too low to measure [73].

net and that of two consecutive Δ-complexes (see Figure 5.5 and Figure 12.7) [13]. However, in practice only the 1D helical part of the **srs**-net has been observed in such structures [13,17,75,76].

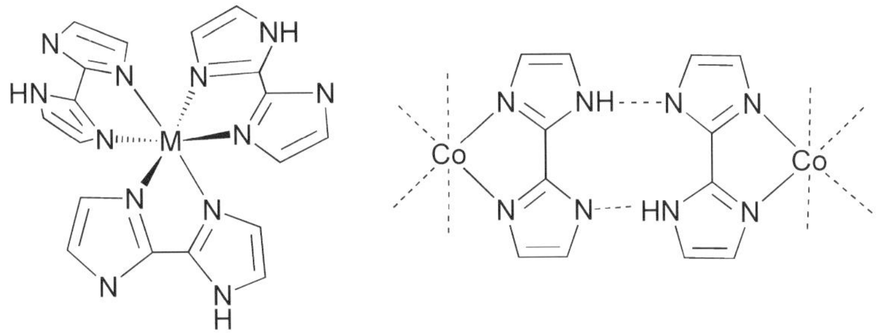

Figure 12.7 Enantiomerically pure Δ(or Λ)-[Co(III)(2,2'-biimidazolato)$_3$] (left) are well predisposed to give the (10,3)-a or **srs** net by hydrogen bonds (right) but this synthetic approach has so far only yielded helices in the solid state [13,17,75,76].

Case 2 giving a racemic mixture of nets seem discouraging at a first glance since even if it were possible to differentiate between the enantiomers by sight (sometimes this may happen), the task of separating them would be far too tedious. However, subsequent batches may be obtained enantiomerically pure by seeding with the desired enantiomer, a practice sometime used in the pharmaceutical industry to ensure that the right polymorph (different crystal forms of the same compound) of a drug substance is formed [77].

In case 3 we obtain enantiomerically pure products from a racemic mixture, seemingly violating standard textbook phrases such as "chirality has to come from somewhere" [78]. However, in view of what was said about seeding earlier this is not all that surprising. If the crystallisation starts from one seed only it is quite natural that the chirality of each batch will be determined by the chirality of this seed.

To ensure if we have a racemic mixture or not, all the crystals in a batch need to be investigated, which is most conveniently done with solid state Circular Dichroism (CD).

12.3.2. Interpenetration

An general problem with the **srs** net is interpenetration (see Chapter 11). Class I interpenetration (translation only) reduces the free volume, but might stabilise the structure and enhance the surface area. The important thing is that the chirality stays the same. Figure 11.7 shows two interpenetrated **srs** nets with the same chirality still having large channels and Figure 11.8 shows four interpenetrated nets of the same type completely blocking the channels.

More problematic is that the **srs** nets often give interpenetrated structures of class II containing a racemic mixture of nets, see Figures 11.9-11.11. Apart from the series of compounds based on [Ni$_3$(1,3,5-benzenetricarboxylate)$_2$] the majority of interpenetrated **srs** structures are racemic [79]. It is worth noting that Wells anticipated both types of interpenetration of the **srs** nets in his book [80].

This problem with the **srs** net is likely due to that it is a self dual net (see section 10.2.4 and Figure 11.4) and thus easily accommodates copies of itself.

12.3.3. Some chiral nets other than srs

It might therefore be interesting to turn to some other chiral nets that are simple enough (low genus) but that are not self dual. Figure 12.8 shows some of them that we have already encountered. Perhaps they will make suitable synthetic targets?

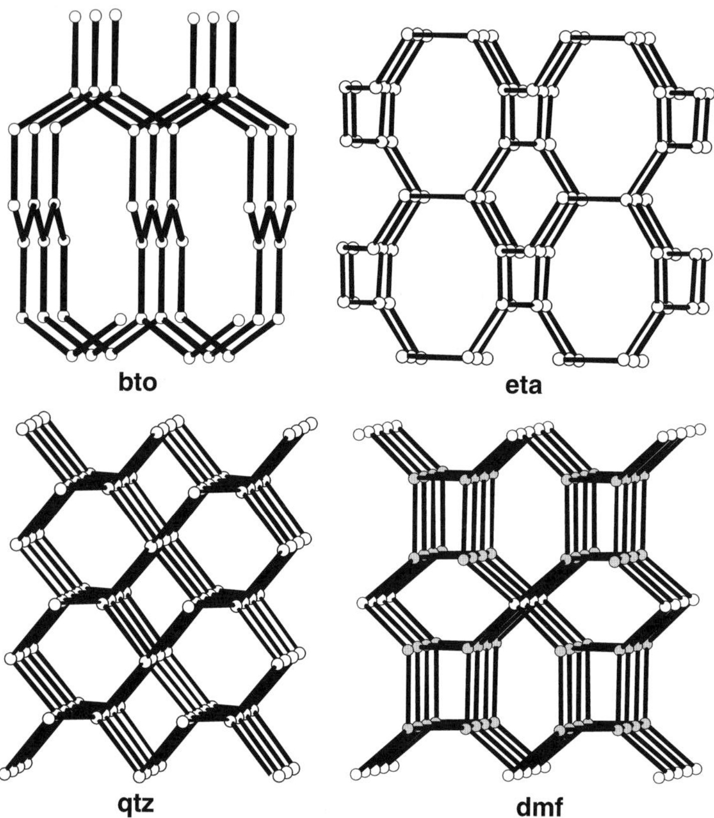

Figure 12.8 Some chiral nets: three-connected **bto** (10,3)-c and **eta** (8,3)-a, four-connected **qtz** (quartz) and three- and four-connected **dmf**.

Of the nets in Figure 12.8 **bto** and **eta** are very rare (Chapter 6), **qtz** (Chapter 4) has three known cases of class I interpenetration [79] and **dmf** (Chapter 8) has a known structure with two nets of different chirality. Regarding the **dmf** case one should note that it is not necessary to assemble all nodes at the same time, one could envisage a scheme where the four-connected nodes are first connected (covalently) to enantiomerically pure *tris*-chelating complexes[4] and then these are assembled in the last step, see Figure 12.9.

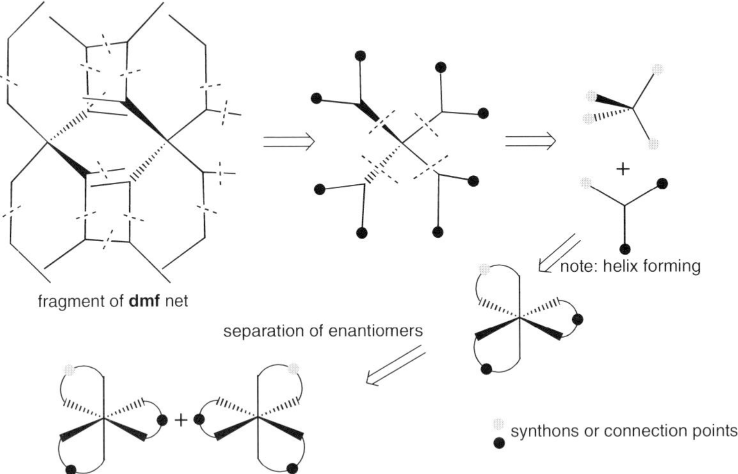

Figure 12.9. Schematic retrosynthetic analysis of the chiral **dmf** net. Grey and black dots represent different synthons (covalent, coordination or hydrogen bonding) holding the net together.

Finally, we should note that there are only a few studies on the effects of either type of chiral nets (intrinsic and induced) on chiral processes, thus for the moment we cannot say if either type has any distinctive advantage.

12.4. Some notes on polymorphism and supramolecular isomerism

In the preceding chapters we have seen many different mixtures of ions, ligands and other molecules giving specific nets. The reader may have, as have we, occasionally wondered why just this specific net was produced, and if, under other circumstances, another net may not have been formed? This is the field of *polymorphism*, [82] and it is of immense importance in the industrial manufacture of solid compounds since the principal method of purification is crystallisation and it is, for many reasons, unacceptable to get a different crystal form than the one specified.

---

[4] Methods for such preparations exist in the coordination chemist's toolbox ever since Alfred Werner and co-workers separation of the enantiomers of $[Co(ethylenediamine)_2Cl_2]^+$ in 1911 [81].

Polymorphism has been defined as "a solid crystalline phase of a given compound resulting from the possibility of at least two crystalline arrangements of the molecule of that compound in the solid state". The conceptual link between this and what has been called "supramolecular isomerism" has been pointed out by Moulton and Zaworotko: "...since polymorphs can be rationalized on the basis of supramolecular interactions, polymorphism can be regarded as a type of supramolecular isomerism." [83] They made the following distinctions:

- **Structural** The constituents of the nets are the same but they may be, for example, connected in different ways.
- **Conformational** Conformational changes in flexible molecules can generate different, but often related, nets. This is related to conformational polymorphism.
- **Catenatal** It may happen that the same starting materials under different conditions, for example different concentrations, give a compound with different degree of interpenetration.
- **Optical** A chiral net has two optical isomers. However, if it is constructed from chiral molecular building blocks these net-isomers are not polymorphs unless they rapidly interconvert in solution (and even then it is debatable).[5]

From a more practical point of view we have to be aware of two things:

- If you obtain a product that is not your desired compound, remember that this is not the end. Probably it is only one of several possible polymorphs of your system and it is probable well worth to investigate alternative methods of preparation and crystallisation.
- Any crystalline product you get needs to be closely examined for "concomitant polymorphs", that is, two or more polymorphs may appear simultaneously, or at least be present in your sample at the same time. This is one of the reasons you should to follow the advice not to put your crystallisation preparations away in a dark place and forget about them for days, you may obtain one polymorph first and another one later.

In order to check the latter point, the powder X-ray diffractogram could be measured for the bulk sample and compared to a calculated spectrum from single crystal data. Visual inspection with a good microscope (with heating to observe different melting points) may also do, often it is quite obvious if all the crystals are of the same type.

Polymorphism may look like an additional complication, but it could also be seen as a gift from nature allowing us to find out more about the particular

---

[5] Another proposed definition states that all polymorphs of a substance should revert to the same species (or equilibrium mixture) on dissolution, melting or vaporisation.

system we are studying. Organometallic polymorphism was recently review by Braga and Greponi [82].

12.5. Specific properties of nets and their analysis

This section should be either very long or very brief and we opt for the latter. Each class of specific properties has its own set of methods (SQUID magnetometer for magnetism etc) and we will not go into details concerning these.

A general remark for those who, just like us, come from a solution chemistry/small molecule/organic and organometallic syntheses tradition: It is worthwhile to investigate the equipment of neighbouring laboratories more geared towards materials (materials science, physics, polymers) where you might more easily find instruments for thermogravimetry (TG), differential scanning calorimetry (DSC), surface and porosity analysis.

A few words concerning porosity may finally be useful since this may not be completely clear to everyone. If a material is porous (or to be really explicit have "permanent porosity") this means that you can remove whatever it is that sits in the channels and get them completely empty and then put something else in there with only minor perturbations of the structure.

If you want to claim a porous crystalline material it is probably a good idea to:

- Have a crystal structure with filled channels.
- Have another crystal structure or powder diffraction data demonstrating the same structure but with empty channels.
- Furthermore you need to be sure that no re-crystallisation takes place during any of these transformations. AFM (atomic force microscopy) or related methods may be useful.
- Demonstrate the porosity through gas sorption isotherms.

References

[1] J. L. C. Rowsell, O. M. Yaghi, Microporous Mesoporous Mater. 73 (2004) 3.

[2] F. H. Allen, W. D. S. Motherwell, P. R. Raithby, G. P. Shields, R. Taylor, New J. Chem. 23 (1999) 25.

[3] T. Steiner, Angew. Chem. Int. Ed. 41 (2002) 48.

[4] N. C. Seeman, Clin. Chem. 39 (1993) 722.

[5] N. C. Seeman, Angew. Chem. Int. Ed. 37 (1998) 3220.

[6] N. C. Seeman, Nano Letters 1 (2001) 22.

[7] T. Sasaki, M. Lieberman, in: Comprehensive Supramolecular Chemistry, J.-M. Lehn (Ed.) 4Pergamon Press, Oxford, 1996, pp. 193.

[8] A. Carbone, N. C. Seeman, in: Aspects of Molecular Computing,Lecture Notes in Computer Science 2950, 2004, pp. 61.

[9] P. J. Paukstelis, J. Nowakowski, J. J. Birktoft, N. C. Seeman, Chemistry & Biology 11 (2004) 1119.

[10] A. Chworos, I. Severcan, A. Y. Koyfman, P. Weinkam, E. Oroudjev, H. G. Hansma, L. Jaeger, Science 306 (2004) 2068.

[11] R. Martin, L. Waldmann, D. L. Kaplan, Biopolymers 70 (2003) 435.

[12] I. L. Karle, J. Mol. Struct. 474 (1999) 103.

[13] L. Öhrström, K. Larsson, Dalton Trans. (2004) 347.

[14] B. F. Abrahams, S. R. Batten, H. Hamit, B. F. Hoskins, R. Robson, J. Chem. Soc., Chem. Commun. (1996) 1313.

[15] O. M. Yaghi, G. Li, Angew. Chem. Int. Ed. 34 (1995) 207.

[16] S. R. Halper, S. M. Cohen, Inorg. Chem. 44 (2005) 486.

[17] L. Öhrström, K. Larsson, S. Borg, S. T. Norberg, Chem. Eur. J. 7 (2001) 4805.

[18] V. R. Pedireddi, S. Varughese, Inorg. Chem. 43 (2004) 450.

[19] K. Kobayashi, A. Sato, S. Sakamoto, K. Yamaguchi, J. Am. Chem. Soc. 125 (2003) 3035.

[20] B. L. Chen, F. R. Fronczek, A. W. Maverick, Chem. Commun. (2003) 2166.

[21] A. J. Blake, N. R. Champness, P. A. Cooke, J. E. B. Nicolson, Chem. Commun. (2000) 665.

[22] F. Pointillart , C. Train, M. Gruselle, F. Villain, H. W. Schmalle, D. Talbot, P. Gredin, S. Decurtins, M. Verdaguer, Chem. Mat. 16 (2004) 832.

[23] S. Decurtins, H. W. Schmalle, P. Schneuwly, H. R. Oswald, Inorg. Chem. 32 (1993) 1888.

[24] S. Decurtins, H. W. Schmalle, R. Pellaux, P. Schneuwly, A. Hauser, Inorg. Chem. 35 (1996) 1451.

[25] E. Coronado, J. R. Galán-Mascarós, C. J. Gómez-García, J. M. Martínez-Agudo, Inorg. Chem. 40 (2001) 1331.

[26] J. Lacour, C. Ginglinger, C. Grivet, G. Bernardinelli, Angew. Chem. Int. Ed. 36 (1997) 608.

[27] D. T. de Lill, N. S. Gunning, C. L. Cahill, Inorg. Chem. 44 (2005) 258.

[28] V. V. Komarchuk, V. V. Ponomarova, H. Krautscheid, K. V. Domasevitch, Z. Anor. Allgem. Chem. 630 (2004) 1413.

[29] Z. M. Wang, B. Zhang, T. Otsuka, K. Inoue, H. Kobayashi, M. Kurmoo, Dalton Trans. (2004) 2209.

[30] M. L. Tong, J. Wang, S. Hu, S. R. Batten, Inorg. Chem. Comm. 8 (2005) 48.

[31] M. L. Tong, J. Ru, Y. M. Wu, X. M. Chen, H. C. Chang, K. Mochizuki, S. Kitagawa, New J. Chem. 27 (2003) 779.

[32] B. F. Abrahams, A. Hawley, M. G. Haywood, T. A. Hudson, R. Robson, D. A. Slizys, J. Am. Chem. Soc. 126 (2004) 2894.

[33] B. F. Abrahams, M. G. Haywood, R. Robson, J. Am. Chem. Soc. 127 (2005) 816.

[34] P. M. v. d. Werff, S. R. Batten, P. Jensen, B. Moubaraki, K. S. Murray, Inorg. Chem. 40 (2001) 1718.

[35] P. D. Boyle, Growing Crystals That Will Make Your Crystallographer Happy, 2001, http://www.xray.ncsu.edu/GrowXtal.html

[36] M. Lutz, Tips for Crystal Growing, 2002, http://www.cryst.chem.uu.nl/growing.html

[37] O. M. Yaghi, G. Li, H. Li, Chem. Mat. 9 (1997) 1074.

[38] O. M. Yaghi, L. Guangming, T. L. Groy, J. Solid State Chem. 117 (1995) 256.

[39] C. Livage, C. Egger, G. Ferey, Chem. Mat. 13 (2001) 410.

[40] P. M. Forster, A. R. Burbank, C. Livage, G. Ferey, A. K. Cheetham, Chem. Commun. (2004) 368.

[41] J. P. Zhang, S. L. Zheng, X. C. Huang, X. M. Chen, Angew. Chem. Int. Ed. 43 (2004) 206.

[42] S. O. H. Gutschke, A. M. Z. Slawin, P. T. Wood, J. Chem. Soc., Chem. Commun. (1995) 2197.

[43] J. P. Zhang, Y. Y. Lin, X. C. Huang, X. M. Chen, J. Am. Chem. Soc. 127 (2005) 5495.

[44] L. H. Xie, S. X. Liu, B. Gao, C. D. Zhang, C. Y. Sun, D. H. Li, Z. M. Su, Chem. Commun. (2005) 2402.

[45] F. T. Xie, L. M. Duan, X. Y. Chen, P. Cheng, J. Q. Xu, H. Ding, T. G. Wang, Inorg. Chem. Comm. 8 (2005) 274.

[46] D. R. Xiao, H. Y. An, E. B. Wang, L. Xu, C. W. Hu, J. Mol. Struct. 733 (2005) 69.

[47] M. L. Tong, J. Wang, S. Hu, J. Solid State Chem. 178 (2005) 1518.

[48] J. Y. Lu, Z. H. Ge, Inorg. Chim. Acta 358 (2005) 828.

[49] T. Loiseau, H. Muguerra, G. Ferey, M. Haouas, F. Taulelle, J. Solid State Chem. 178 (2005) 621.

[50] M. Kurmoo, C. Estournes, Y. Oka, H. Kumagai, K. Inouey, Inorg. Chem. 44 (2005) 217.

[51] S. Hu, M. L. Tong, Dalton Trans. (2005) 1165.

[52] Z. He, E. Q. Gao, Z. M. Wang, C. H. Yan, M. Kurmoo, Inorg. Chem. 44 (2005) 862.

[53] D. Y. Kong, A. Clearfield, Cryst. Growth Des. 5 (2005) 1263.

[54] P. Lidstrom, J. Tierney, B. Wathey, J. Westman, Tetrahedron 57 (2001) 9225.

[55] C. O. Kappe, Angew. Chem. Int. Ed. 43 (2004) 6250.

[56] G. Kaupp, Curr. Opin. Solid State Mat. Sci. 6 (2002) 131.

[57] D. Braga, F. Grepioni, Angew. Chem. Int. Ed. 43 (2004) 4002.

[58] D. Braga, G. Cojazzi, L. Maini, M. Polito, F. Grepioni, Chem. Commun. (1999) 1949.

[59] D. Braga, S. L. Giaffreda, F. Grepioni, M. Polito, Crystengcomm 6 (2004) 458.

[60] B. Kesanli, W. B. Lin, Coord. Chem. Rev. 246 (2003) 305.

[61] D. Bradshaw, J. B. Claridge, E. J. Cussen, T. J. Prior, M. J. Rosseinsky, Acc. Chem. Res. 38 (2005) 273.

[62] M. J. Rosseinsky, Microporous Mesoporous Mater. 73 (2004) 15.

[63] W. Lin, H. L. Ngo, in: Chemistry of Nanostructured Materials, P. Yang (Ed.) World Scientific Publishing Co, Singapore, 2003, pp. 261.

[64] B. Kesanli, W. Lin, Coord. Chem. Rev. 246 (2003) 305.

[65] H. L. Ngo, W. Lin, J. Am. Chem. Soc. 124 (2002) 14298.

[66] J. S. Seo, D. Whang, H. Lee, S. I. Jun, J. Oh, Y. J. Jeon, K. Kim, Nature 404 (2000) 982.

[67] A. Hu, H. L. Ngo, W. Lin, J. Am. Chem. Soc. 125 (2003) 11490.

[68] A. Hu, H. L. Ngo, W. Lin, Angew. Chem. Int. Ed. 42 (2003) 6000.

[69] B. F. Abrahams, M. Moylan, S. D. Orchard, R. Robson, Angew. Chem. Int. Ed. 42 (2003) 1848.

[70] N. Guillou, C. Livage, M. Drillon, G. Ferey, Angew. Chem. Int. Ed. 42 (2003) 5314.

[71] D. Laliberte, T. Maris, J. D. Wuest, Can. J. Chem. 82 (2004) 386.

[72] M. Vestergren, A. Johansson, A. Lennartson, M. Hakansson, Mendeleev Commun. (2004) 258.

[73] K. Mislow, Collect. Czech. Chem. Commun. 68 (2003) 849.

[74] C. J. Kepert, T. J. Prior, M. J. Rosseinsky, J. Am. Chem. Soc. 122 (2000) 5158.

[75] M. Tadokoro, H. Kanno, T. Kitajima, H. Shimada-Umemoto, N. Nakanishi, K. Isobe, K. Nakasuji, Proc. Nat. Acad. Sci. USA 99 (2002) 4950

[76] K. Larsson, Thesis: Chirality in crystal engineering, Chalmers Tekniska Högskola, Göteborg, 2003.

[77] W. Beckmann, Organic Process Research & Development 4 (2000) 372.

[78] J. McMurry, Organic Chemistry, 6 ed. Brooks-Cole, New York, 2003

[79] V. A. Blatov, L. Carlucci, G. Ciani, D. M. Proserpio, Crystengcomm 6 (2004) 377.

[80] A. F. Wells, Three-dimensional nets and polyhedra, John Wiley & Sons, New York, 1977.

[81] A. J. Ihde, The Development of Modern Chemistry, Dover Publications, New York, 1984.

[82] J. M. Bernstein, Polymorphism in molecular crystals, Clarendon Press, Oxford, 2002.

[83] B. Moulton, M. J. Zaworotko, Chem. Rev. 101 (2001) 1629.

Chapter 13

Computational tools

So far, we have dealt with net analysis on a simple basis where only the nodes have been displayed. In reality, finding and assigning a 3D-net in a crystal structure can be far from easy. Fortunately, there is help available in the form of computer programs. In this chapter, we will first try to give some tips on how to use some standard visualisation software to check for nets and in the second part we will introduce two newer tools capable of searching and analysing nets in a crystal structure. Bear in mind that this is a short introduction on how we have used the software and it is not intended to be a complete guide. All programs discussed in this chapter come with user guides and you will probably find your own approach in using them.

Two example structures, one coordination polymer containing [Zn(isonicotinato)$_2$] (CSD refcode VACFUB)[1] and a hydrogen bonded net of 3,3',5,5'-tetramethyl-4,4'-bipyrazolyl (CSD refcode UDAYUT) [2], will be employed to explain how to use these programs. The 2D and 3D views of both structures can be found in Figures 13.1-2. Since all programs are not suited to analyse both types of nets, both examples are not used with all software.

13.1. Regular visualisation tools for net discovery and analysis

We will have a look at some widely used free and commercial software for crystallography and molecular visualisation and analysis: Mercury, Diamond, CrystalMaker and RPluto. This section should not be seen as a complete review of these programs, if you are in the process of choosing such software ensure that you consult other sources and review other properties as well.

Figure 13.1A The coordination polymer [Zn(isonicotinato)$_2$] [1]

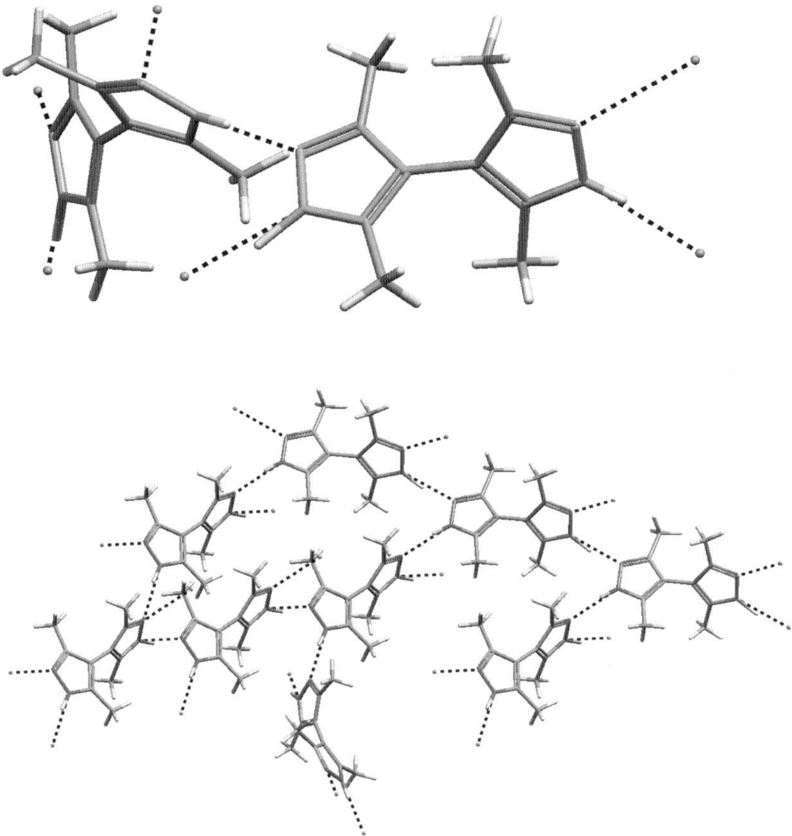

Figure 13.4. Short contacts in structure UDAYUT, in this case hydrogen bonds. Note that Mercury distinguishes between hydrogen bonds and short contacts by using different colours.

13.1.2. Diamond

This is a commercial visualisation software developed by Crystal Impact for Windows systems (free demo available) [4].

Just as Mercury, this program allows the user to expand the net one molecule at a time for a first impression of the connectivity. In addition, it is possible to simplify the net by changing molecules to nodes. This also makes it possible to manually count the rings and determine the topology, and in the following paragraphs we will outline how to accomplish this. It should be noted that the following procedure can be used with many other visualisation software with some adaptation.

The path from molecule to a simplified view of the net can be summarized as:

- Examine the single molecule and its short contacts.
- Expand the structure through these contacts and identify the nodes and their connectivity.
- Add centroids by selecting a suitable number of atoms at the identified node.
- Check the distances between the nodes to be connected.
- If you are lucky, there are no other nodes from interpenetrating nets within the distance of two nodes belonging to the same net. In this case, set the appropriate connectivity range and expand the structure.
- Before deleting the bridge atoms and reducing the structure to a net, always plot the simplified node-only network over the top of the real structure to check that the simplified net does indeed correspond to the real one.

If you are not so lucky, there are nodes from different nets within the distances (or having exactly the same distances). To solve this, there two different methods can be use. The first is to add extra two-connected nodes between the first set of nodes and then check the distances again. This has the drawback that it is impossible to get a C10 value. The other solution involves changing the original unit cell and thus the distance between the nodes. For unit cells of higher symmetry, it is often necessary to release the restraints on the the cell lengths.

When the nodes for one net are connected check for other, interpenetrating, nets in the structure. This is done by filling the unit cell with nodes and expanding. At this point, start using lots of different colours for the individual nets.

Assign the correct net to the structure by comparing the net to schematic views in this or other books [5-7] or using the Reticular Chemistry Structure Resource [8]. If you are unsure of the net, count rings. An example on how to count rings can be found in Chapter 4.

Coming back to our example structures, we have determined the nodes already; four-connected for VACFUB and three-connected for UDAYUT. Figure 13.5 shows a part of VACFUB with the Zn atoms highlighted. Since these are in the centre of the node, we'll use them instead of making new ones. Figure 13.6 shows the connected nodes (Zn atoms) without the ligands. The distance is 8.81 Å between all nodes in one net and there is no need to create extra dummy atoms. The finished image is shown in Figure 13.7.

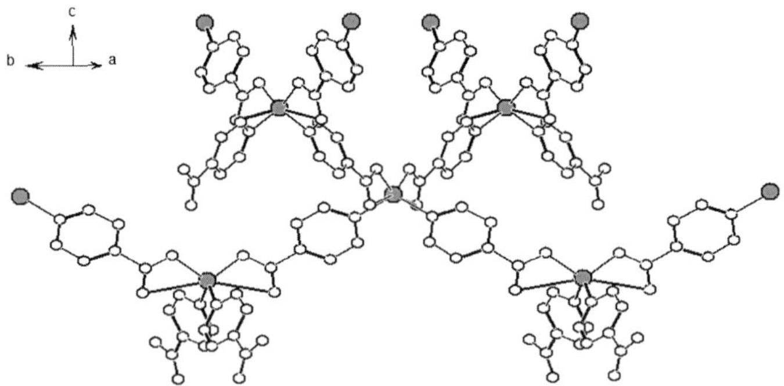

Figure 13.5. A part of VACFUB with the Zn atoms highlighted

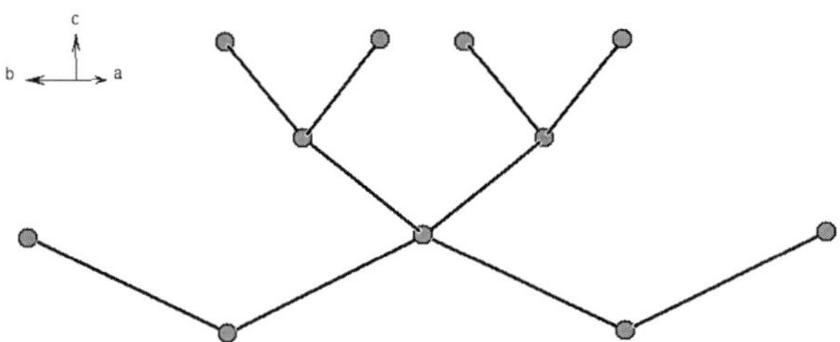

Figure 13.6. Same view of VACFUB as in Figure 13.5 but with only the Zn nodes shown.

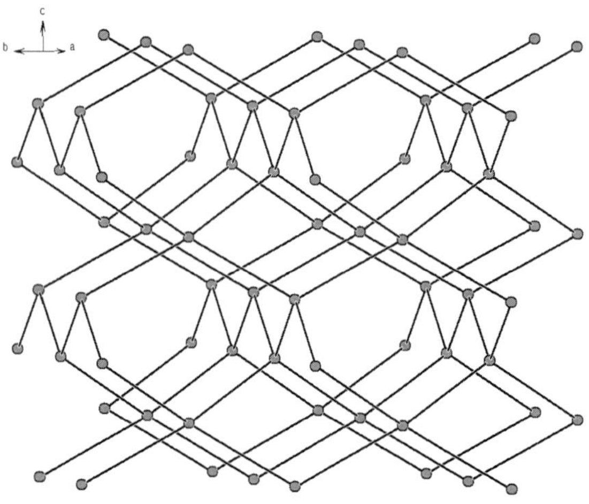

Figure 13.7 The net in VACFUB expanded.

13.1.3. Crystal Maker

CrystalMaker is a commercial program from CrystalMaker Inc. for "building, displaying and manipulating all kinds of crystal and molecular structures" [9]. It runs on MacOS X and Windows. It can be used in much the same way as Diamond. One of the advantages of CrystalMaker, is the ability to put a distance range for the bonds. Thus if there are 'false' inter-node distances less than the true inter-node distances, then these can be eliminated by raising the minimum in the bond range.

Both Diamond and CrystalMaker have been used to generate the pictures in this book.

13.1.4. RPluto

RPluto is also available as freeware (for non commercial use) at the CCDC website. It is a Unix/Linux only application and it is used to find hydrogen bonded structures.

To analyse a structure just click on the HBOND- and then the GSET button. A clickable list will appear and when a particular motif is selected, RPluto will show the result graphically.

13.2. Software packages that can search for a nets

13.2.1. OLEX

OLEX, written by Oleg Dolomanov, is freeware for academic and non-commercial use [10, 11]. It is an excellent tool for finding nets in coordination networks and although not as straightforward, it is also possible to analyse hydrogen bonded networks.

The procedure for finding nets is often very simple for coordination networks. The net in our example structure VACFUB the procedure is as follows:

- Import the structure (cif-file).
- Remove fragments which are not a part of the net by selecting them in the Fragments window and pressing the "del" key. Which fragments to remove are sometimes easier to figure out by a preliminary examination of the structure in Mercury. In our case, the oxygen (O5) and oxygen (O6) are solvent water molecules and not part of the net and therefore removed.
- Expand the structure by clicking on the Generate... button. Since this is the first time the analysis is made, it is a good idea to use a low number of unit cells (1-2 cells) and then examine the resulting connectivity.
- If everything looks OK, select Construct Net in the Network menu and examine the nodes in the red network. There should be no extra nodes.

- Go back to the asymmetric unit using the Fragment --> Uniq menu command.
- Use Generate again and create a larger net. Start with 3 or 4 cells in all directions. Create the net and then select Evaluate Topology in the Network menu. Open a text editor of your choice (Notepad for example) and paste the contents on the clipboard. The result when using 4 cells is $5 \cdot 5 \cdot 5 \cdot 6 \cdot 7 \cdot 7_2$. Since OLEX only checks the network topology for the created net we cannot be sure that the result is correct. Therefore is a good idea to generate larger nets using a higher number of unit cell ranges in the creation of the net. When the symbols converge on one value, we have (probably) found the correct network topology.
- Always checked manually by visual comparison of the suspected topology obtained from OLEX with the ideal net from RCSR or other sources.

The procedure for our example can be found in Table 13.1 and the final result is Short symbol $6 \cdot 8^2$ and Long symbol $6 \cdot 6 \cdot 6_2 \cdot 6_2 \cdot 8_7 \cdot 8_7$. A quick look in Chapter 7 reveals that this is a quartz, **qtz** net. The corresponding OLEX output is shown in Figure 13.9

Table 13.1. The Short and long symbol for structure VACFUB using different cell ranges.

Cell range	Short symbol	Long symbol
4	$5^3 \cdot 6 \cdot 7^2$	$5 \cdot 5 \cdot 5 \cdot 6 \cdot 7 \cdot 7_2$
5	$6^3 \cdot 8^3$	$6_2 \cdot 6_2 \cdot 6 \cdot 8_4$
6	$6^4 \cdot 8^2$	$6 \cdot 6 \cdot 6_2 \cdot 6_2 \cdot 8_7 \cdot 8_7$
7	$6^4 \cdot 8^2$	$6 \cdot 6 \cdot 6_2 \cdot 6_2 \cdot 8_7 \cdot 8_7$

```
Topological analysis for: VACFUB
Topological Terms for:      short          long
  Zn1 (Zinc, count: 746)            6(4).8(2)
     6.6.6(2).6(2).8(7).8(7)
```

Figure 13.9 Output from the topological analysis in of VACFUB with OLEX.

As mentioned in the beginning of this section, the analysis of hydrogen bonded nets is not as straightforward. The hydrogen bonds involved in the formation of the net have to be identified manually and converted to covalent bonds in the software. Starting out as before by importing our example structure UDAYUT, note the atom labels involved in the net, see Figure 13.8. Then one procedure to find the net is:

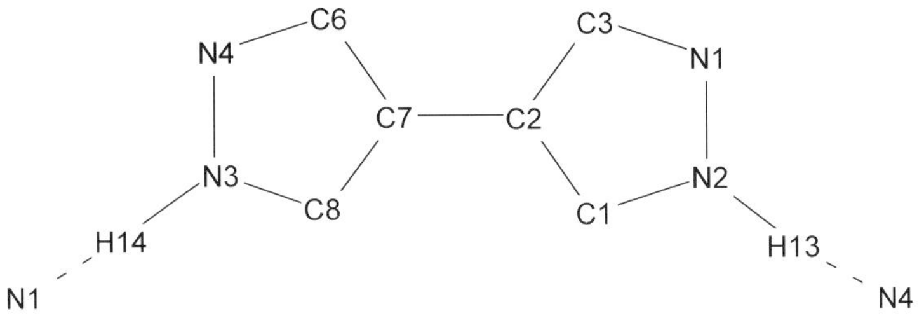

Figure 13.8 Atom labels in the UDAYUT structure, methyl groups have been removed for clarity.

- In the initial tries with Generate, many bonds are formed between the hydrogen atoms on the methyl groups, forming an unwanted bridge between our building blocks. Therefore the input file is edited and the hydrogen atoms at the methyl groups are removed.
- Now, the generation of multiple unit cells works better and the molecules are created without any intermolecular bonds. So how do we locate the hydrogen bonds?
- By using the Short Interaction Analysis on the Analysis menu, a window is shown where different interactions can be selected. In our case, intermolecular N-H is selected and OK is clicked.
- By selection Bond --> Short Interactions, a list is shown with all created bonds. To make our selection easier, we sort the list by labels. In the sorted list, the interactions responsible for the propagation of the nets are highlighted, H13⋯N4 and H14⋯N3 our case. Then, by right clicking in the list, Convert To Covalent is selected.
- Then, Network --> Construct Net is applied and the resulting red net is inspected just as in the former structure. In this case, we can see that there are extra nodes in the net and these have to be collapsed before the topology is evaluated.
- This is also done on the Bond menu with the Network Bonds command. This shows a similar list as for the short interactions, but now all bonds in the red network are shown. It is also convenient to sort this list by atom labels.
- In the list select the three bonds in one of the five membered rings, right click and select Collapse Nodes. The first set of bonds is N1-N2, N1-C2 and N2-C2. Repeat for the second set N4-N3, N4-C7 and N3-N7. Now the extra nodes should be replaced by two.
- Select Network --> Evaluate Topology and paste the result in a text editor. If enough unit cells were used in the creation of the net, the result is $10 \cdot 10^2 \cdot 10^2$.

As we of course remember from Chapter 3 this is a **bto** or (10,3)-c net.

13.2.2. TOPOS

TOPOS is a commercial software package written by V.A. Blatov and A.P. Shevchenko [12, 13]. Among its many features, the analysis of networks is the one we will use. This program also has the ability to find interpenetrating nets and assign the proper interpenetration symbol

The general procedure when analysing a net is: inspection, calculate the connectivity for all atoms using AutoCN and finally calculate the topology using ADS. It should be noted that the default calculated Schläfli symbol uses circuits instead of rings (this can be changed, see below) and the listed C10 value does not include the starting node.

Since this program contains a lot of features and options in network analysis, we will only give a short introduction:

- Import the file using the Import command located in the Database menu. Create a new TOPOS file with the filename of your choice and type any number except the default value of "0" in the following popup window.
- The structure can now be inspected using the built in program IsoCryst.
- After the initial inspection, AutoCN is started. In the options window, under the matrix tab, use Sectors as Method and check the Spec. Cont. and Dist.+Rsdsin under Matrix Data. Close the options window and select Run. Check the connectivity for the atoms, especially coordinating metals
- If there are unwanted bonds or if some bonds are not created, this can be fixed by right clicking on the compound in the list and select Adj. Matrix followed by Edit. On the Adj. Matrix tab you will find a list with all the atoms and by clicking on them you will get a complete list of the bond as well as all the short contacts. By right clicking on a bond or short contact, it is possible to change the type.
- When satisfied, open the ADS program. In the options window, under the topology tab, check Schläfli and Interpenetration. Then, depending on the type of structure, check the appropriate boxes under Bond types. If your structure consist of only coordination bonds only check the At. option under Valence and turn off the others. In the case of a hydrogen bonded net, use At. under Valence as well as Mol. under H Bonds. The last option means that there are discrete molecules connected by hydrogen bonds. To include the Long symbol in the output list, add a number in the Max. Ring box. A typical value would be 12 or 14, but it depends on the ring size in the net. Note: these calculations may be very time consuming.
- Close the option window and select Run. In the following window, select the central atoms representing the nodes. Finally inspect the result of the calculation, the Schläfli and vertex symbols will be located at the bottom of the output listing.

Some of the other features of TOPOS includes the generation of a new structure, containing only the nodes of the net. This is a very nice method to illustrate interpenetrating nets with different colours.

```
####################################
1;RefCode:VACFUB:C12 H12 N2 O6 Zn1
Author(s): Jinyu Sun,Linhong Weng,Yaming Zhou,Jinxi Chen,Zhenxia
Chen,Zhicheng Liu,Dongyuan Zhao
Journal: Angew.Chem.,Int.Ed. Year: 2002 Volume: 41 Number:
Pages: 4471
####################################

Topology for Zn1
------------------
Atom Zn1 links by bridge ligands and has
Common vertex with                                R(A-A)        f
Zn 1     0.8333     0.6667    -0.6182   ( 1 0-1)  8.813A        1
Zn 1     0.3333     0.1667     0.7151   ( 1 1 0)  8.813A        1
Zn 1     0.8333    -0.3333    -0.6182   ( 1-1-1)  8.813A        1
Zn 1     1.3333     0.1667     0.7151   ( 2 1 0)  8.813A        1
-------------------------
Structural group analysis
-------------------------


-------------------------
Structural group No 1
-------------------------
Structure consists of 3D framework with ZnO4N2C12
There are 2 interpenetrated nets
FIV: Full interpenetration vectors
-----------------------------------
[0,0,1] (6.26A)
-----------------------------------
PIC: [0,0,2][0,1,0][1,0,0] (PICVR=2)

Zt=2; Zn=1
Class Ia  Z=2

Coordination sequences
----------------------
Zn1: 1  2   3   4    5    6    7    8    9    10
Num  4 12  30  52   80  116  156  204  258   318
Cum  4 16  46  98  178  294  450  654  912  1230
----------------------

Vertex symbols for selected sublattice
--------------------------------------
Zn1 Schlafli symbol:{6^4;8^2}
With circuits:[6.6.6(2).6(2).8(9).8(9)]
With rings:   [6.6.6(2).6(2).8(7).8(7)]
All rings (up to 14):
[(6,8a(2),8b(2),8c,8d,8e(2)).(6,8a(2),8b(2),8c,8d,8e(2)).(6(2),8a
,8b(2),8c,8d).(6(2),8a,8c,8d,8e(2)).(8a,8b,8c(2),8d(2),8e).(8a,8b
,8c(2),8d(2),8e)]
--------------------------------------
Total Schlafli symbol: {6^4;8^2}

Topological type: Quartz  qtz {6^4;8^2} - VS
[6.6.6(2).6(2).8(7).8(7)]

Elapsed time: 7.29 sec.
```

Figure 13.10 Output from the topological analysis in TOPOS

References

[1] J. Sun, L. Weng, Y. Zhou, J. Chen, Z. Chen, Z. Liu, D. Zhao, Angew.Chem.,Int.Ed. 41 (2002).

[2] I. Boldog, E. B. Rusanov, A. N. Chernega, J. Sieler, K. V. Domasevitch, Angew. Chem. Int. Ed. 40 (2001) 3435.

[3] CCDC 2001-2004, http://www.ccdc.cam.ac.uk/mercury/

[4] Diamond - Crystal and Molecular Structure Visualization, Crystal Impact - K. Brandenburg & H. Putz GbR, Postfach 1251, D-53002 Bonn, http://www.crystalimpact.com/

[5] M. O'Keeffe, B. G. Hyde, Crystal Structures I: Patterns and Symmetry, ed., Mineral Soc. Am., Washington, 1996.

[6] A. F. Wells, Three-dimensional nets and polyhedra, ed., John Wiley & Sons, New York, 1977.

[7] A. F. Wells, Further Studies of Three-Dimensional Nets, ed., Polycrystal book service, Pittsburgh, 1979.

[8] M. O'Keeffe, O. Yaghi, M., Reticular Chemistry Structure Resource, Tucson, Arizona State University, 2005, http://okeeffe-ws1.la.asu.edu/RCSR/home.htm

[9] CrystalMaker, CrystalMaker Software Ltd. Yarnton, Oxfordshire, OX5 1PF, UK 2005, http://www.crystalmaker.com/

[10] O. V. Dolomanov, A. J. Blake, N. R. Champness, M. Schroder, Journal of Applied Crystallography 36 (2003) 1283.

[11] O. V. Dolomanov, OLEX, http://www.ccp14.ac.uk/ccp/web-mirrors/lcells/index.htm

[12] V. A. Blatov, 2004, http://www.topos.ssu.samara.ru/

[13] V. A. Blatov, A. P. Shevchenko, V. N. Serezhkin, J. Appl. Cryst. 33 (2000) 1193.

Appendix A

Ideal Nets

The following table contains all the ideal nets that can be found in this book. Each entry contains all information on the unit cell and node coordinates for generating the net. In addition, the short and long symbol as well as the c10 value can also be found. Each entry should be "decoded" as follows:

RSCR symbol	*Space group (number)* *x, y, z coordinates*	*Unit cell edges* *Short symbol*	*Unit cell angles* *Long symbol c10*

The parameters are set so that the distance between the nodes is 1 Å. However, a few of the nets (cds,tcb...) have extra node-node distances of 1Å, so be careful when connecting the nodes.

acs	P6$_3$/mmc (#194)	1.4142/1.4142/1.1547	90/90/120
	0.3333/0.6667/0.25	$4^9.6^6$ $4\cdot4\cdot4\cdot4\cdot4\cdot4\cdot4_2\cdot4_2\cdot4_2\cdot6_4\cdot6_4\cdot6_4\cdot$	
	6$_4$·6$_4$·6$_4$		1751

asv	P4/mmm (#123)	3.0477/3.0477/2.1545	90/90/90
	0/0.5/0.5	$6^4.10^2$ $6\cdot6\cdot6_2\cdot6_2\cdot12_8\cdot12_8$	787
	0/0.232/0.2321	$4^3.6^3$ $4\cdot6\cdot4\cdot6\cdot4\cdot6$	787

bbf	Pmna (#53)	2.8285/1.732/1.6333	90/90/90
	0/0/0	$6^4.8^2$ $6_2\cdot6_2\cdot6_2\cdot6_2\cdot8_4\cdot8_4$	1225
	0.25/0.3333/0.25	6^6 $6\cdot6\cdot6\cdot6\cdot6_2\cdot6_2$	1223

bbm	C2/c (#15)	2.2387/2.8846/3.6542	90/114.166/90
	0/0.3002/0.25	$3.6^2.7^3$ $3\cdot7\cdot6\cdot6\cdot7\cdot7$	975
	0.0996/0/0.65	$3.6^2.7^3$ $3\cdot6\cdot6\cdot7\cdot7\cdot8_2$	979

bcu   Im - 3m (#229)   1.1547/1.1547/1.1547   90/90/90
       0/0/0 $4^{24}.6.^4$ 4·4·4·4·4·4·4·4·4·4·4·4·4·4_3·4_3·
       4_3·4_3·4_3·4_3·4_3·4_3·4_3·4_3·4_3·4_3·*·*·*·*   2331

bcu-1 Im - 3m(#229)   3.9834/3.9834/3.9834   90/90/90
       0.3551/0/0.1775 $4^6.6^4$ 4·4·4·4·4·4·6·6·6_2·6_2 1176

bnn   P6/mmm(#191)   1.7321/1.7322/1   90/90/120
       0.3333/0.6667/0 $4^6.6^4$ 4·4·4·4·4·4·6·6·6·*   1176

bor   P - 43m (#215)   2.4495/2.4495/2.4495   90/90/90
       0.1666/0.1666/0.1666 6^3 6·6·6   822
       0.5/0/0 $6^2.8^4$ 6_2·6_2·8·8·8·8   817

bsn   I4$_1$/amd (#141)   1.9367/1.9367/1   90/90/90
       0/0.75/0.125 $4^8.5^4.6^3$ 4·4·4·4·4·4·4·4·5_3·5_3·5_3·5_3·
       ··*   1941

bto   P6$_2$22 (#180)   1.7321/1.7321/4.5   90/90/120
       0.5/0/0.111 10^3 10·10_2·10_2 730

cab   Pm -3 m (#221)   2.4142/2.4142/2.4142   90/90/90
       0.2929/0/0 $3^4.4^2.8^4$ 3·3·3·3·4·4·8·8·8·8   882

cag   Cmca (#64)   2.8701/3.2727/2.7658   90/90/90
       0.1742/0.1338/0.0872 4.6^5 4·6_2·6·6·6·6   969

cds   P4$_2$/mmc (#131)   1/1/2   90/90/90
       0/0/0 $6^5.8$ 6·6·6·6·6_2·*   1489

coe   C2/c (#15)   2.3585/4.1223/2.2511   90/119.98/90
       0.1592/0.0921/0.0874 $4^2.6.8^2.9$ 4·8·4·9_7·6·8   1322
       0.0157/0.3237/0.0366 $4^2.6^3.8$ 4·6·4·6·8·9_2 1315

cor   R - 3c (#167)   2.4674/2.4674/7.0344   90/90/120
       0.2939/0/0.25 $4^3.6^3$ $4·4·4_2·6_4·6_2·6_2$ 1371
       0/0/0.3479 $4^6.6^9$ $4·4·4·4·4·4·6_2·6_2·6_2·6_3·6_3·6_3·$
       ..*   1306

crb   I4/mmm (#139)   2.8687/2.8687/1.5786   90/90/90
       0.1743/0.1743/0 4.6^5 $4·6_2·6·6·6·6$ 959

ctn   I - 43d (#220)   3.7033/3.7033/3.7033   90/90/90
       0.2083/0.2083/0.2083 8^3 $8_5·8_5·8_5$ 906
       0.375/0/0.25 8^6 $8_3·8_3·8_3·8_3·8_4·8_4$ 879

ctn-d I - 43d (#220)   4.7115/4.7115/4.7115   90/90/90
       0.2689/0/0.25 10^3 $10_4·10_4·10_4$ 602
       0.173/0.173/0.173 10^3 $10_4·10_4·10_4$ 608

dia   Fd - 3m (#227)   2.3094/2.3094/2.3094   90/90/90
       0.125/0.125/0.125 6^3 $6_2·6_2·6_2·6_2·6_2·6_2$ 981

dia-a Fd - 3m (#227)   5.138/5.138/5.138   90/90/90
       0.0562/0.0562/0.0562 $3^3.12^3$ $3·12_2·3·12_2·3·12_2$ 497

dia-b F - 43m (#216)   2.3094/2.3094/2.3094   90/90/90
       0/0/0 6^3 $6_2·6_2·6_2·6_2·6_2·6_2$ 981
       0.25/0.25/0.25 6^3 $6_2·6_2·6_2·6_2·6_2·6_2$ 981

dia-e[1] Fd - 3m (#227)   2.8286/2.8286/2.8286   90/90/90
       0/0/0 $3^6.6^6.7$ $3·3·3·3·3·3·6·6·6·6·6·6·*·*·*$
                                                                          1941

dia-f I 4_1/amd (#141)   3.9489/3.9489/4.6504   90/90/90
       0/0.4035/0.9304 4.14^2 $4·14_{12}·14_{12}$ 352

---

[1] Also known as the crs net

dia-g   I $4_1$22 (#98)   3.5999/3.5999/5.6805   90/90/90
        0.0411/0.1327/0.0779 4.14^2 $4 \cdot 14_{12} \cdot 14_{12}$ 350

dia-j   Fd - 3m (#227)   5.6568/5.6568/5.6568   90/90/90
        0.25/0.25/0.125 $3^5.4^4.5^4.6^2$ $3 \cdot 3 \cdot 3 \cdot 3 \cdot 3 \cdot 4 \cdot 4 \cdot 4 \cdot 4 \cdot 6 \cdot 6 \cdot 12_5 \cdot$
        $12_5 \cdot 12_{13} \cdot 12_{13}$ 1359

dmc   P2_1/c (#14)   2.1419/2.8038/3.0878   90/115.044/90
        0.3918/0.6779/0.822 4.8^2 $4 \cdot 8_3 \cdot 8_3$ 725
        0.2372/0.5696/0.0695 4.8^5 $4 \cdot 8_2 \cdot 8_2 \cdot 8_3 \cdot 8_2 \cdot 8_3$ 735

dmd   P4_2/mmc (#141)   2.4748/2.4748/2.7018   90/90/90
        0/0/0.25 $4^2.10^4$ $4 \cdot 4 \cdot 10_4 \cdot 10_4 \cdot 10_4 \cdot 10_4$ 551
        0/0.298/0 4.10^2 $4 \cdot 10_4 \cdot 10_4$ 542

dme   C2/m (#12)   5.8044/2.4009/2.9074   90/142.143/90
        0.1116/0.2917/0.2819 $4.5^2.6.7.8$ $4 \cdot 5 \cdot 5 \cdot 6 \cdot 8 \cdot 10$ 824
        0.5467/0.5/0.4116 5.8^2 $5 \cdot 8 \cdot 8$ 809

dmf   C222 (#21)   4.1798/1.8691/1.8955   90/90/90
        0/0/0 10^6 $10_2 \cdot 10_2 \cdot 10_3 \cdot 10_3 \cdot 10_4 \cdot 10_4$
                                                     997
        0.1327/0.3020/0.6774 10^3 $10_3 \cdot 10_4 \cdot 10_6$ 984

eta   P$6_2$22 (#180)   3.1018/3.1018/2.1864   90/90/120
        0.4069/0.8138/0 8^3 $8 \cdot 8 \cdot 8_2$ 508

etb   R - 3m (#166)   5.3731/5.3731/2.1861   90/90/120
        0.4069/0/0 8^3 $8 \cdot 8 \cdot 8_2$ 496

fcu   Fm - 3m (#225)   1.4142/1.4142/1.4142   90/90/90
        0/0/0 $3^{24}.4^{36}.5^6$ $3 \cdot 3 \cdot 3 \cdot 3 \cdot 3 \cdot 3 \cdot 3 \cdot 3 \cdot 3 \cdot 3 \cdot 3 \cdot 3 \cdot$
$3 \cdot 3 \cdot 3 \cdot 3 \cdot 3 \cdot 3 \cdot 3 \cdot 3 \cdot 3 \cdot 3 \cdot 3 \cdot 3 \cdot 3 \cdot 4 \cdot 4 \cdot 4 \cdot 4 \cdot 4 \cdot 4 \cdot 4 \cdot 4 \cdot 4 \cdot 4 \cdot 4 \cdot * \cdot * \cdot * \cdot * \cdot * \cdot * \cdot * \cdot * \cdot * \cdot * \cdot$
$* \cdot * \cdot * \cdot * \cdot * \cdot * \cdot * \cdot * \cdot * \cdot * \cdot * \cdot * \cdot * \cdot * \cdot * \cdot * \cdot *$ 3871

flu	Fm - 3m (#225)	2.3094.3094/2.3094		90/90/90
	0.25/0.25/0.25	4^6	4·4·4·4·4·4	1531
	0/0/0	$4^{12}.6^{12}.8^4$	4·4·4·4·4·4·4·4·4·4·	

4·6_2·6_2·6_2·6_2·6_2·6_2·6_2·6_2·6_2·6_2·6_2·6_2·*·*·*·* 1401

fsg	Pmmm (#47)	1/1/2		90/90/90
	0/0/0.5	$4^4.6^2$	4·4·4·4·6_2·*	1489
	0/0/0	$4^8.6^7$	4·4·4·4·4·4·4·6·6·6·6·*·*·*	
				1561

fsh	R - 3m (#166)	1.5601/1.5601/5.6062		90/90/120
	0/0/0.4108	$4^3.6^3$	4·6_2·4·6_2·4·6_2	1151
	0/0/0	$4^6.6^6.8^3$	4·4·4·4·4·4·6_2·6_2·6_2·6_2·6_2	
				1159

ftw	Pm - 3m (#221)	1.4142/1.4142/1.4142		90/90/90
	0.5/0/0.5	$4^4.6^2$	4_3·4_3·4_3·4_3·6_{22}·6_{22}	3549
	0/0/0	$4^{36}.6^{30}$	4·4·4·4·4·4·4·4·4·4·4·4·4·4·4·	

4·4·4·4·4·4·4·4·4·4·4·4·4·4·4·4·4·4·4·4·4·4·4·8_8·8_8·8_8·8_8·8_8·8_8·8_8·8_8·8_8·
8_8·8_8·8_8·8_8·8_8·8_8·8_8·8_8·8_8·8_8·8_8·8_8·8_8·8_8·8_8·*·*·*·*·* 3271

| gis | I 4_1/amd (#141) | 3.3340/3.3340/2.9819 | | 0/90/90 |
| | 0.15/0.4/0.875 | $4^3.6^2.8$ | 4·4·4·8_2·8·8 | 726 |

gra	P6_3/mmc (#194)	1.7321/1.7321/2		90/90/120
	0.3333/0.6667/0.25	6^3	6_3·6_3·6_3	1164
	0/0/0.25	$6^9.8$	6·6·6·6_2·6_2·6_2·6_2·6_2·6_2	
				1176

| gsi | Ia - 3 (#206) | 2.7879/2.7879/2.7879 | | 90/90/90 |
| | 0.1036/0.1036/0.1036 | 6^6 | 6·6_2·6·6_2·6·6_2 | 1166 |

hms	P - 6m2 (#187)	1.7321/1.7321/1		90/90/120
	0/0/0	6^3	6_3·6_3·6_3	1172
	0.333/0.6667/0	$6^9.8$	6·6·6·6_2·6_2·6_2·6_2·6_2·6_2·*	
				1176

noc	$P6_422$ (#181)	4.7188/4.7188/3.8937	90/90/120
	0.0955/0.3478/0.9136	4.20^2 $4{\cdot}20_3{\cdot}20_{24}$	307
	0.1224/0.5612/0.8333	$4^2.6$ $4{\cdot}4{\cdot}20_{24}$	284
nod	Cccm (#66)	4/2.8286/3.9998	90/90/90
	0.25/0.25/0.875	$8^2.10$ $8{\cdot}8{\cdot}10_3$	528
	0.125/0/0.75	$8^2.10$ $8{\cdot}8{\cdot}10_3$	524
noe	C2/m (#12)	5.0323/3.4011/4.4185	90/139.358/90
	0.223/0.2893/0.0839	$7^2.8$ $7{\cdot}7{\cdot}8$	529
	0/0.147/0	$7^2.8$ $7{\cdot}7{\cdot}8$	531
	0.1157/0/0.6737	7.12^2 $7{\cdot}12_3{\cdot}12_3$	528
nof	Cccm (#66)	4.9098/3.7839/2.8825	90/90/90
	0.3121/0.3547/0	6.10^2 $6{\cdot}10{\cdot}10$	448
	0.8982/0/0.25	$6^2.10$ $6{\cdot}6{\cdot}10_2$	448
nog	$P2_1/c$ (#14)	4.3022/1.6802/5.9182	90/136.874/90
	0.0884/0.0167/0.2839	$5^2.6^2.7.8$ $5{\cdot}5{\cdot}6{\cdot}8{\cdot}6{\cdot}8_2$	1217
	0.2138/0.7033/0.4798	$5^3.6.7.8$ $5{\cdot}5{\cdot}5{\cdot}6{\cdot}8{\cdot}8_6$	1208
	0.0797/0.2033/0.4716	$4.5^3.6.7$ $4{\cdot}6{\cdot}5{\cdot}5{\cdot}5{\cdot}8_3$	1210
	0.165/0.2025/0.6804	$4.5^2.7^2.8$ $4{\cdot}8_3{\cdot}5{\cdot}5_2{\cdot}8_2{\cdot}8_2$	1220
	0.4801/0.2012/0.7842	$5^4.6.8$ $5{\cdot}5{\cdot}5{\cdot}5{\cdot}8{\cdot}8_2$	1209
noh	Pa - 3 (#205)	4.6016/4.6016/4.6016	90/90/90
	0.1906/0.6906/0.8094	10^3 $10_3{\cdot}10_3{\cdot}10_3$	507
	0.6362/0.0712/0.6363	6.10^2 $6{\cdot}10_3{\cdot}10_3$	505
noj	$P4_322$ (#95)	3.7568/3.7568/1.7491	90/90/90
	0.8307/0/0.75	8^3 $8{\cdot}8{\cdot}8_2$	507
	0.5645/0/0	8^3 $8{\cdot}8{\cdot}8$	499
nol	Pnma (#62)	3.12/3.9275/2.9135	90/90/90
	0.1218/0.0334/0.398	$3^2.4.6.7.8$ $3{\cdot}6{\cdot}3{\cdot}8{\cdot}10_4{\cdot}*$	821
	0.1728/0.8773/0.6635	$3.6^2.7.8^2$ $3{\cdot}6{\cdot}6{\cdot}8{\cdot}8{\cdot}10_7$	822

nom   P6$_1$22 (#178)   2.9197/2.9197/3.1463   90/90/120
      0.4638/0.4638/0.1667   4^2.6^2.8^2 4·4·6$_2$·6$_2$·12$_{33}$·12$_{59}$ 875
      0.6203/0.3797/0.9167   4^2.6^4 4·4·6·6·6$_2$·6$_2$ 879

noq   C2/c (#15)   2.3179/3.9409/2.4715   90/113.848/90
      0.415/0.3881/0.4065   4.5^2.6^2.7   4·5·5·6·6·8   960
      0/0.2674/0.25   4^2.5.6^2.8   4·4·5·8$_2$·6·6   961

nor   C2/m (#12)   2.9914/4.5674/1.7203   90/141.032/90
      0/0/0.5   5^2.6^2.8^2 5·5·6$_2$·6$_2$·8$_2$·8$_2$ 1147
      0.3644/0.1569/0.0667   5^2.6^4 5·6·5·6·6·6   1147

nos   P2$_1$/c (#14)   3.3267/3.3222/6.6504   90/107.546/90
      0.1487/0.711/0.128   8^3 8·8·8   516
      0.3579/0.926/0.1262   8^3 8·8·8$_2$ 516
      0.6695/0.88/0.1733   8^2.10   8·8$_2$·10$_2$ 521
      0.8481/0.8018/0.0827   8^2.10   8·8$_2$·10$_2$ 519
      0.7596/0.929/0.3289   8^3 8·8·8   525
      0.7526/0.6993/0.425   8^3 8·8·8$_2$ 523

not   P2$_1$/c (#14)   2.1448/3.4996/1.9997   90/104.685/90
      0.4865/0.5956/0.1816   4^2.6.7.8   4·7$_2$·4·8·8$_2$·*   1325
      0.217/0.1951/0.5916   4^2.6.7^2.8   4·7·4·8·7·7   1304
      0.064/0.5397/0.2866   4^2.6.7^2.8   4·7·4·8·7·8$_2$ 1321

nou   P4/mbm (#127)   2.7321/2.7321/1.4142   90/90/90
      0/0/0   6^4.8^2 6$_2$·6$_2$·6$_2$·6$_2$·8$_2$·8$_2$ 1169
      0.5/0/0.5   4^2.8^4 4·4·8$_2$·8$_2$·8$_8$·8$_8$ 1163
      0.183/0.317/0   4.6^5 4·6$_2$·6·6·6·6   1155

nov   Cmca (#64)   1.8924/2.9675/1.5952   90/90/90
      0/0.1484/0.1486   4^4.6^6 4·4·4·4·6·6·6$_3$·6$_5$·6$_5$·6$_5$ 1428

nta   R - 3m (#166)   4.4214/4.4214/5.9831   90/90/120
      0.3869/0/0   9^3 9$_2$·9$_2$·9$_2$ 576
      0.1536/0.8464/0.7515   9^3 9$_2$·9$_2$·9$_2$ 573

ntb	P 4$_2$/nmc(#137)	4.4214/4.4214/3.4546		90/90/90
	0.25/0.3631/0.1228	9^3	$9_2 \cdot 9_2 \cdot 9_2$	586
	0.0566/0.4803/0.1228	9^3	$9_2 \cdot 9_2 \cdot 9_2$	562
pcb	Im - 3m (#229)	3.1548/3.1548/3.1548		90/90/90
	0.1585/0.1585/0.1585	$4^3 . 8^3$	$4 \cdot 8_2 \cdot 4 \cdot 8_2 \cdot 4 \cdot 8_2$	787
pcu	Pm - 3m (#221)	1/1/1		90/90/90
	0/0/0	$4^{12} . 6^3$	$4 \cdot 4 \cdot 4 \cdot 4 \cdot 4 \cdot 4 \cdot 4 \cdot 4 \cdot 4 \cdot 4 \cdot 4 \cdot 4 \cdot {*} \cdot {*} \cdot {*}$	
				1561
pcl	Cmcm (#63)	3.2578/3.1261/2.8385		90/90/90
	0.1535/0.3548/0.0738	$4^2 . 6^3 . 8$	$4 \cdot 6 \cdot 4 \cdot 6 \cdot 6 \cdot 8_3$	864
pto	Pm - 3n (#223)	2.8284/2.8284/2.8284		90/90/90
	0.25/0.25/0.25	8^3	$8_5 \cdot 8_5 \cdot 8_5$	914
	0.25/0/0.5	8^6	$8_2 \cdot 8_2 \cdot 8_4 \cdot 8_4 \cdot 8_4 \cdot 8_4$	893
pts	P4$_2$/mmc (#131)	1.6331/1.6331/2.3093		90/90/90
	0/0.5/0	$4^2 . 8^4$	$4 \cdot 4 \cdot 8_2 \cdot 8_2 \cdot 8_4 \cdot 8_4$	979
	0/0/0.25	$4^2 . 8^4$	$4 \cdot 4 \cdot 8_7 \cdot 8_7 \cdot 8_7 \cdot 8_7$	975
ptt	Cccm (#66)	4.6535/1.5165/3.0847		90/90/90
	0.1401/0/0.75	$4 . 6^3 . 8^2$	$4 \cdot 6 \cdot 6 \cdot 6 \cdot 8_6 \cdot 8_6$	1196
	0/0.5/0.75	$4^2 . 6^2 . 8^2$	$4 \cdot 4 \cdot 6 \cdot 6 \cdot 8_2 \cdot 8_2$	1189
	0.25/0.25/0.5	$6^2 . 8^4$	$6_2 \cdot 6_2 \cdot 8_4 \cdot 8_4 \cdot 8_{11} \cdot 8_{11}$	1211
pyr	Pa - 3 (#205)	2.4495/2.4495/2.4495		90/90/90
	0.3333/0.3333/0.3333	6^3	$6_3 \cdot 6_3 \cdot 6_3$	1443
	0/0/0	$6^{12} . 8^3$	$6 \cdot 6 \cdot 6 \cdot 6 \cdot 6 \cdot 6 \cdot 6_2 \cdot 6_2 \cdot 6_2 \cdot 6_2 \cdot 6_2 \cdot 6_2 \cdot$	
	$* . * . *$			1371
pyr-e	Pa - 3 (#205)	3.1206/3.1206/3.1206		90/90/90
	0.1945/0.4034/0.4354	$3^5 . 4^2 . 6^5 . 7^3$	$3 \cdot 3 \cdot 3 \cdot 3 \cdot 3 \cdot 4 \cdot 4 \cdot 6 \cdot 6 \cdot 6 \cdot 6 \cdot 6_2 \cdot$	
	$8 \cdot 8 \cdot *$			2014

qom   P - 31c (#163)   3.464/3.464/2.8287   90/90/120
      0/0/0.25 6^3 $6_3 \cdot 6_3 \cdot 6_3$ 1479
      0.3333/0.6667/0.25 6^3 $6_3 \cdot 6_3 \cdot 6_3$ 1491
      0.1111/0.3889/0.9167 6^3 $6_3 \cdot 6_3 \cdot 6_3$ 1485
      0.3333/0.6667/0.75 $6^{12}.8^3$ $6 \cdot 6 \cdot 6 \cdot 6 \cdot 6 \cdot 6 \cdot 6 \cdot 6 \cdot 6 \cdot 6_3 \cdot 6_3 \cdot 6_3 \cdot$
      ..*   1425
      0.1667/0.8333/0.25 $6^{12}.8^3$ $6 \cdot 6 \cdot 6 \cdot 6 \cdot 6 \cdot 6 \cdot 6_2 \cdot 6_2 \cdot 6_2 \cdot 6_2 \cdot 6_2 \cdot 6_2 \cdot$
      ..*   1425

qtz   P6$_2$22 (#180)   1.633/1.633/1.7321   90/90/120
      0.5/0/0 $6^4.8^2$ $6 \cdot 6 \cdot 6_2 \cdot 6_2 \cdot 8_7 \cdot 8_7$ 1231

qzd   P6$_2$22 (#180)   1/1/3   90/90/120
      0/0/0 $7^5.9$ $7_2 \cdot 7_3 \cdot 7_3 \cdot 7_3 \cdot 7_3 \cdot *$ 2079

rhr-a Im - 3m (#229)   8.2930/8.2930/8.2930   90/90/90
       0.3701/0.2848/0.0603   4.8.10 $4 \cdot 8 \cdot 12$ 265

rtl   P4$_2$/mnm (#136)   2.3571/2.3571/1.4906   90/90/90
       0.3/0.3/0 4.6^2 $4 \cdot 6_2 \cdot 6_2$ 1210
       0/0/0 $4^2.6^{10}.8^3$ $4 \cdot 4 \cdot 6 \cdot 6 \cdot 6 \cdot 6 \cdot 6 \cdot 6 \cdot 6 \cdot 6 \cdot 6_2 \cdot 6_2 \cdot$
       ..*   1121

rtw   P4/mbm (#127)   2.5779/2.5779/2   90/90/90
       0/0/0 $5^4.6^2$ $5 \cdot 5 \cdot 5 \cdot 5 \cdot 6_2 \cdot 6_2$ 1557
       0.3629/0.8629/0 $4^2.5^3.6^5$ $4 \cdot 4 \cdot 5 \cdot 5 \cdot 5 \cdot 6_3 \cdot 6_3 \cdot 6_3 \cdot 6_3 \cdot *$ 1534

sit   Imm2 (#44)   1.3093/2.6186/2.2678   90/90/90
       0.5/0.25/0.1667 4.6^2 $4 \cdot 6_2 \cdot 6_2$ 1214
       0/0/0 $4^2.6^{10}.8^3$ $4 \cdot 4 \cdot 6 \cdot 6 \cdot 6 \cdot 6 \cdot 6 \cdot 6 \cdot 6 \cdot 6 \cdot 6_2 \cdot 6_2 \cdot$
       $8 \cdot 8 \cdot 8_6$ 1121

sln   P6$_3$/m (#176)   4.4146/4.4146/1.6719   90/90/120
       0.3333/0.6667/0.25 8^3 $8_4 \cdot 8_4 \cdot 8_4$ 774
       0.3339/0.0605/0.25 6^3 $6 \cdot 6 \cdot 6$ 770
       0.2040/0.7989/0.25 $6^3.8^3$ $6 \cdot 8_2 \cdot 6 \cdot 8_3 \cdot 6 \cdot 8_3$ 749

smn	R -3c (#167)	3.8761/3.8761/1.2774	90/90/120
	0.1985/0/0.25	$4^8.5^3.6^4$ $4·4·4·4·4·4·4·4·4·5_2·5_2·5_5·6_4·$	
	$6_4·*.*$		2003

| sod | Im - 3m (#229) | 2.8284/2.8284/2.8284 | 90/90/90 |
| | 0.25/0/0.5 | $4^2.6^4$ $4·4·6·6·6·6$ | 791 |

| sqp | I4/mmm (#139) | 1.3333/1.3333/2.6667 | 90/90/90 |
| | 0/0/0.1875 | $4^4.6^6$ $4·4·4·4·6·6·6_5·6_5·6_5·6_5$ | 1366 |

| sra | Imma (#74) | 3.2592/1.6842/2.6331 | 90/90/90 |
| | 0.1534/0.25/0.1024 | $4^2.6^3.8$ $4·6·4·6·6·8_2$ | 833 |

| srs | $I4_132$ (#213) | 2.8284/2.8284/2.8284 | 90/90/90 |
| | 0.125/0.125/0.125 | 10^3 $10_5·10_5·10_5$ | 530 |

tbo	Fm - 3m (#225)	4.889/4.889/4.889	90/90/90
	0.3333/0.3333/0.3333	6^3 $6·6·6$	822
	0.25/0/0.25	$6^2.8^2.10^2$ $6_2·6_2·8_2·8_2·12_2·12_2$	817

| tcb | Pnna (#52) | 1.4636/0.6816/1.8807 | 90/90/90 |
| | 0/0/0 | 8^6 $8_2·8_2·8_5·8_5·8_5·8_5$ | 3113 |

tfa	I - 4m2 (#119)	1.8016/1.8016/3.7370	90/90/90
	0/0.5/0.3838	8^3 $8_4·8_4·8_4$	742
	0/0/0	8^6 $8_2·8_2·8_3·8_3·8_3·8_3$	747

tfc	Cmmm (#65)	4.732/1.8613/1	90/90/90
	0.2113/0/0	8^3 $8·8_3·8_3$	978
	0/0/0	8^6 $8_2·8_2·8_2·8_2·8_2·*$	1009

| ths | $I4_1$/amd (#141) | 1.8856/1.8856/5.3338 | 90/90/90 |
| | 0/0.25/0.9687 | 10^3 $10_2·10_4·10_4$ | 608 |

| twt | $P6_222$ (#180) | 2.475/2.475/2.0265 | 90/90/120 |
| | 0.298/0/0 | 12^3 $12_4·12_7·12_7$ | 790 |

| unh | $P6_122$ (#178) | 3.0614/3.0614/1.3526 | 90/90/120 |
| | 0.4305/0.861/0.25 | $5^4.6^2$ $5 \cdot 5 \cdot 5 \cdot 5_2 \cdot 12 \cdot 12$ | 1029 |

| utk | $P4_122$ (#91) | 2.8187/2.8187/1.7556 | 90/90/90 |
| | 0.3818/0.8242/0.2876 | 10^3 $10 \cdot 10 \cdot 10_3$ | 817 |

| utj | $P4_2/nbc$ (#133) | 3.6314/3.6314/2.5454 | 90/90/90 |
| | 0.0386/0.378/0.9275 | 10^3 $10 \cdot 10 \cdot 10_3$ | 710 |

| utm | $I4_1/acd$ (#142) | 4.9881/4.9881/3.0276 | 90/90/90 |
| | 0.0666/0.4568/0.1009 | 10^3 $10_2 \cdot 10_4 \cdot 10_4$ | 611 |

| utn | $I4_1/acd$ (#142) | 5.1050/5.1050/1.8803 | 90/90/90 |
| | 0.0867/0.4548/0.9859 | 10^3 $10 \cdot 10 \cdot 10_3$ | 892 |

| uto | $I4_1/acd$ (#142) | 3.6331/3.6331/5.0306 | 90/90/90 |
| | 0.1227/0.5342/0.9623 | 10^3 $10 \cdot 10 \cdot 10_3$ | 712 |

| utp | $Pnna$ (#52) | 2.985/2.3848/2.6643 | 90/90/90 |
| | 0.1002/0.0937/0.1249 | 10^3 $10_2 \cdot 10_4 \cdot 10_4$ | 622 |

wfq	$I4_1/amd$ (#141)	2.3724/2.3724/2.1648	90/90/90
	0/0.4821/0.2301	$4^{15}.5^4.6^2$ $4 \cdot 4 \cdot 4 \cdot 4 \cdot 4 \cdot 4 \cdot 4 \cdot 4 \cdot 4 \cdot 4 \cdot 4 \cdot 4_3 \cdot 4_3 \cdot$	
	$4_3 \cdot 4_3 \cdot 5 \cdot 5 \cdot 5_3 \cdot 5_3 \cdot * \cdot *$		2853

| zni | $I4_1/acd$ (#142) | 4.2251/4.22511.9036 9 | 0/90/90 |
| | 0.0686/0.598/0.1028 | $4.6^4.8$ $4 \cdot 6 \cdot 6 \cdot 6_3 \cdot 6_2 \cdot 12_{40}$ | 1478 |

Appendix B

Stereo drawings

Instructions on how to see the stereo pairs:
"Gaze at the stereo pair, keeping your eyes level and cross your eyes slightly. As you know, crossing your eyes makes you see double, so you will see four images. Try to cross your eyes slowly, so that the two images in the center come together. When they converge or fuse, you will see them as a single 3D image. The fused image will appear to lie between two flat images, which you should ignore. When you are viewing correctly, you see three images instead of four." (Adapted from Gale Rhoades excellet web page on stereo viewing. For more tips on how to view the stereo drawings, go to: http://www.usm.maine.edu/~rhodes/0Help/StereoView.html)

acs

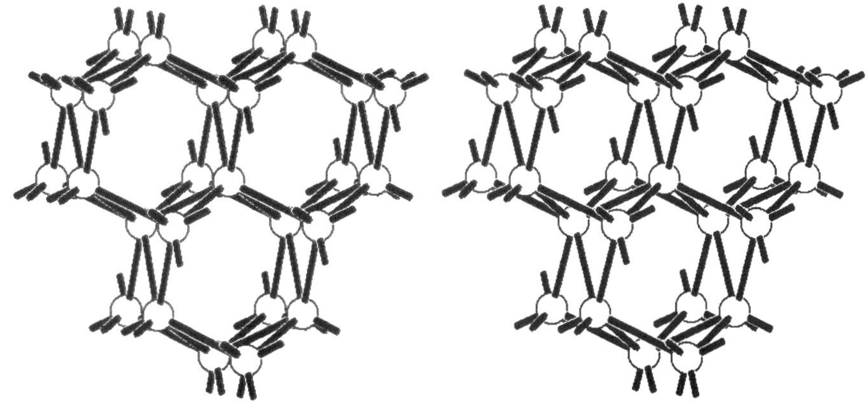

asv

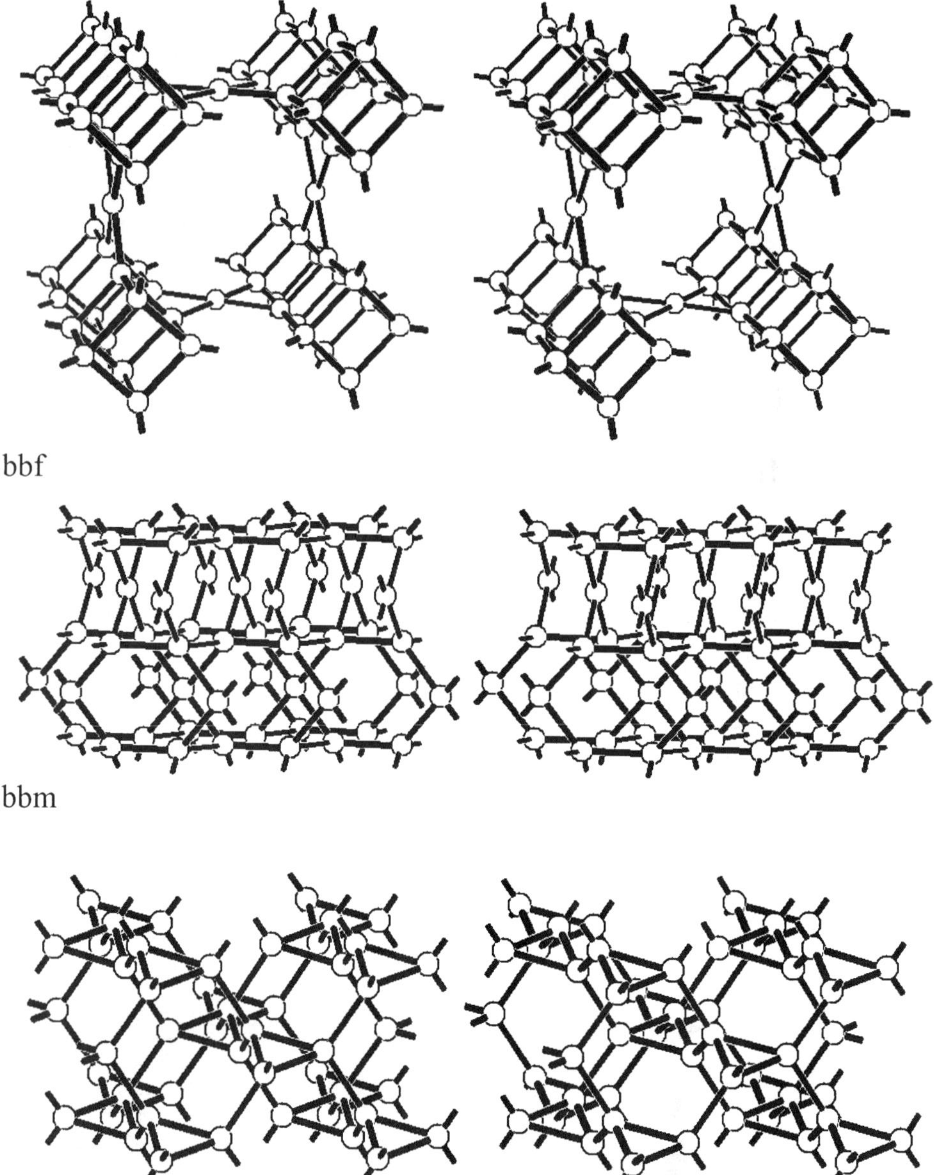

bbf

bbm

bcu

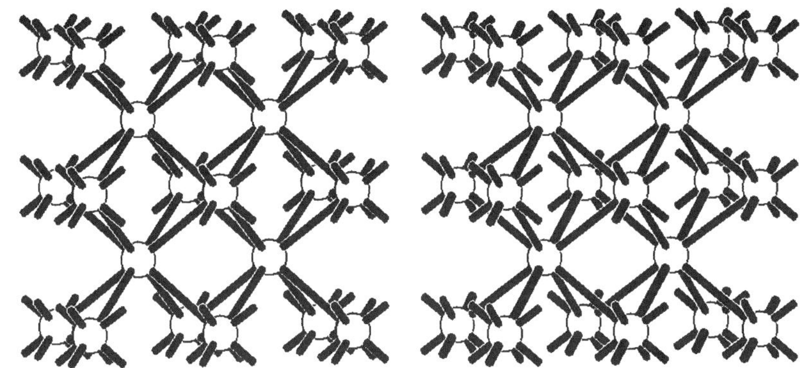

bcu-l

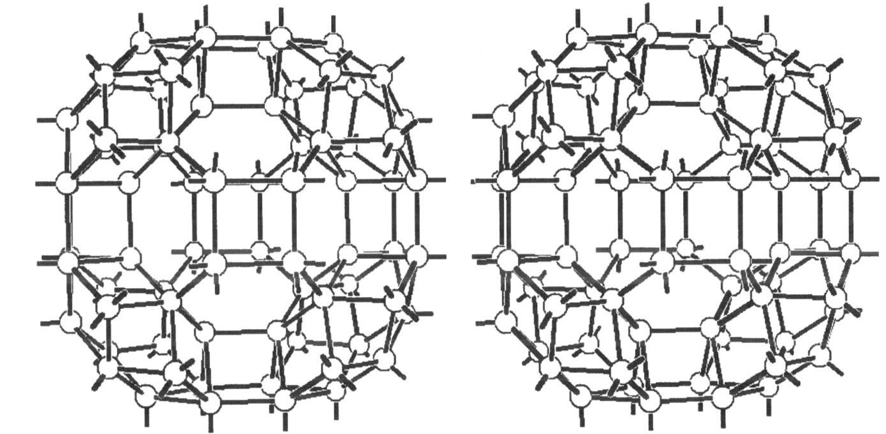

bnn

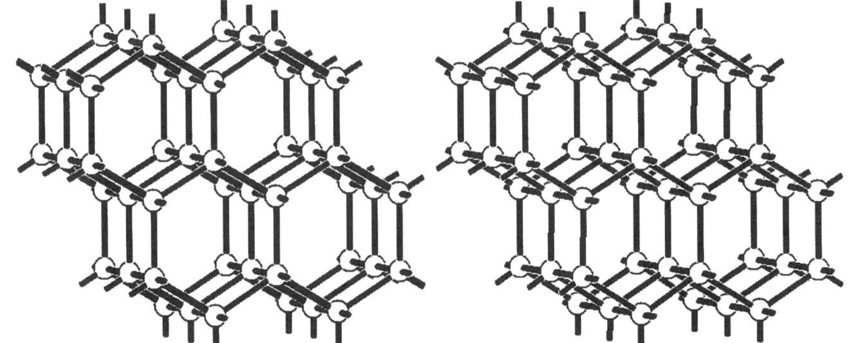

bor

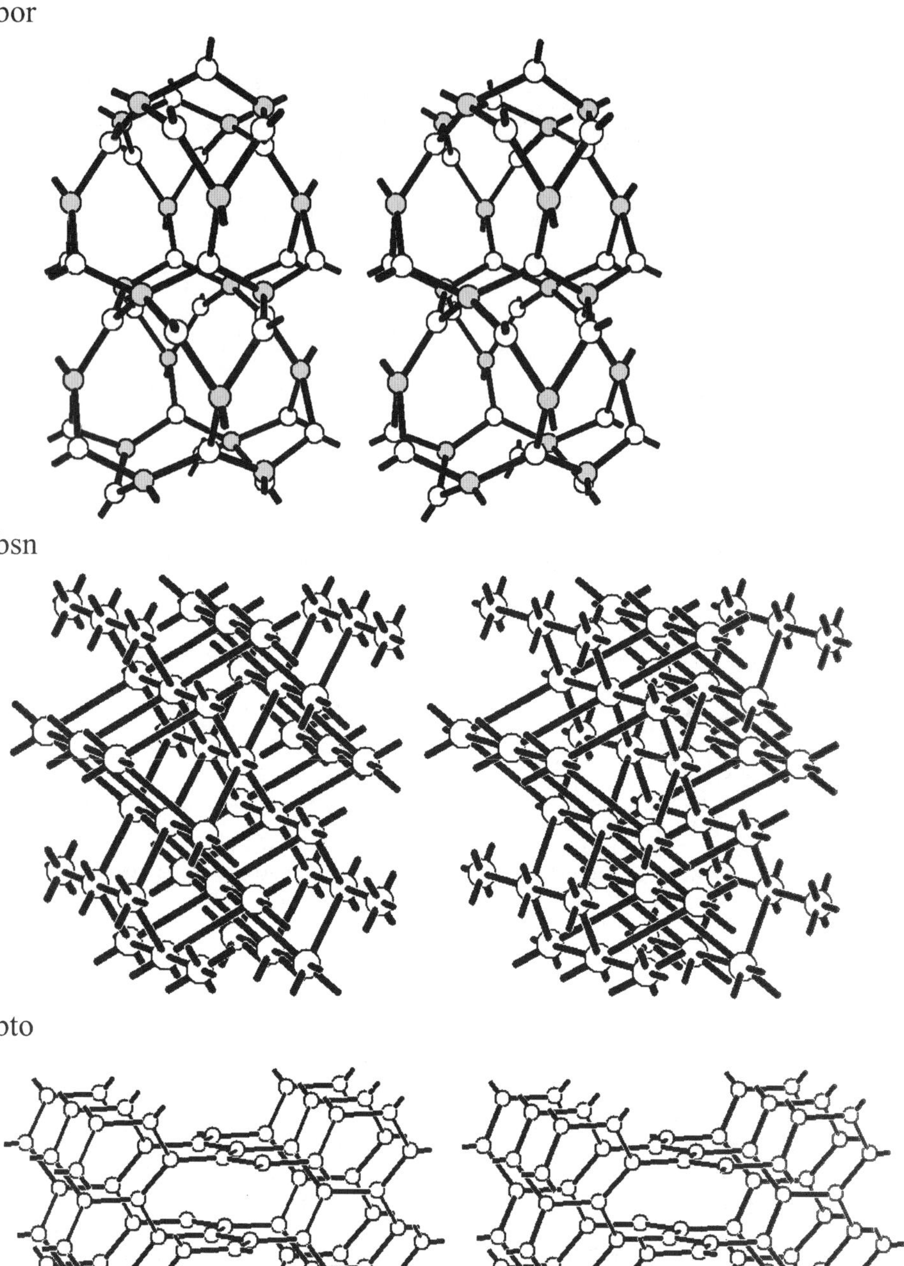

bsn

bto

cab

cag

cds

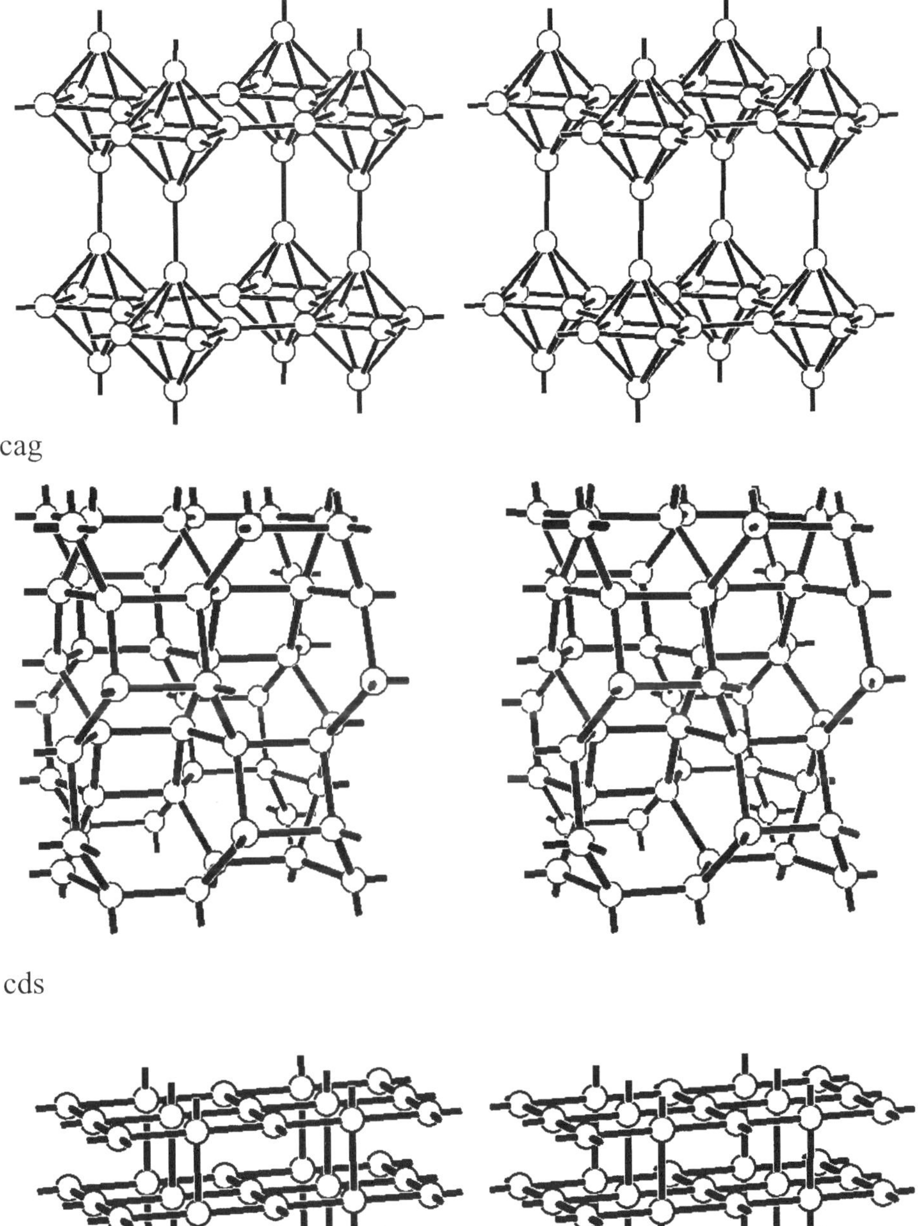

coe

cor

crb

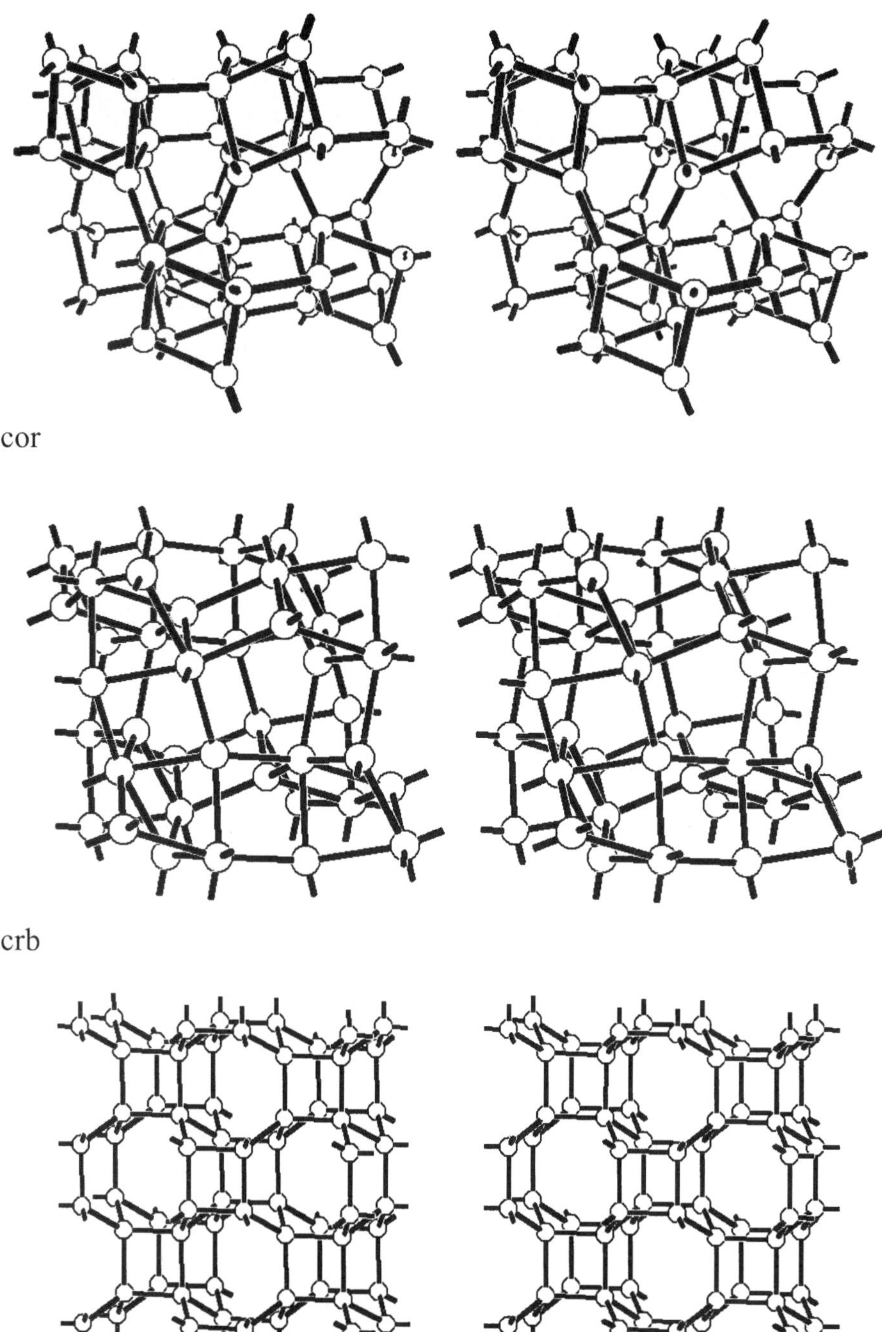

ctn

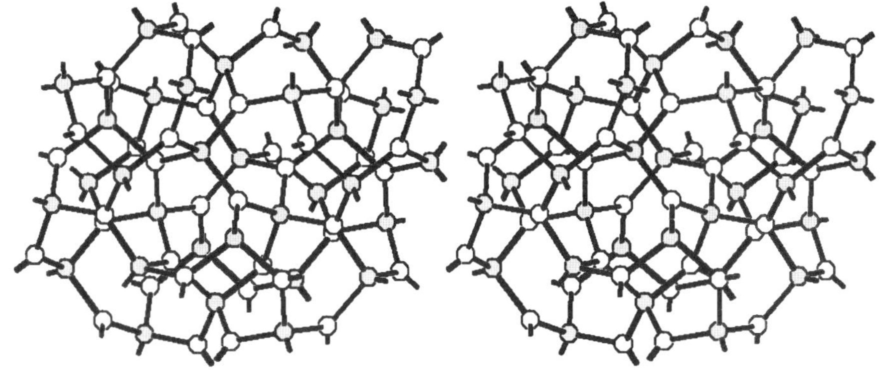

ctn-d

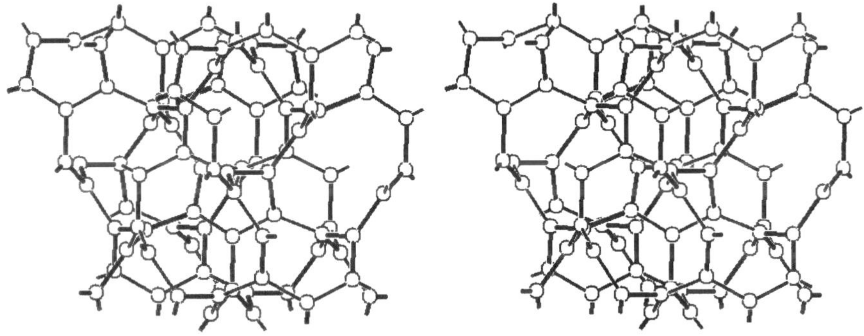

dia

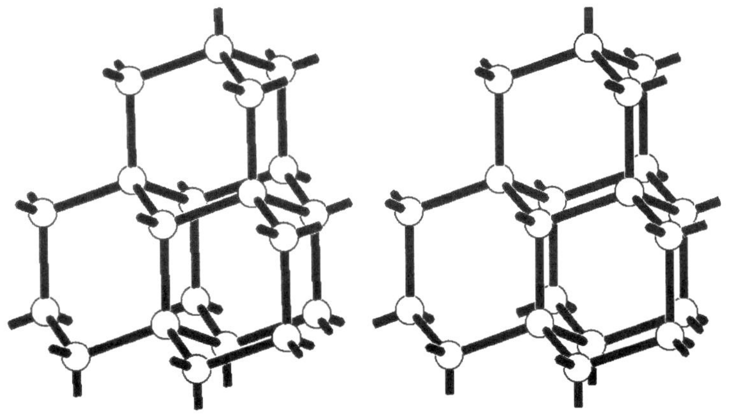

dia-a

dia-b

dia-e

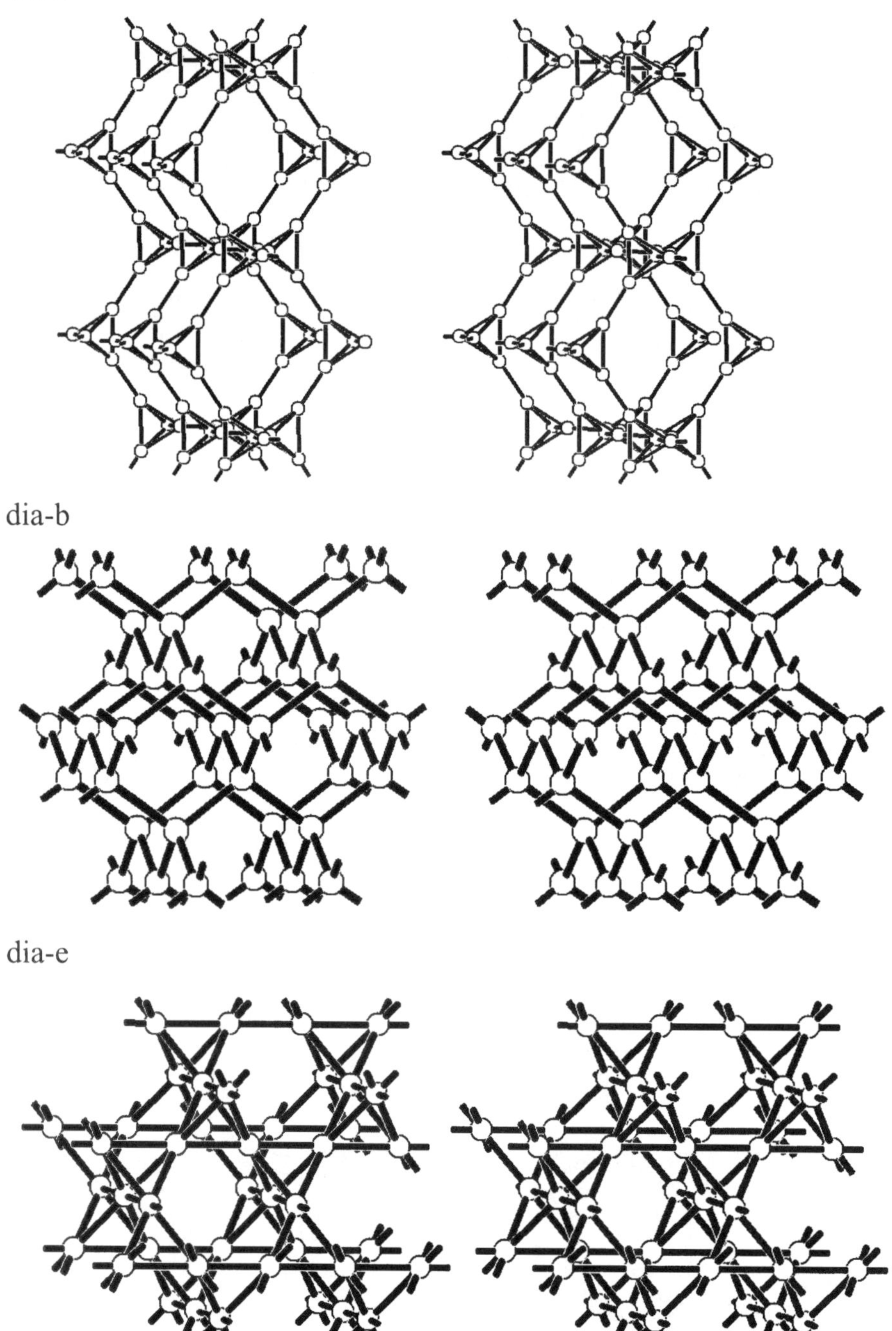

dia-f

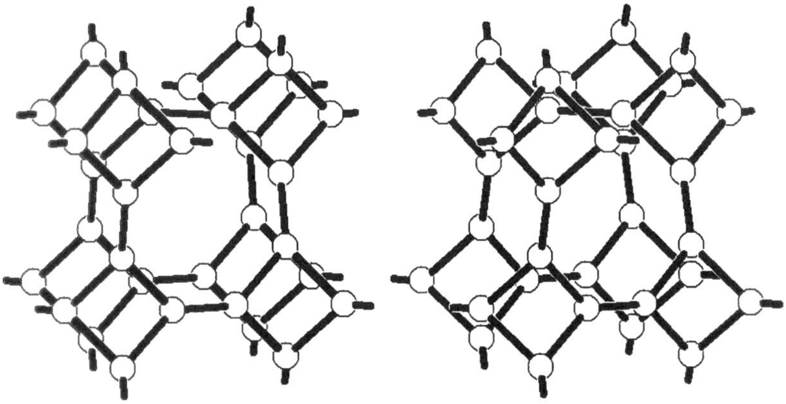

dia-g

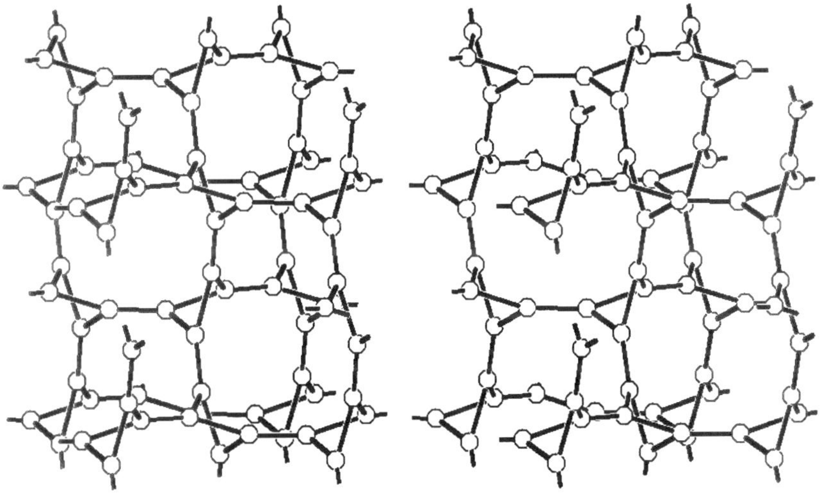

dia-j

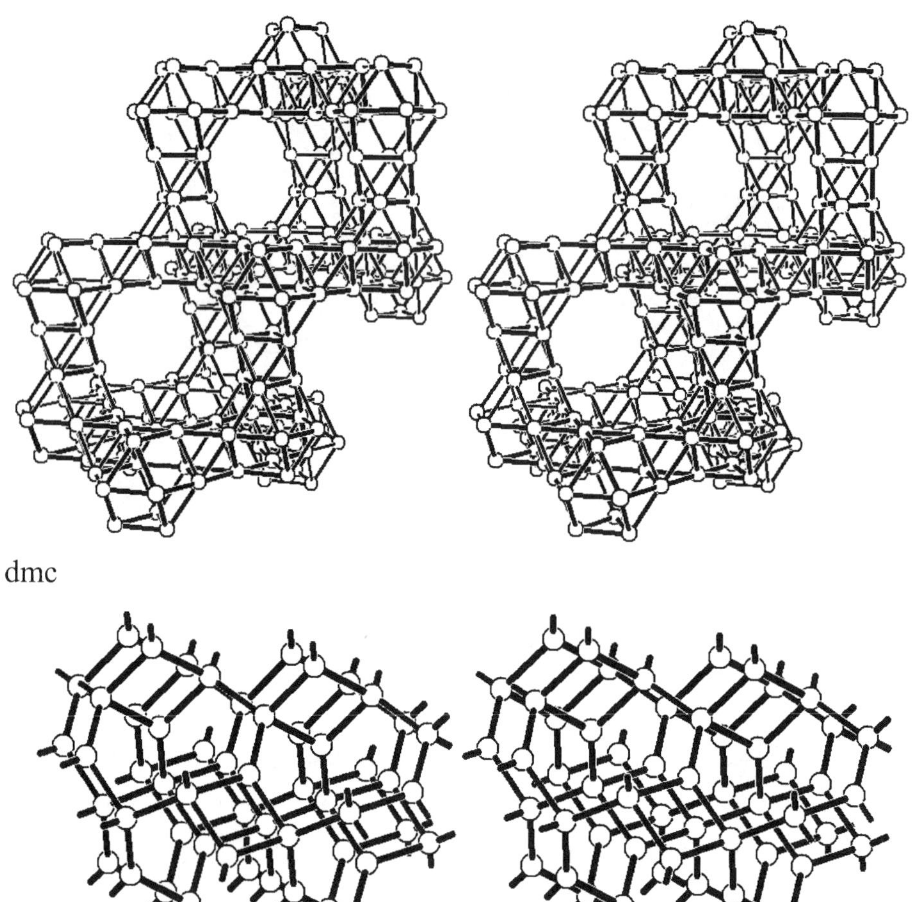

dmc

dmd

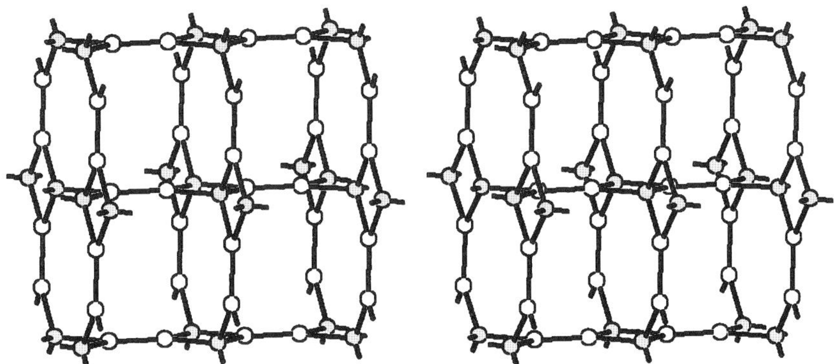

dme

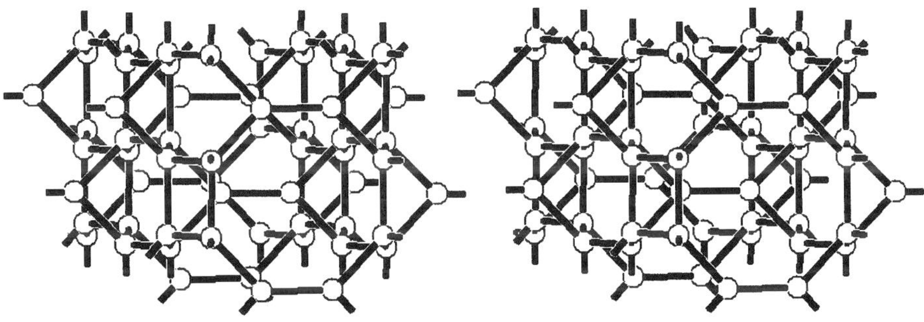

dmf

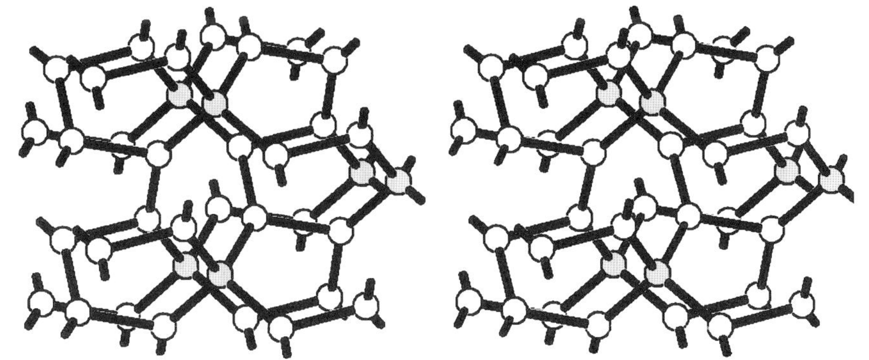

ftw

gis

gra

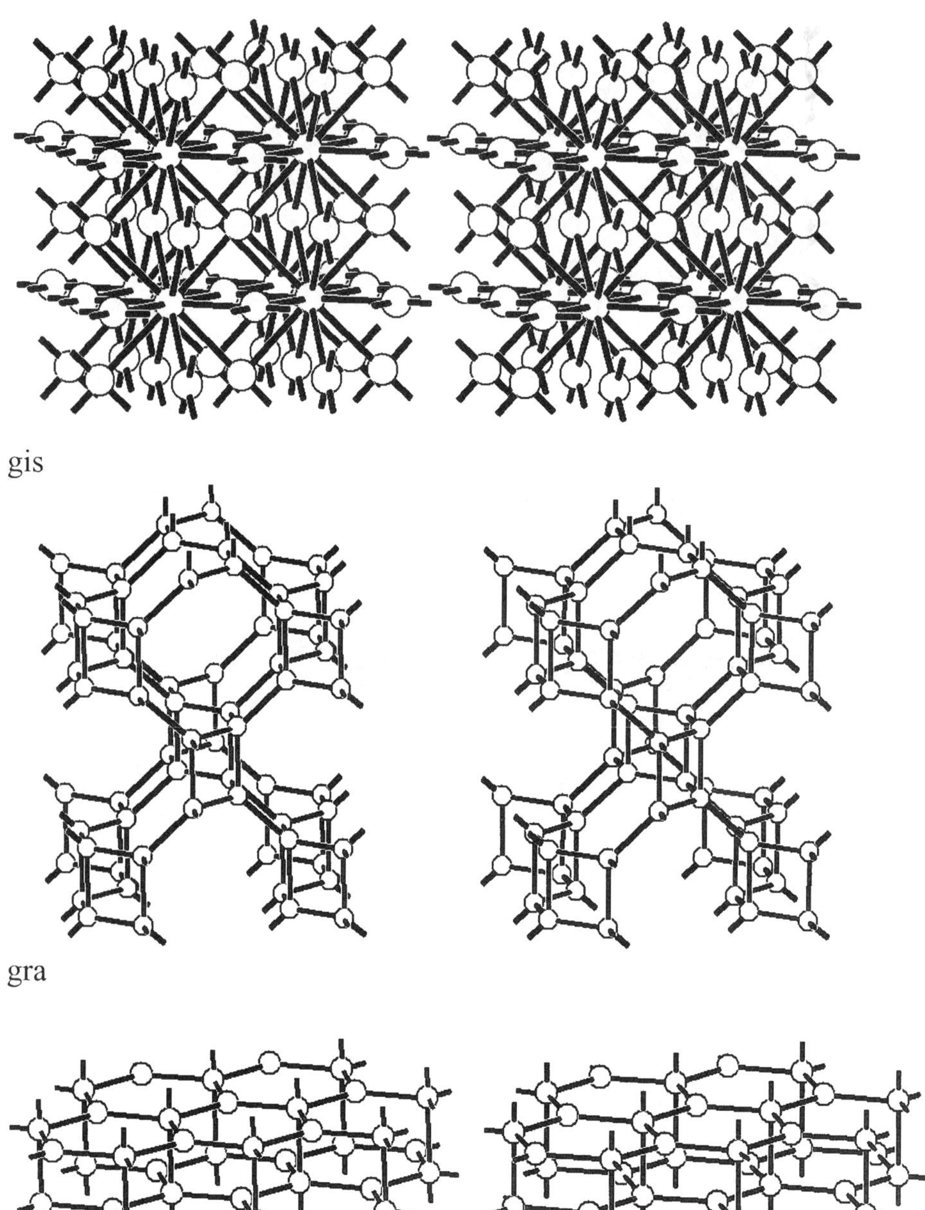

gsi

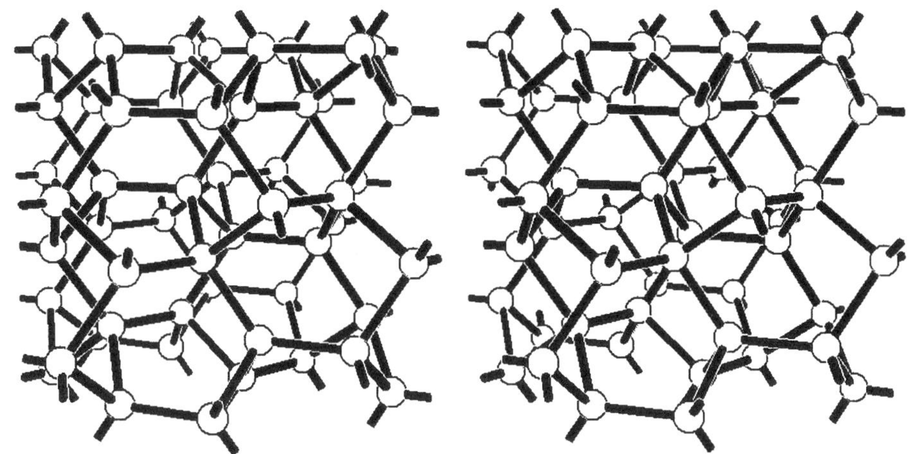

hms

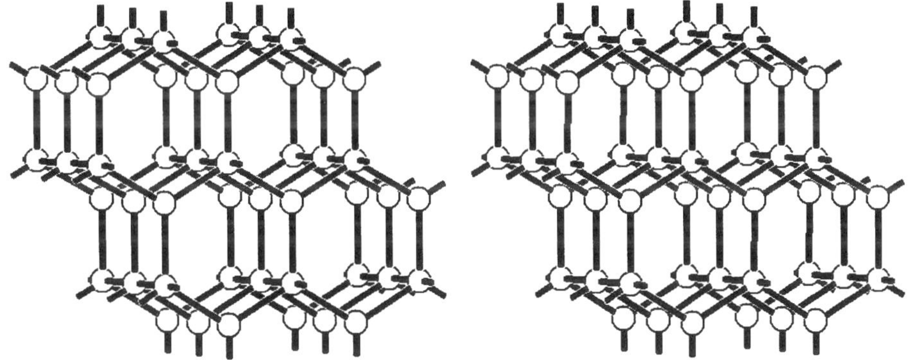

ins

irl

jph

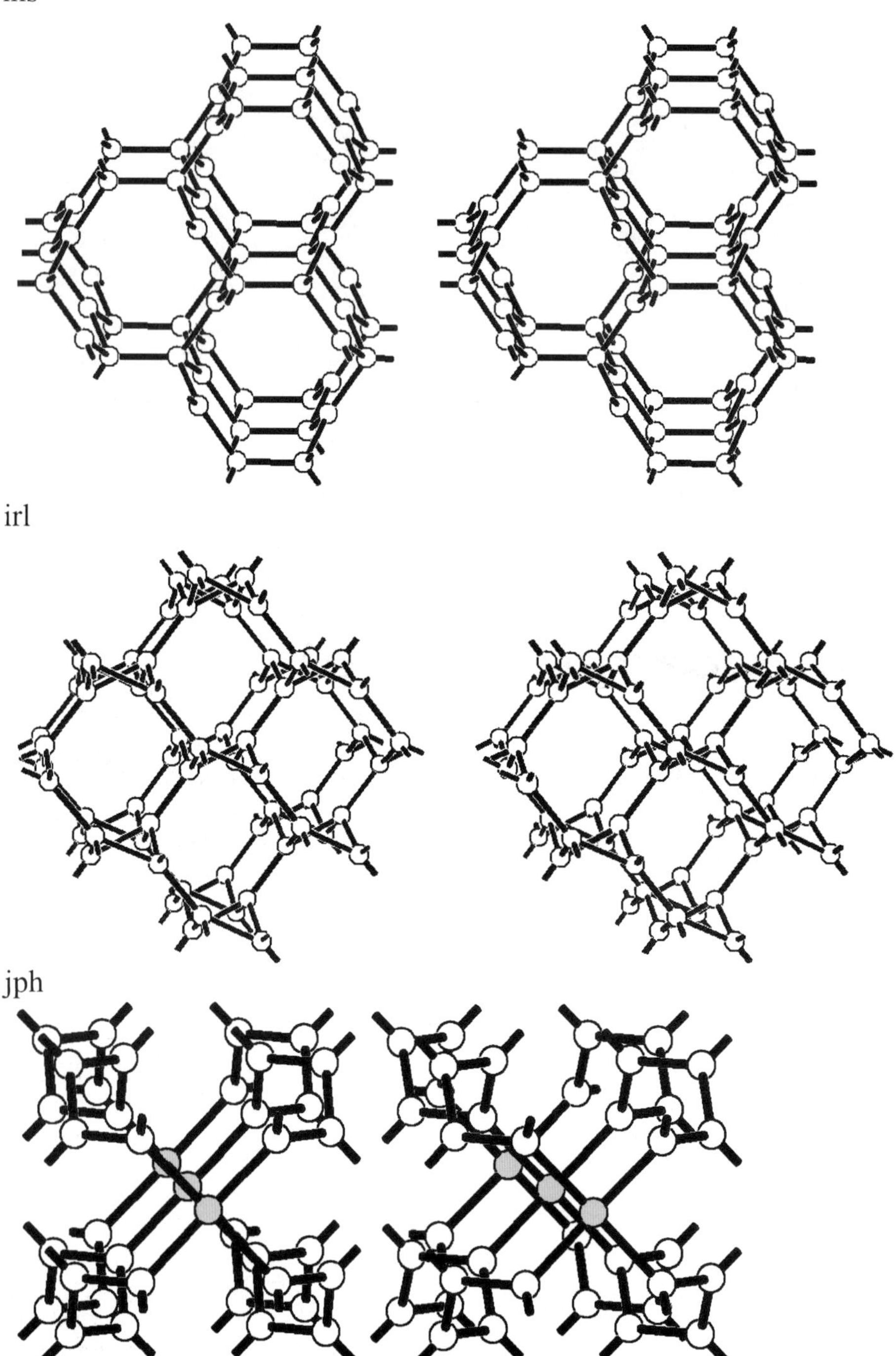

lig

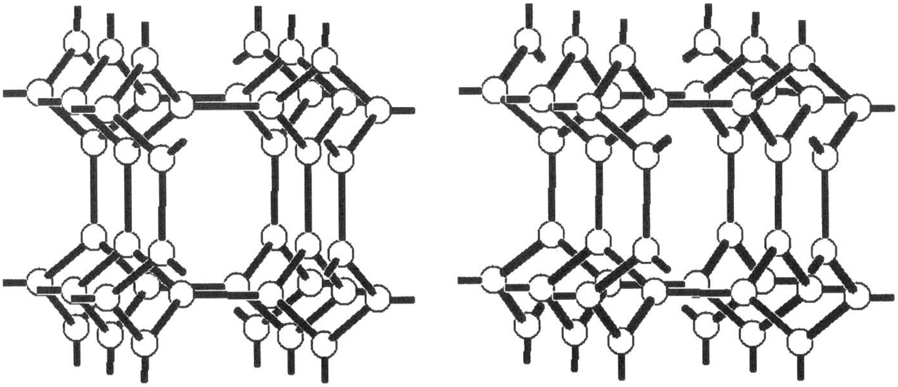

lon

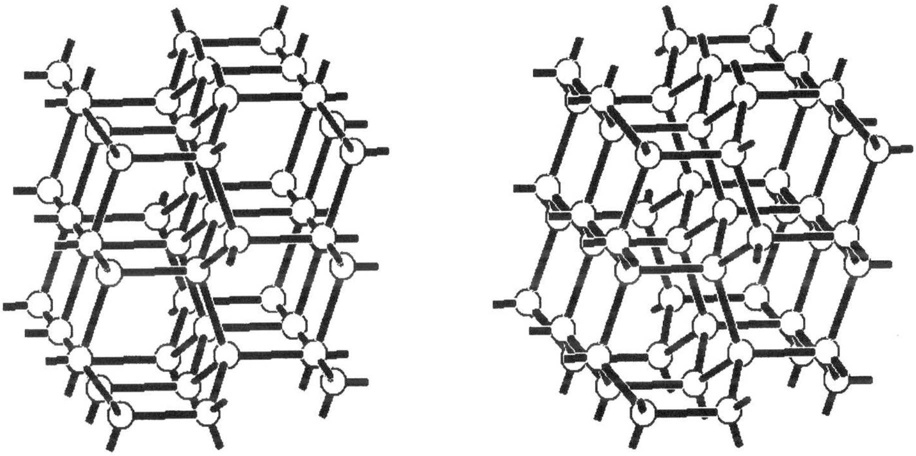

lvt

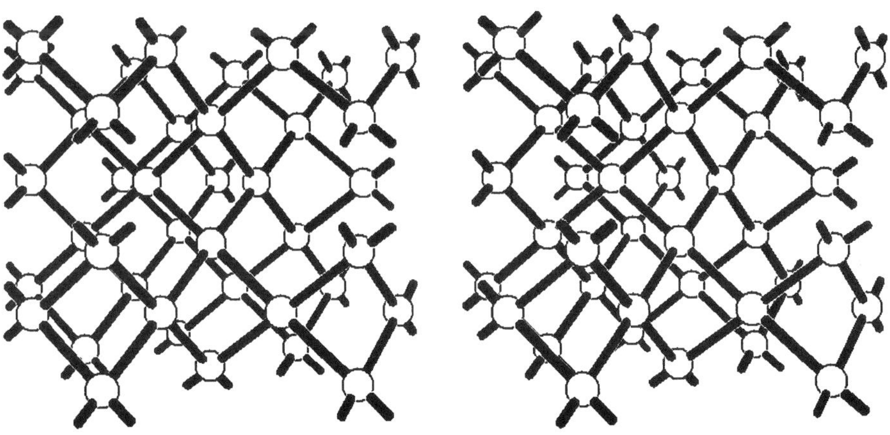

lvt-a

mcf

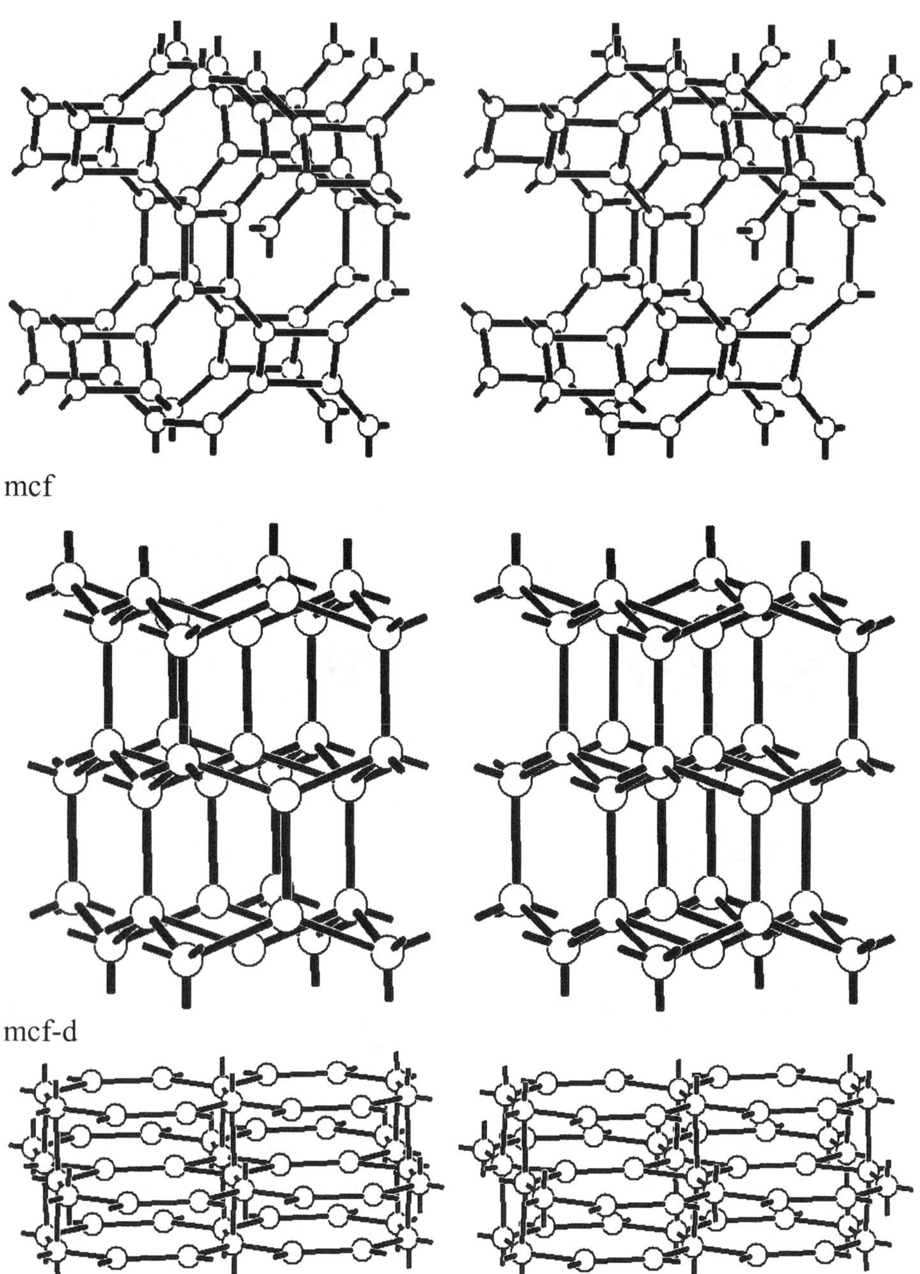

mcf-d

mmt

mog

mot

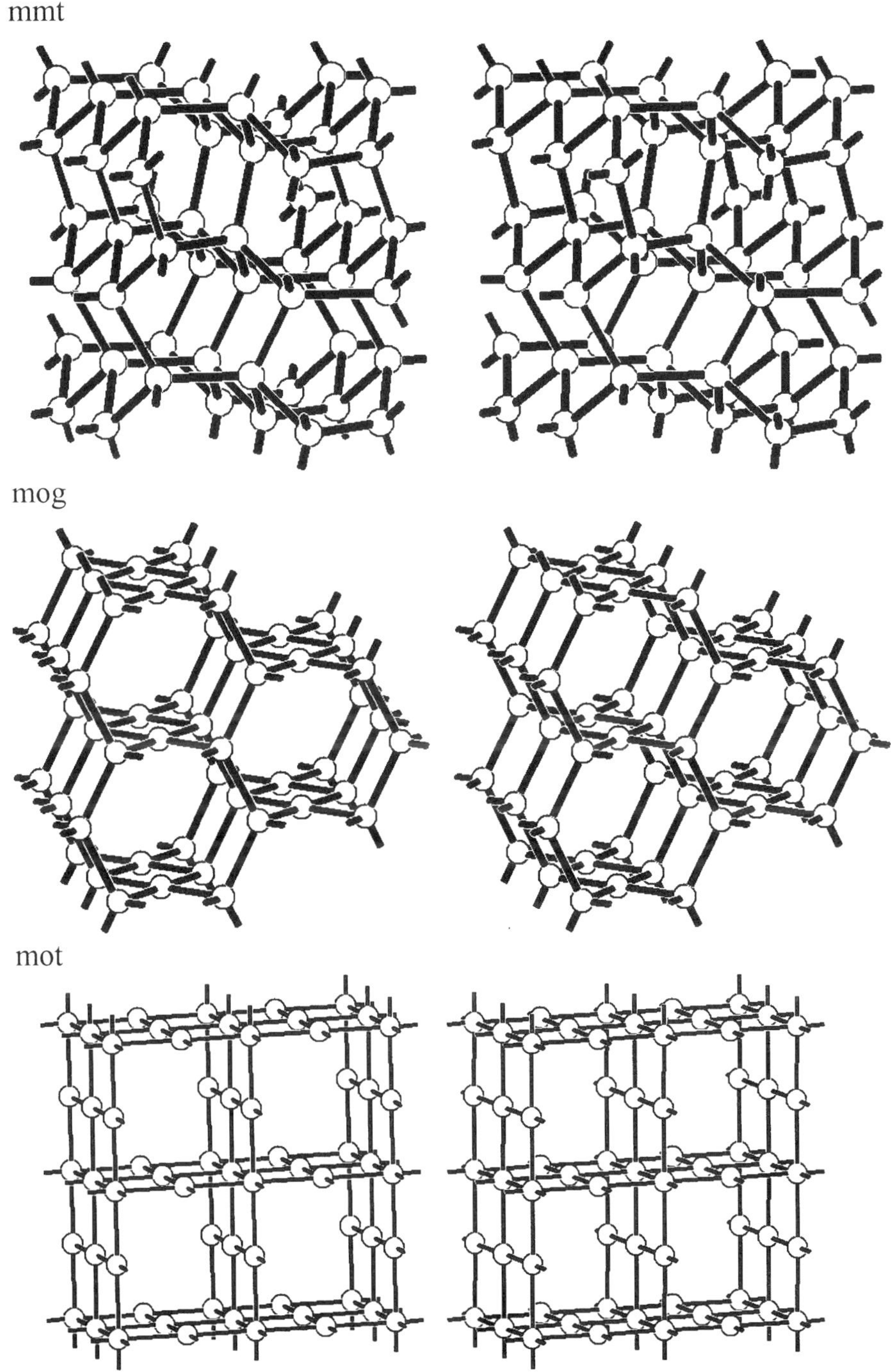

mot-a

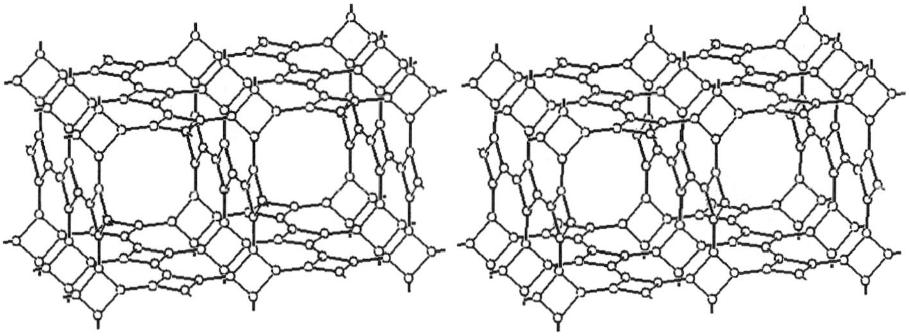

nbo

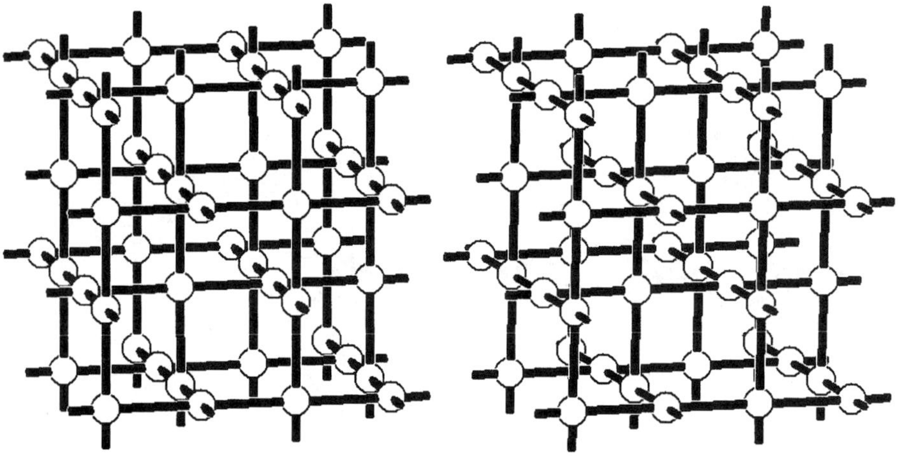

neb

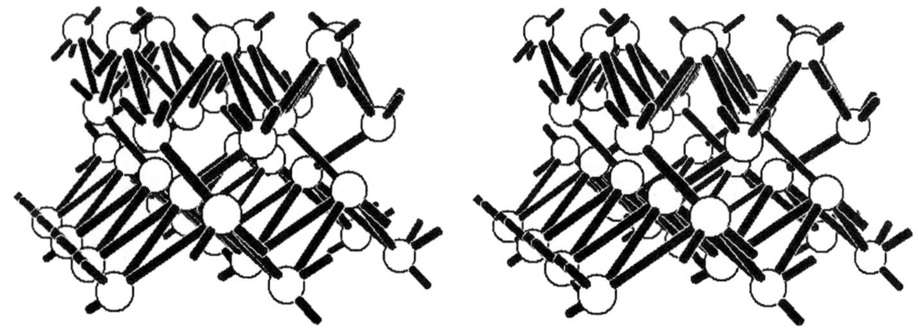

net 3

nia

nob

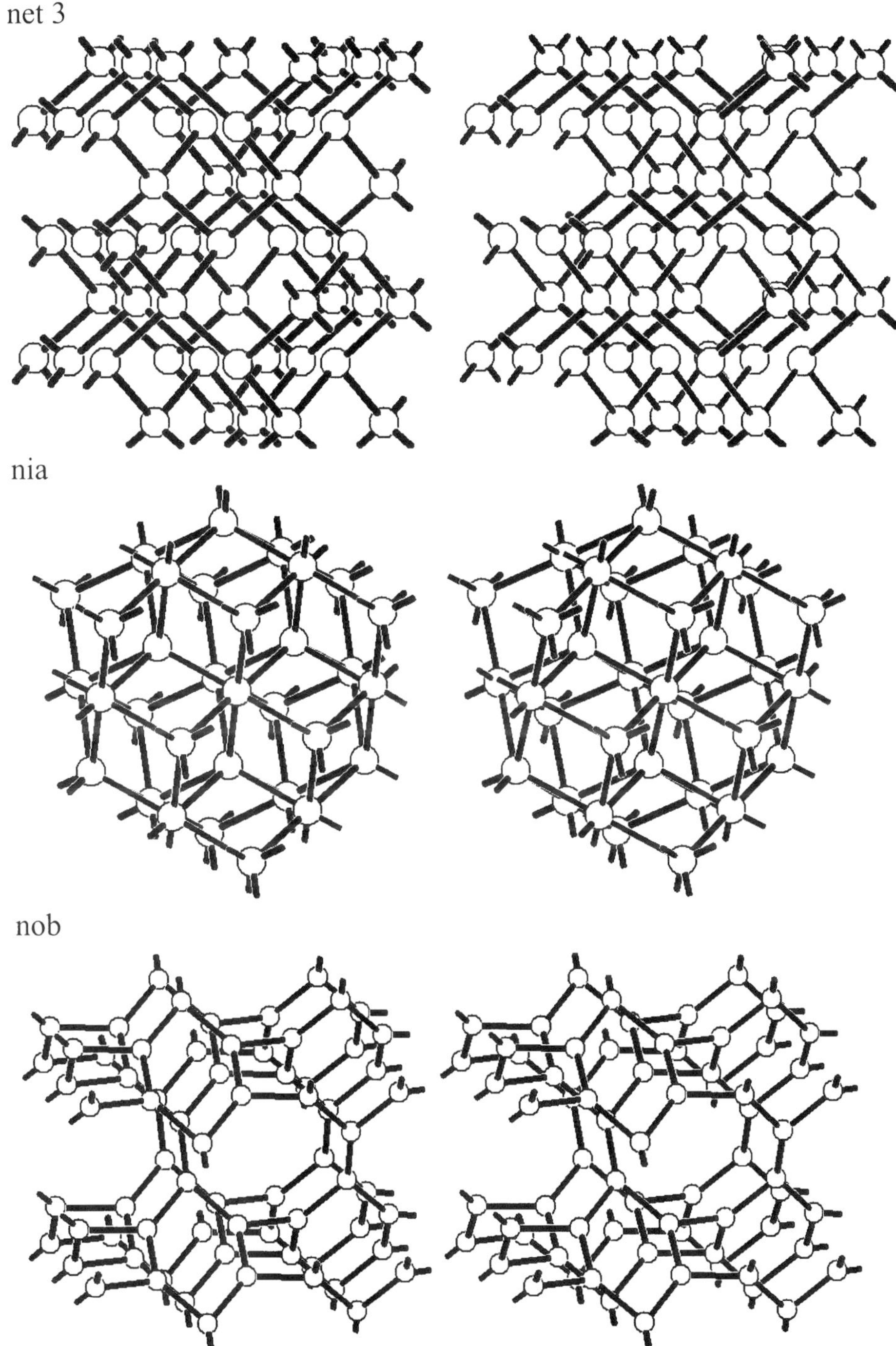

noc

nod

noe

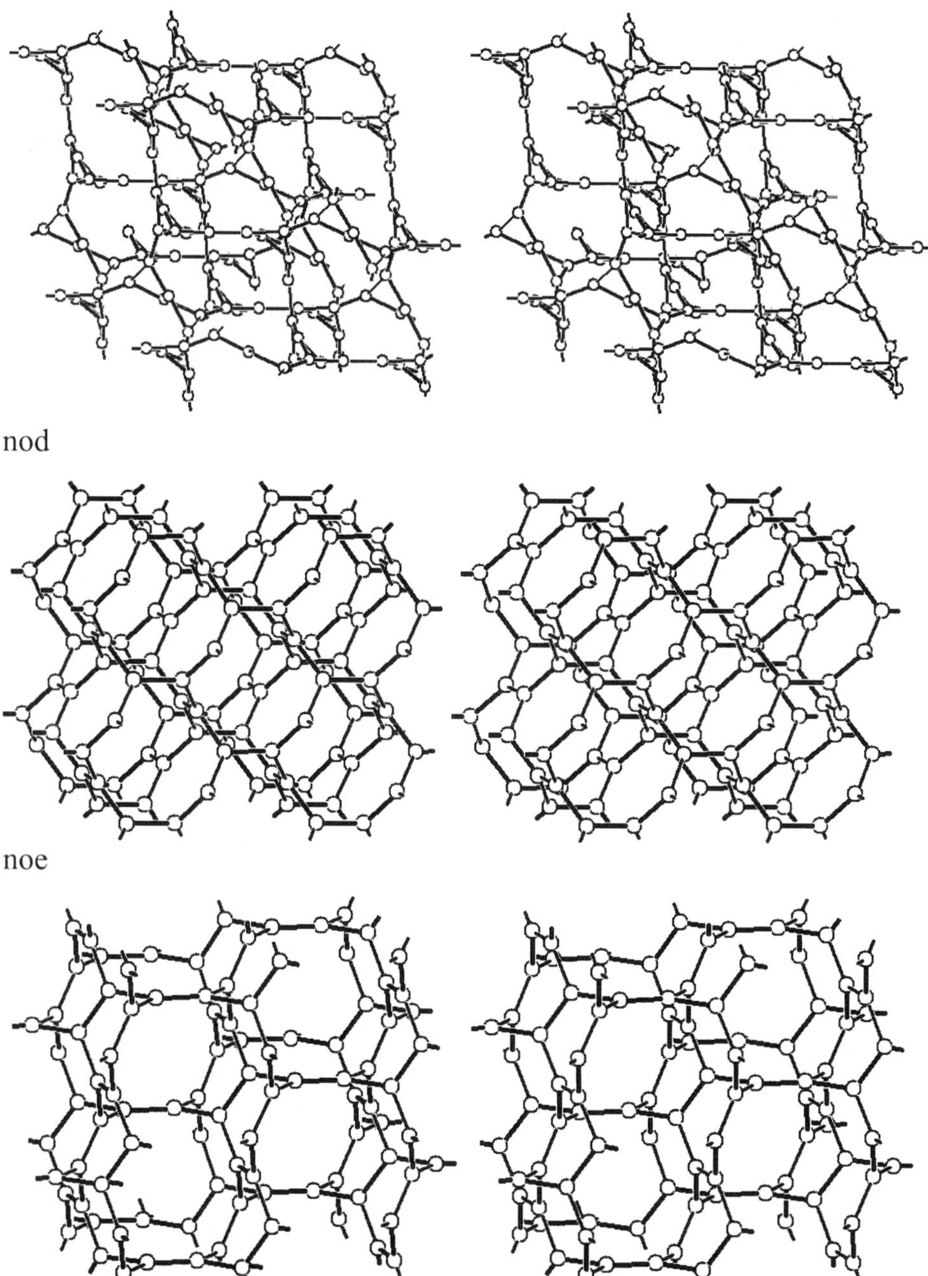

nof

nog

noh

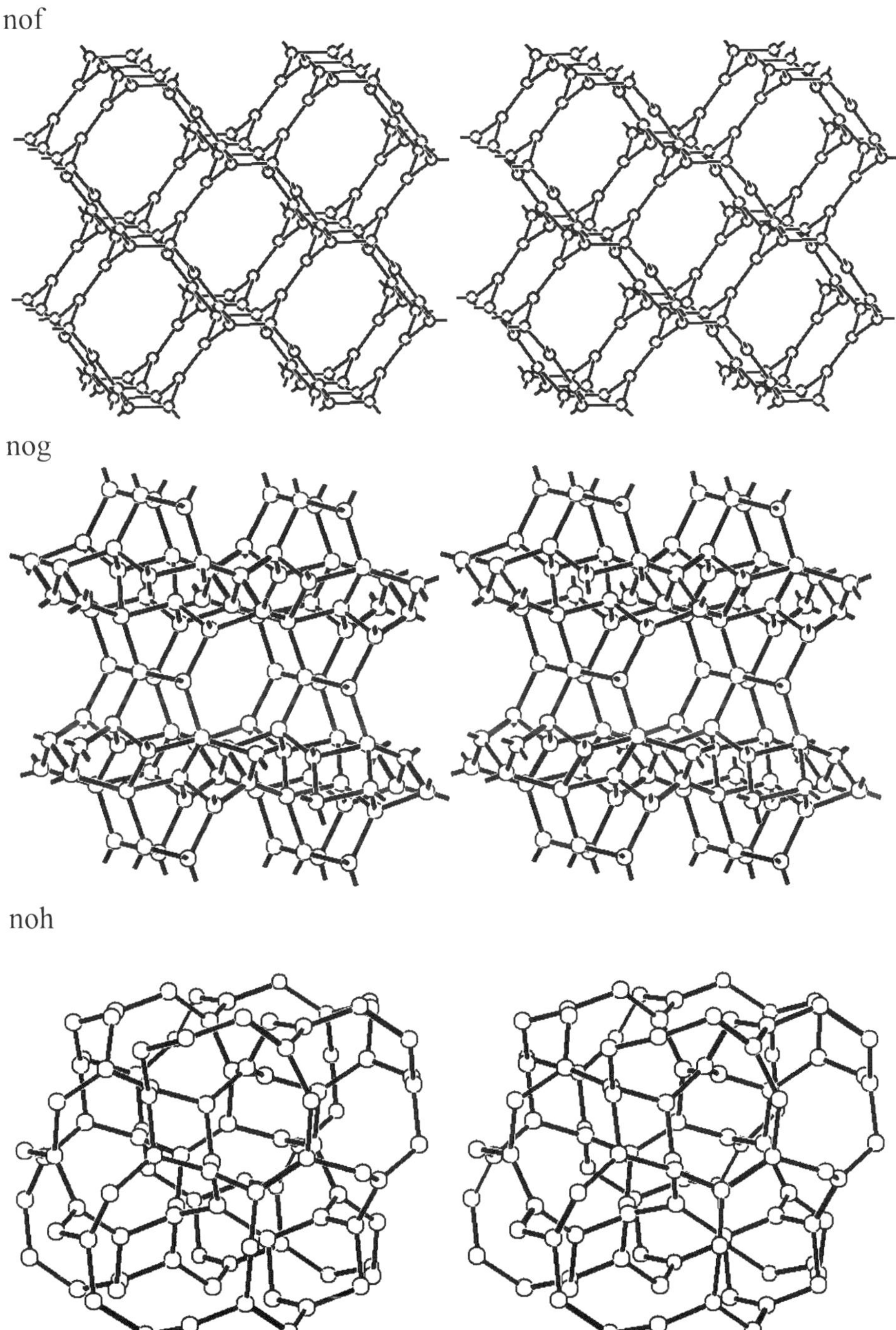

noj

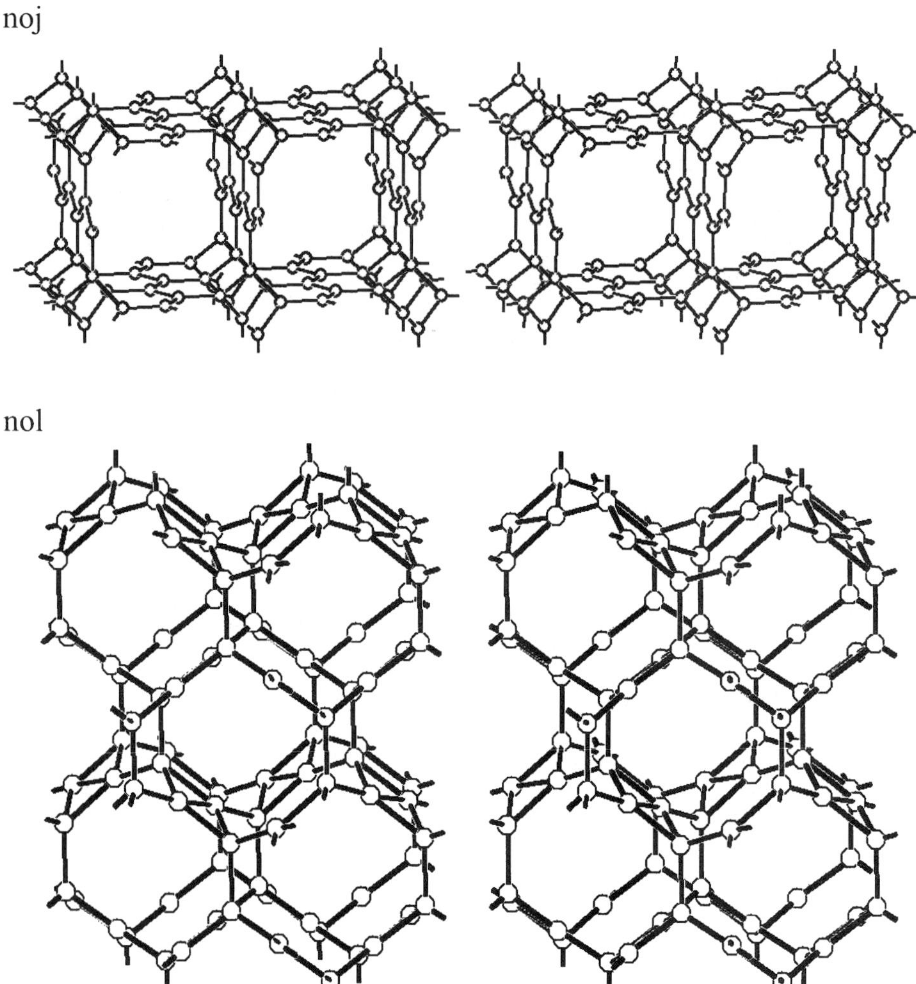

nol

nom

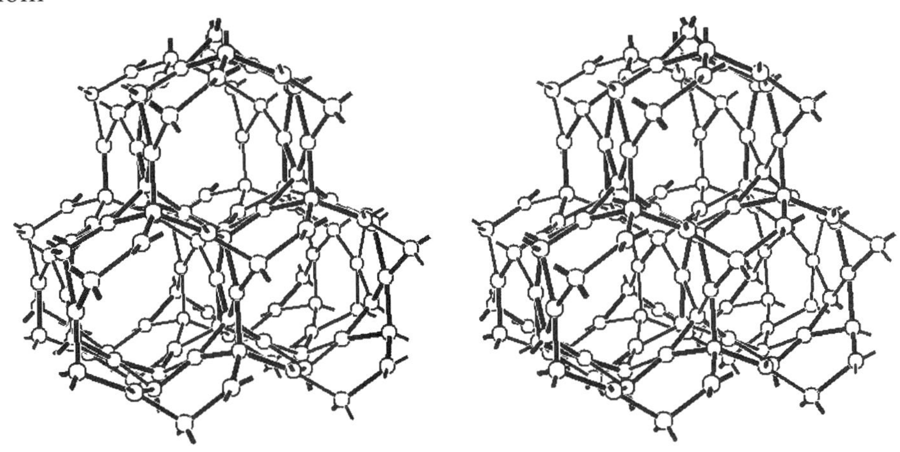

noq

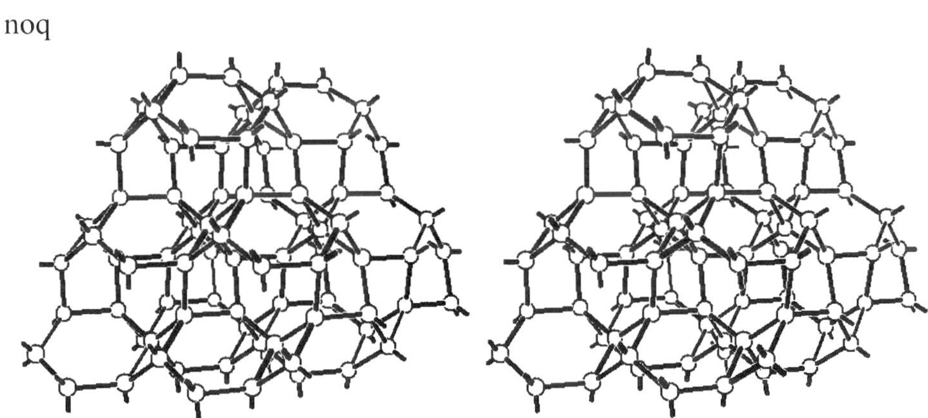

nor

nos

not

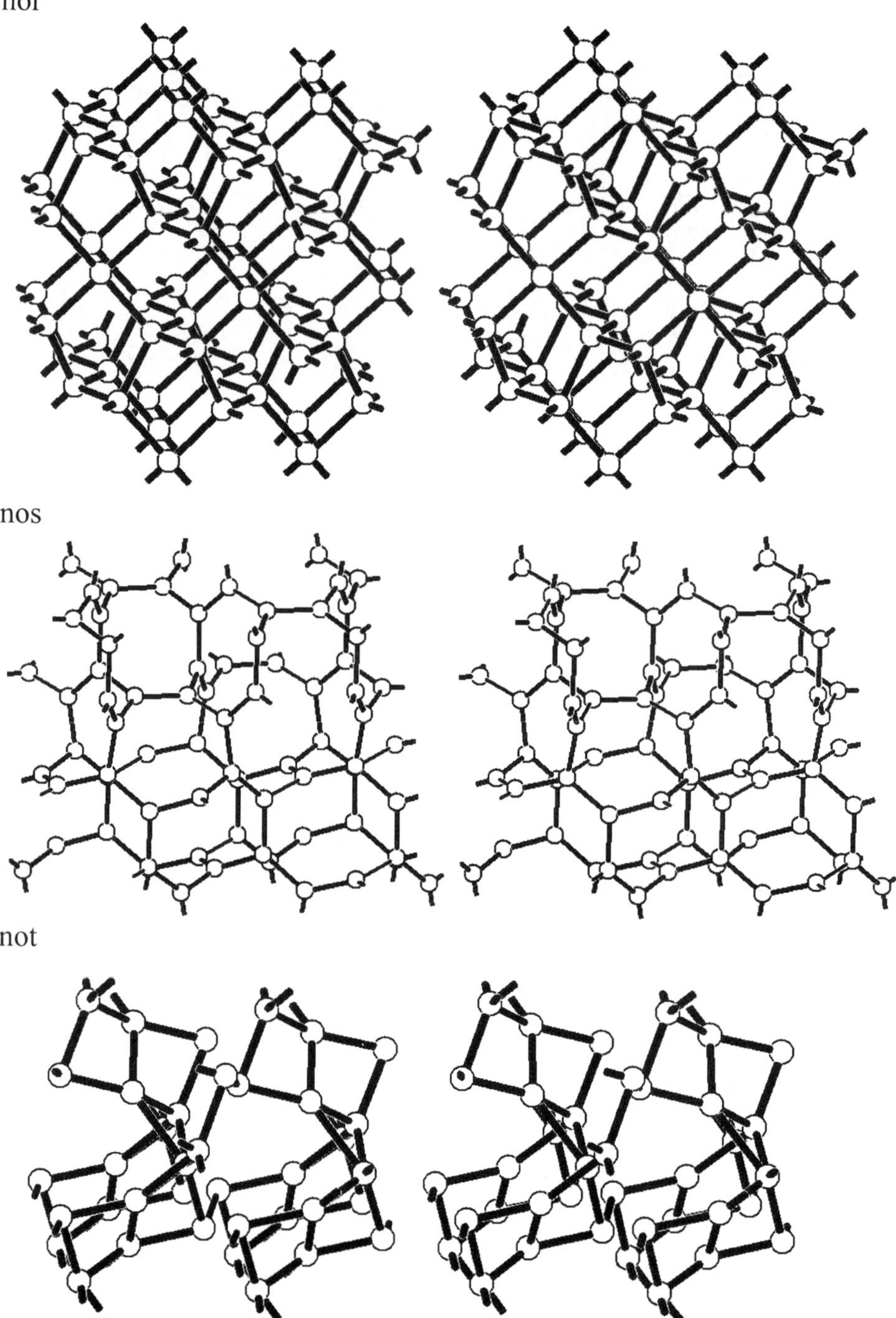

nou

nov

nta

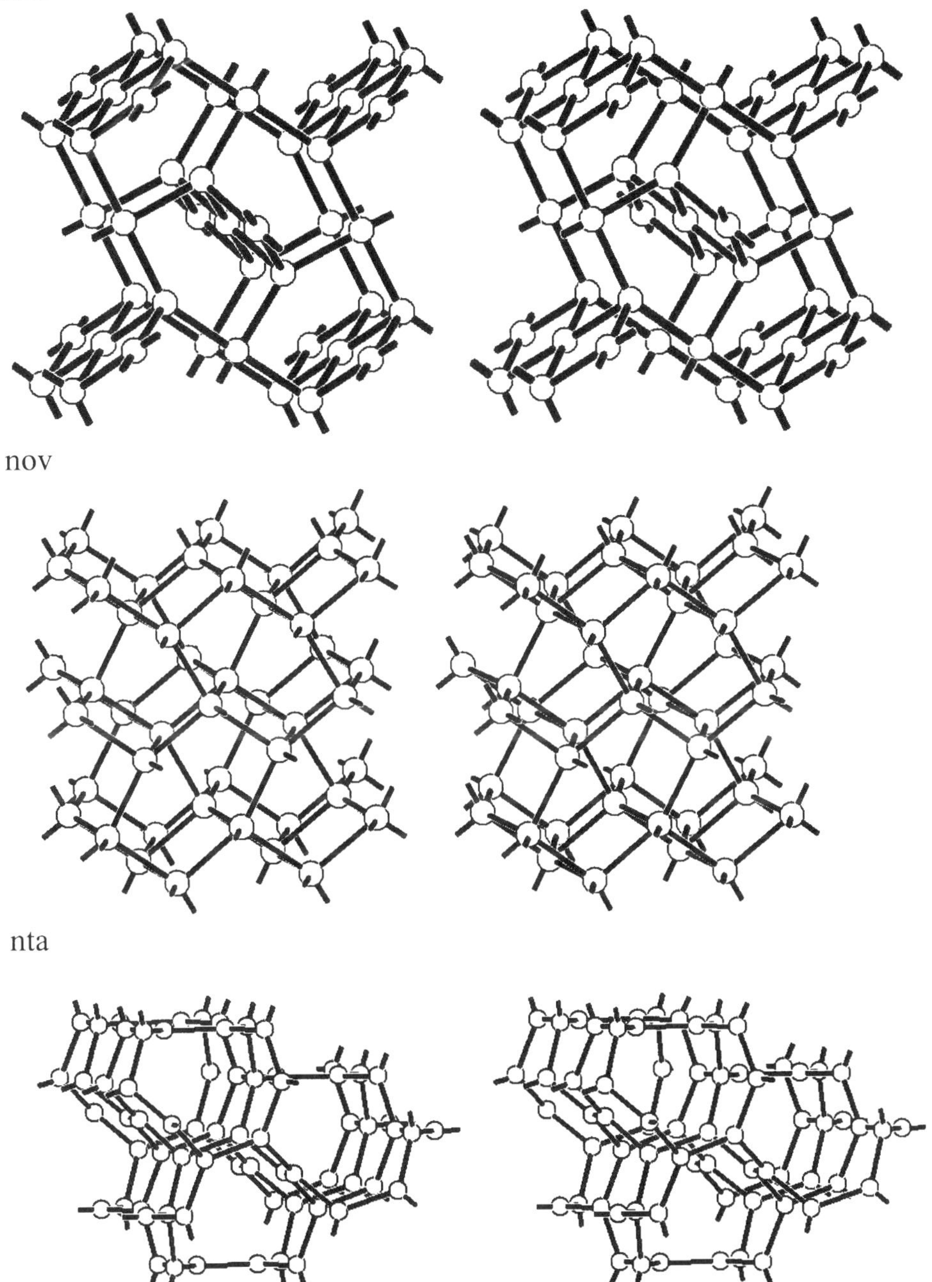

ptt

pyr

pyr-e

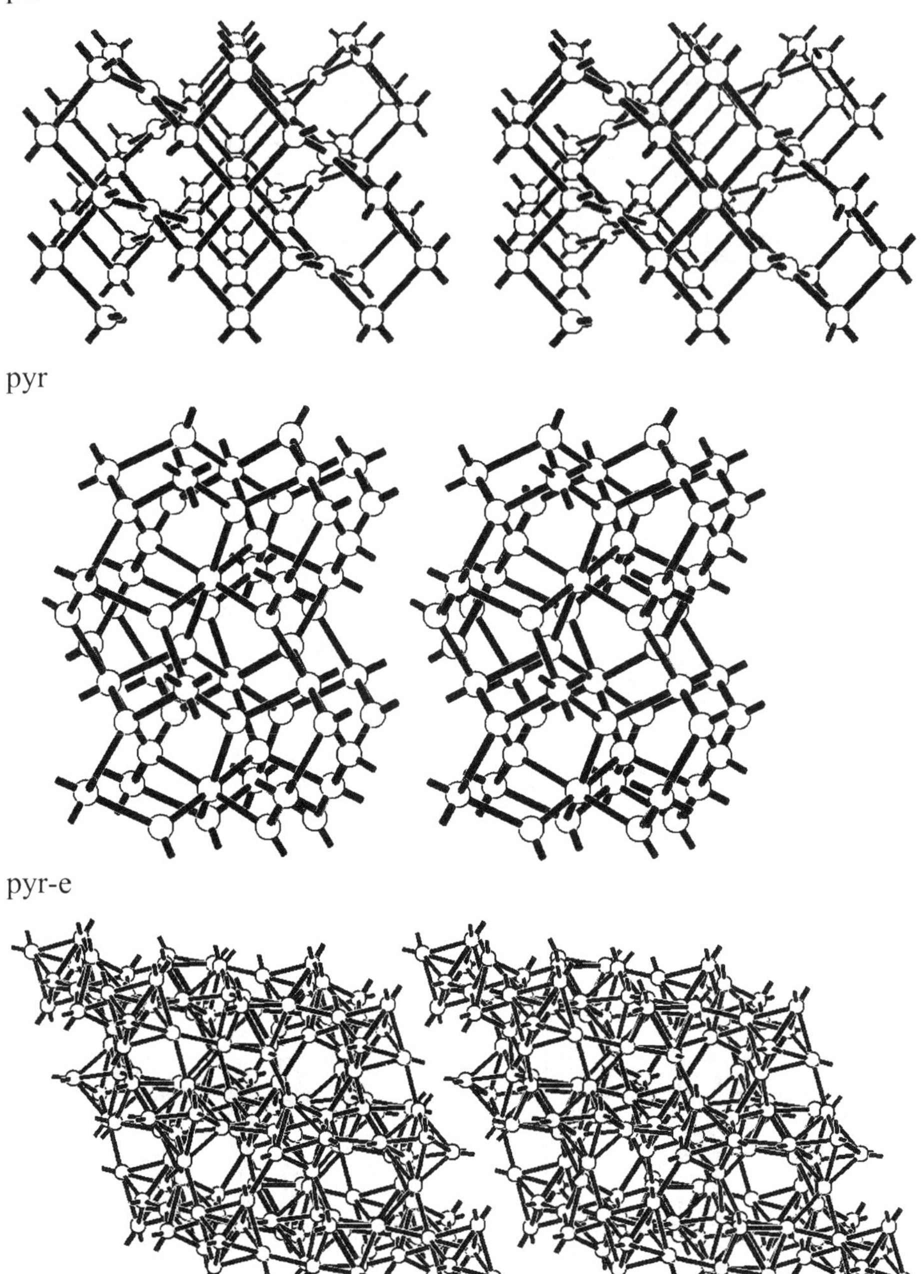

qom

qtz

qzd

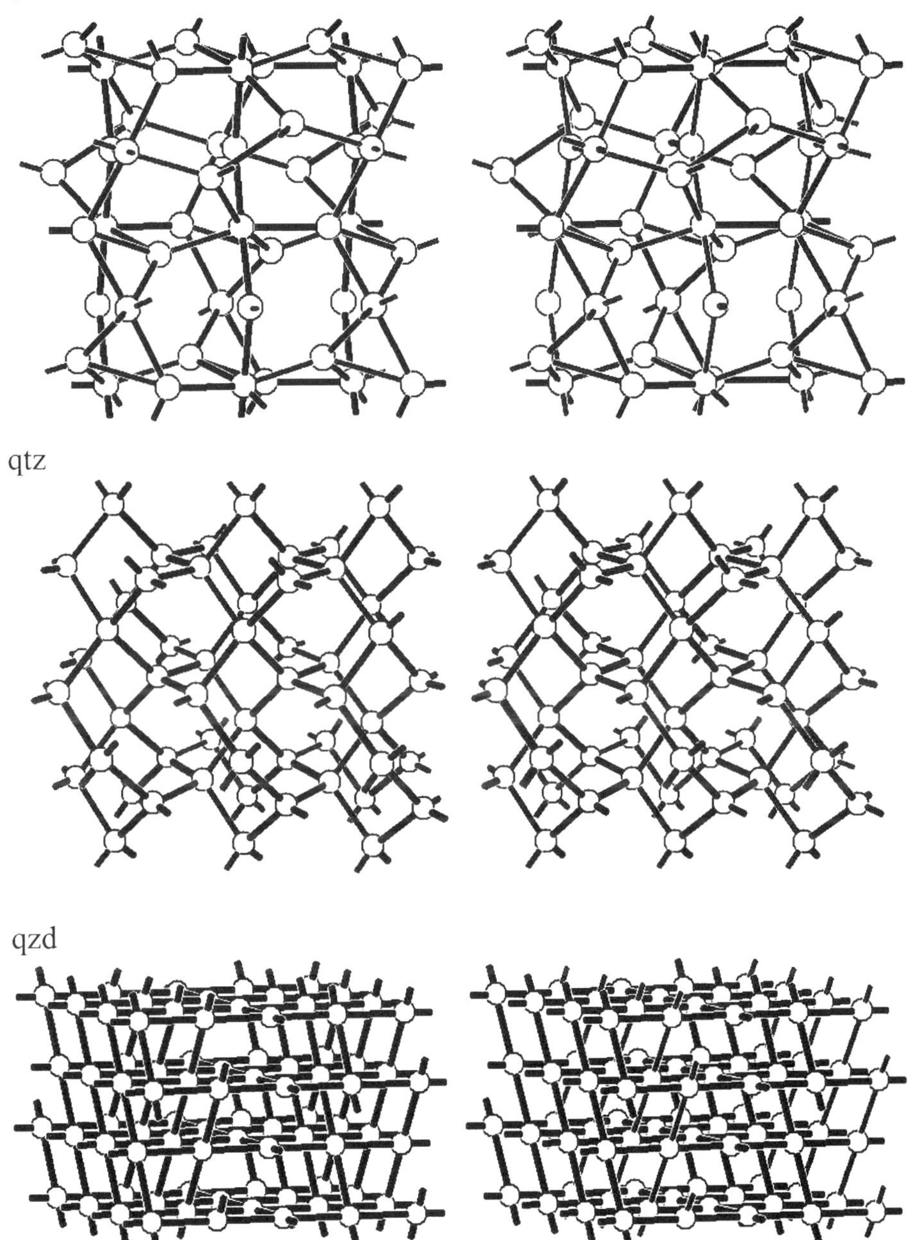

rhr-a

rtl

rtw

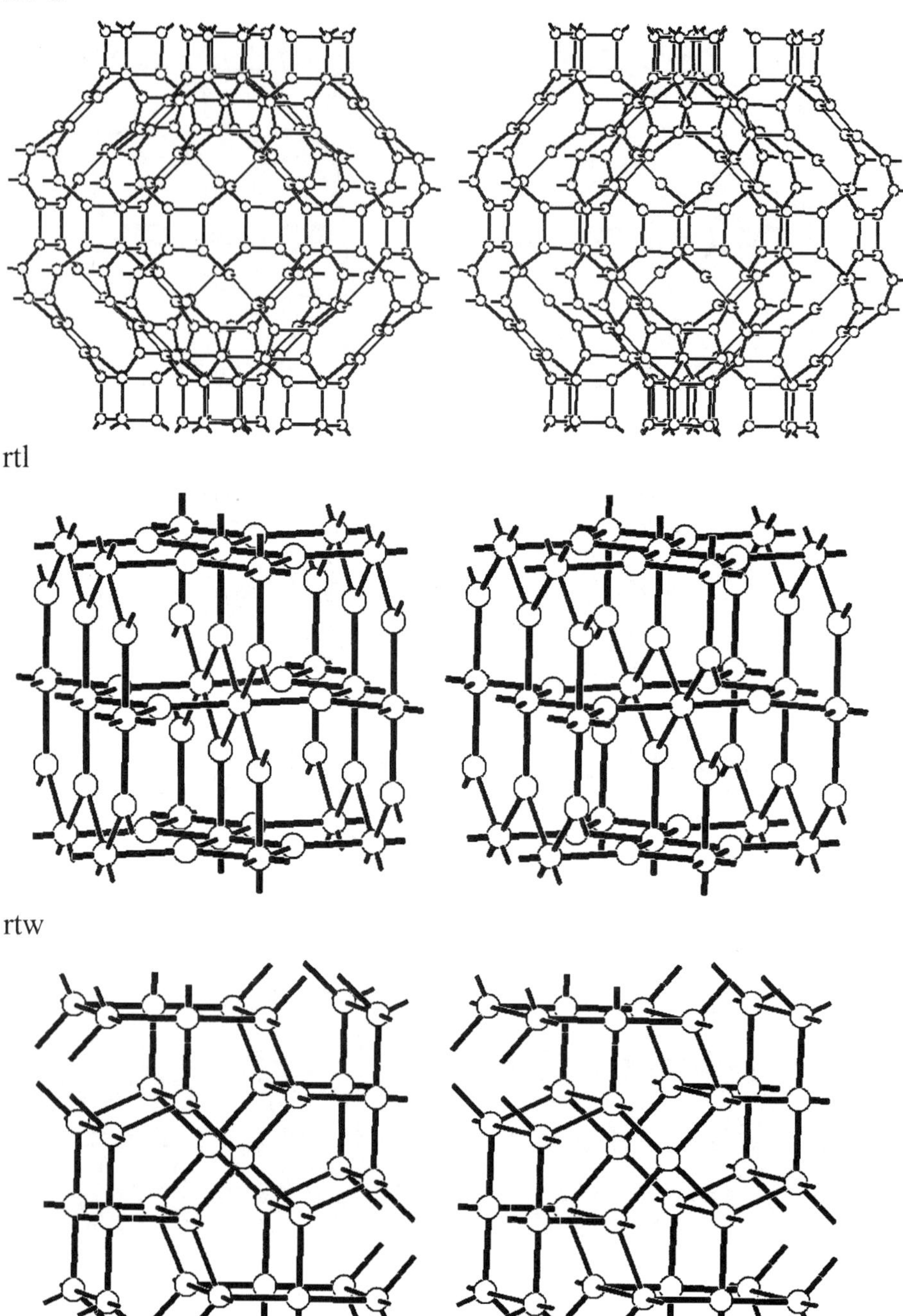

sit

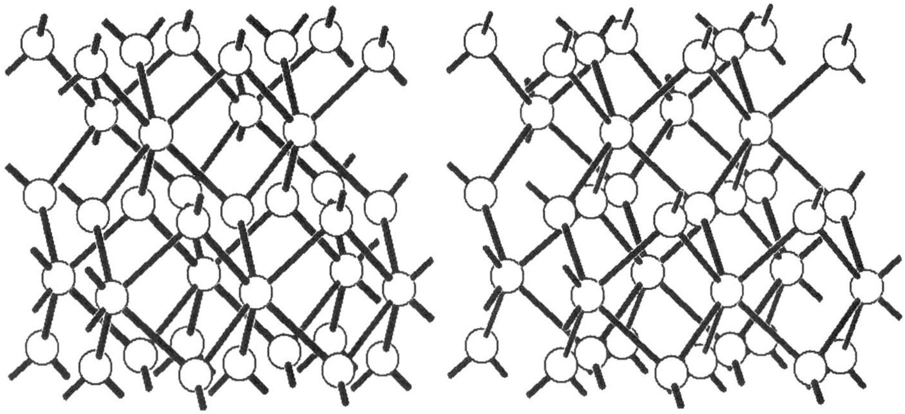

sln

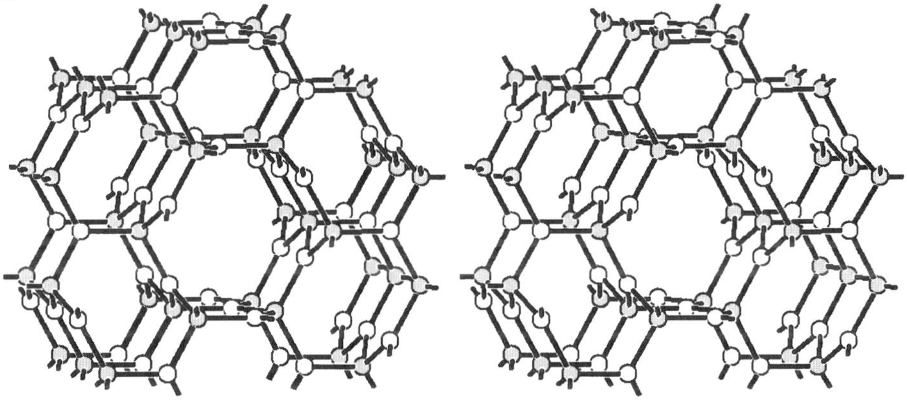

smn

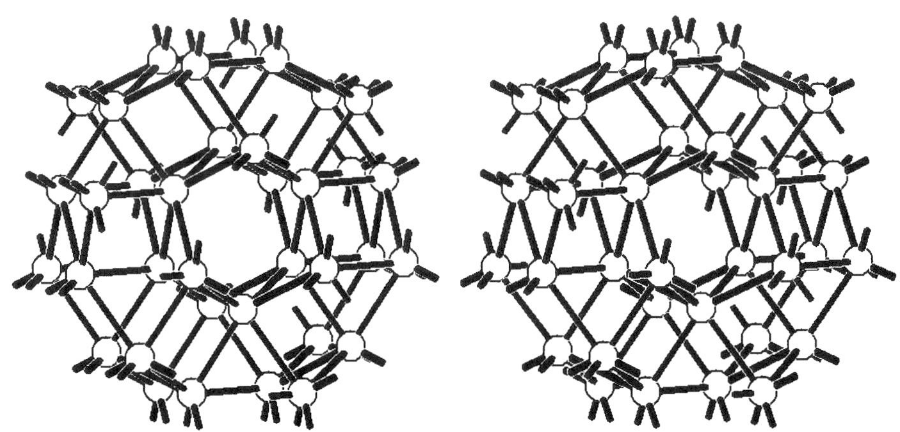

sod

sqp

sra

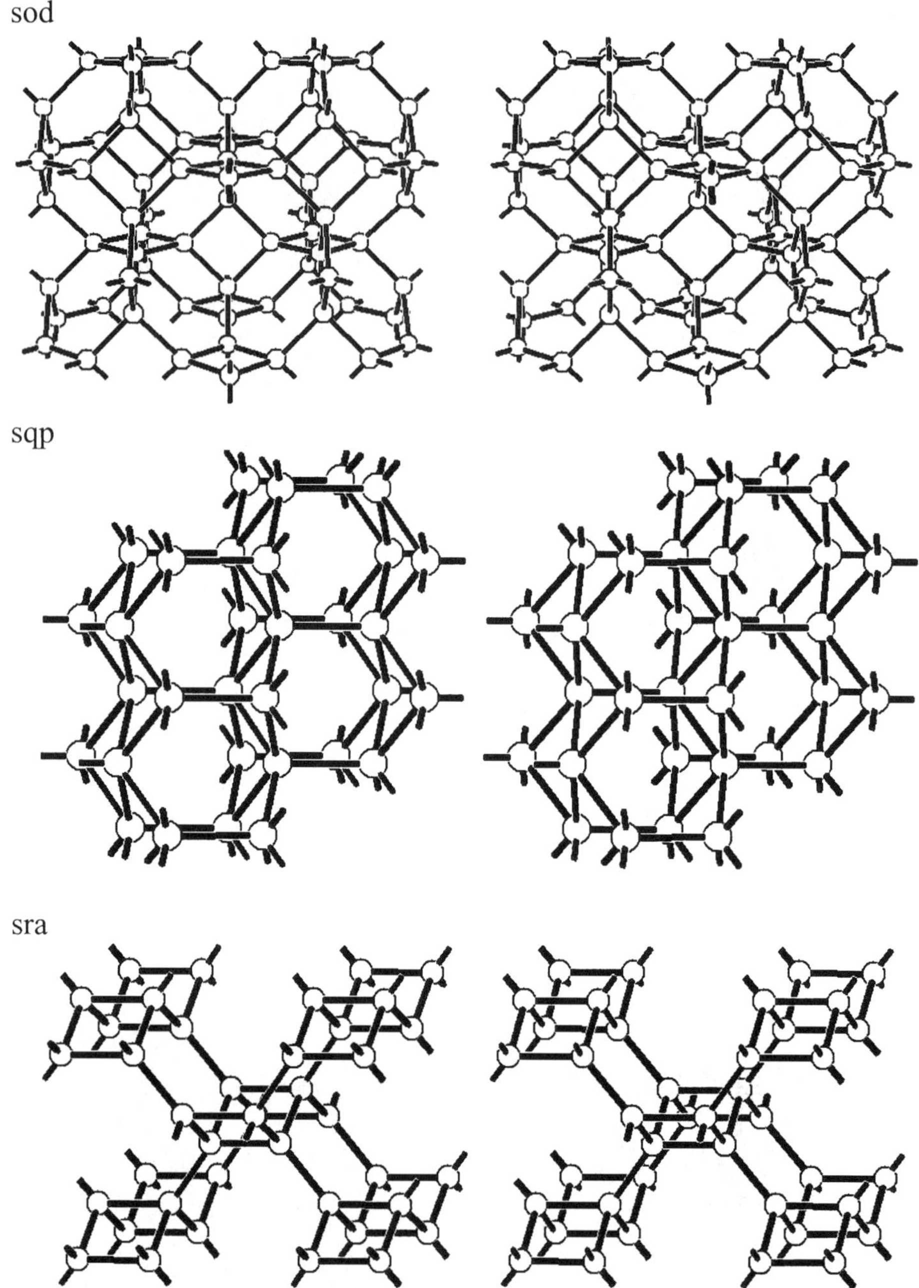

srs

tbo

tcb

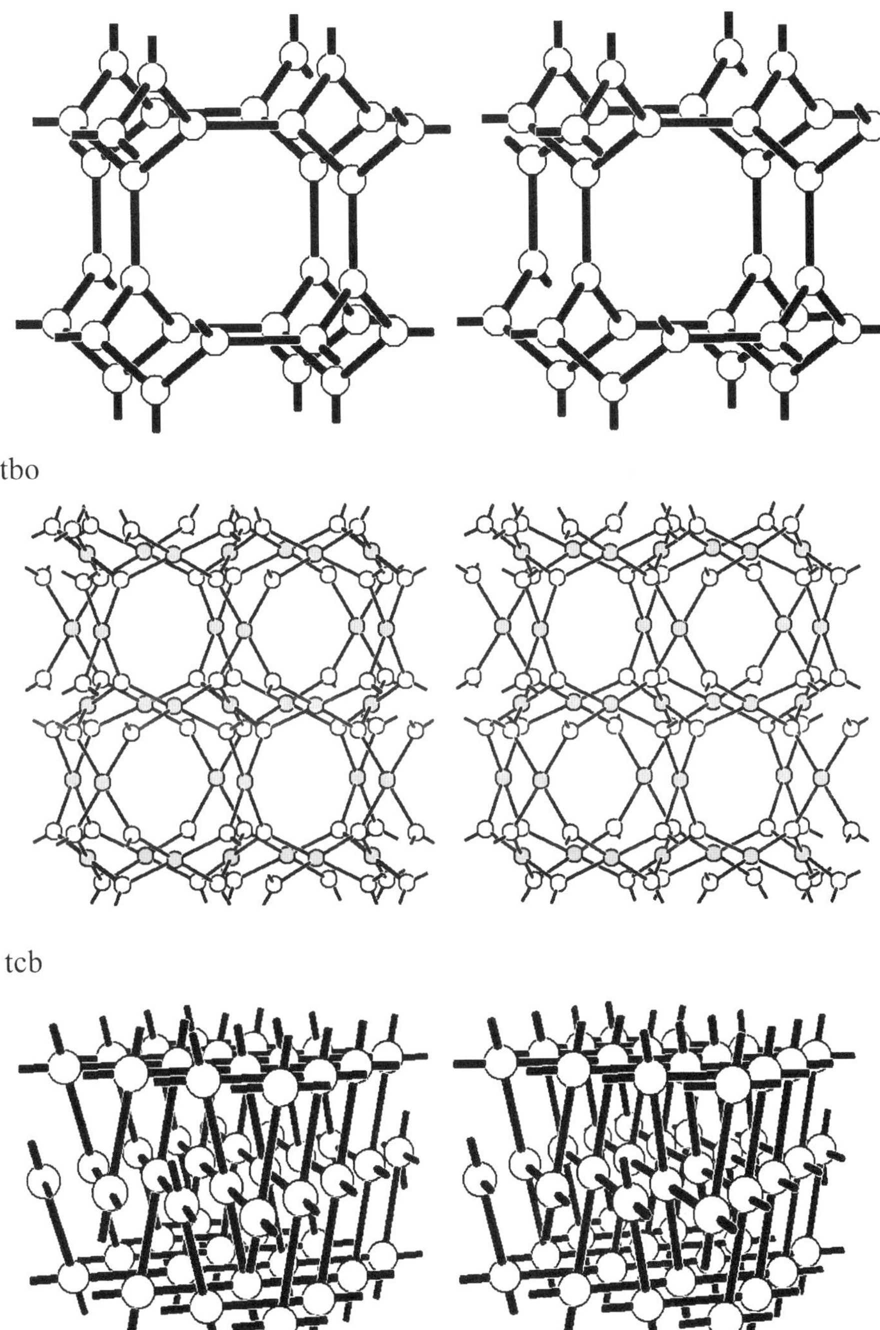

tfa

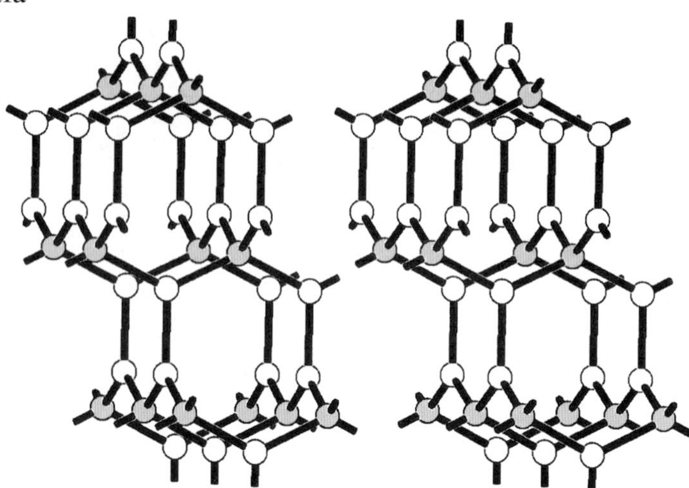

tfc

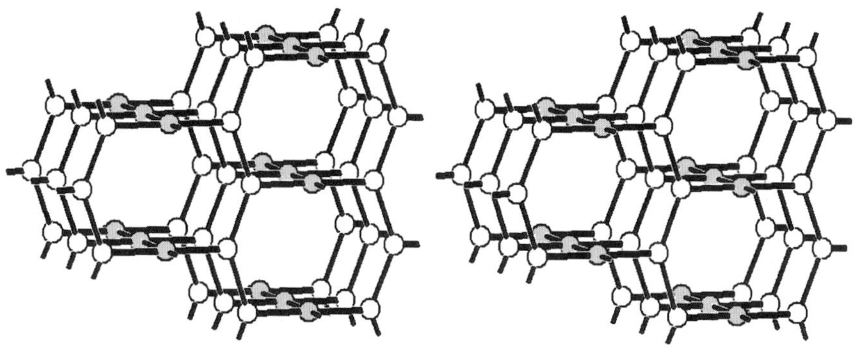

ths

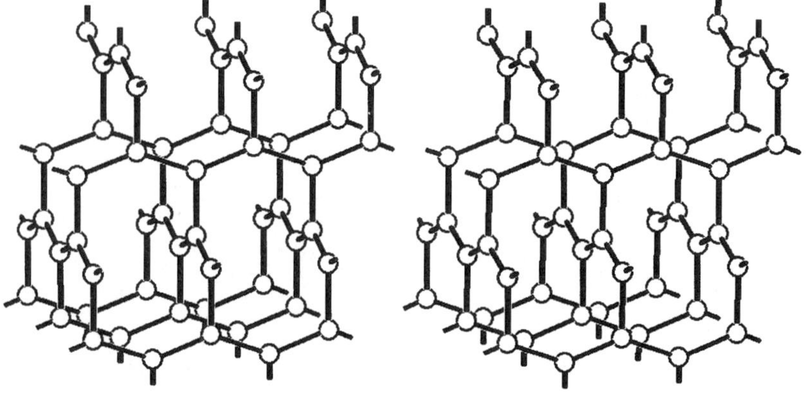

twt

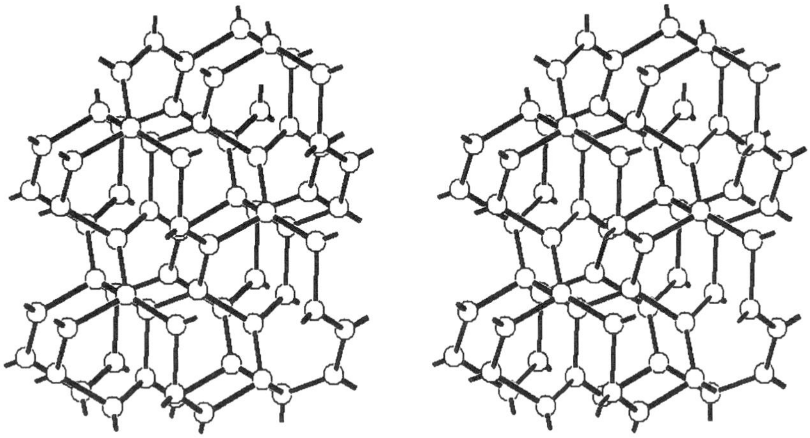

unh

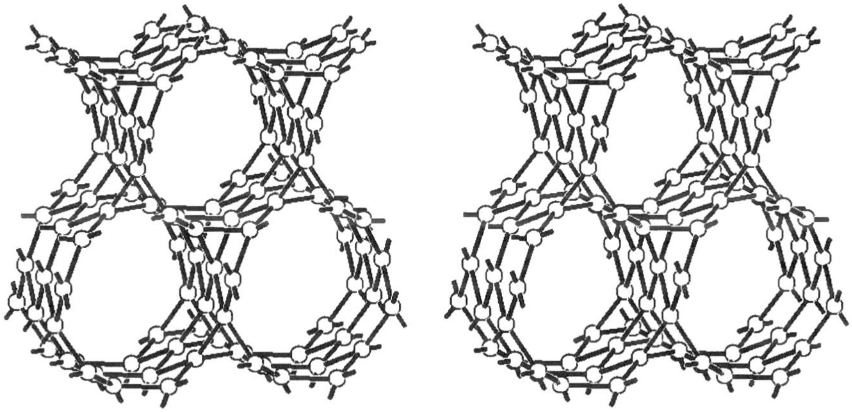

utk

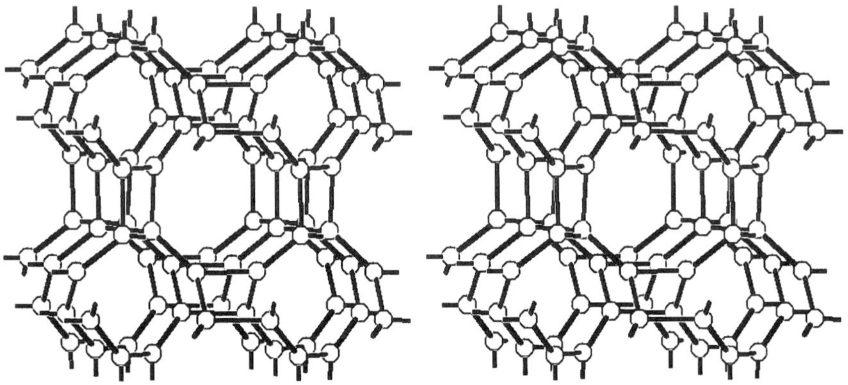

utj

utm

utn

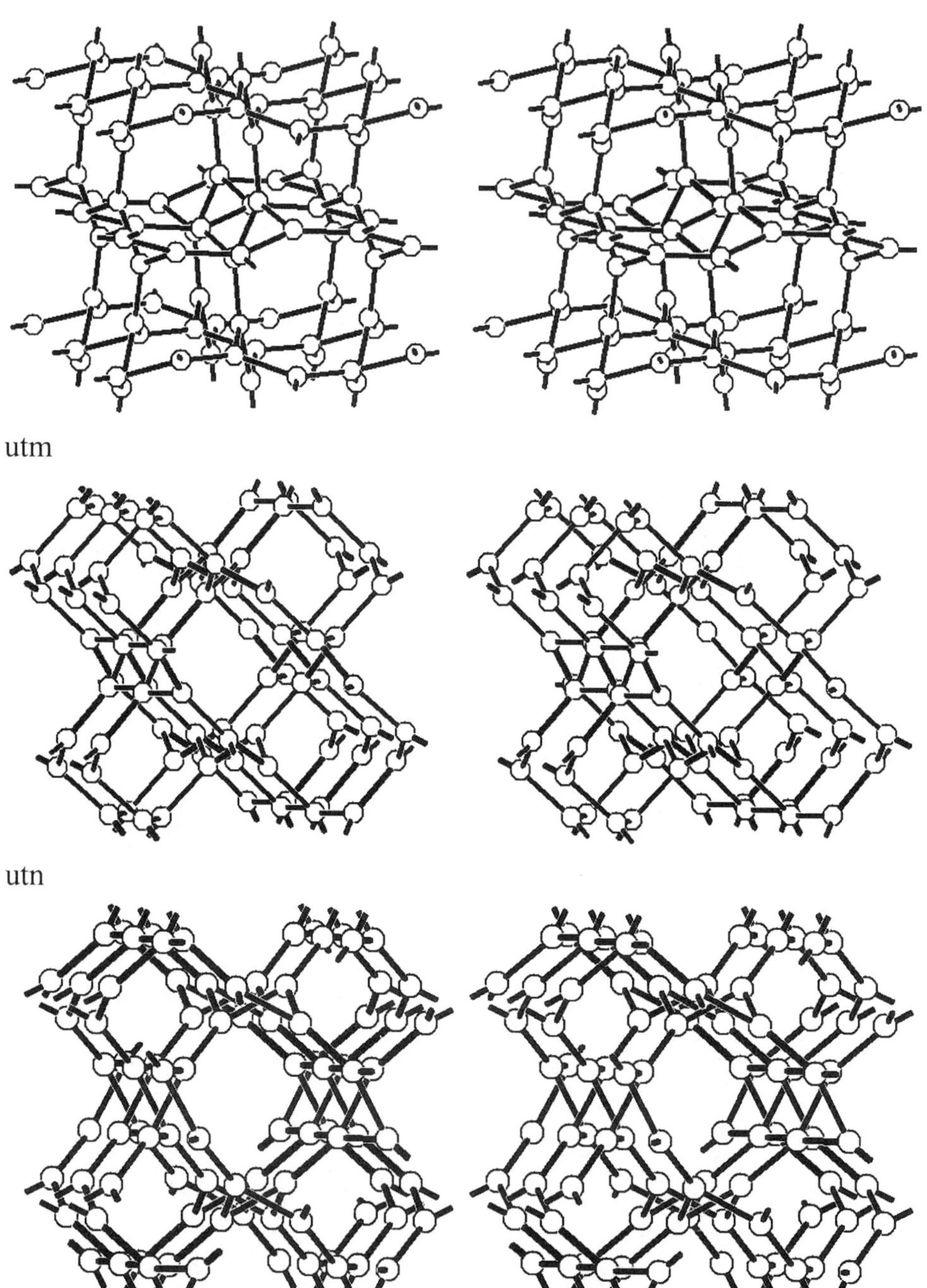

uto

utp

wfq

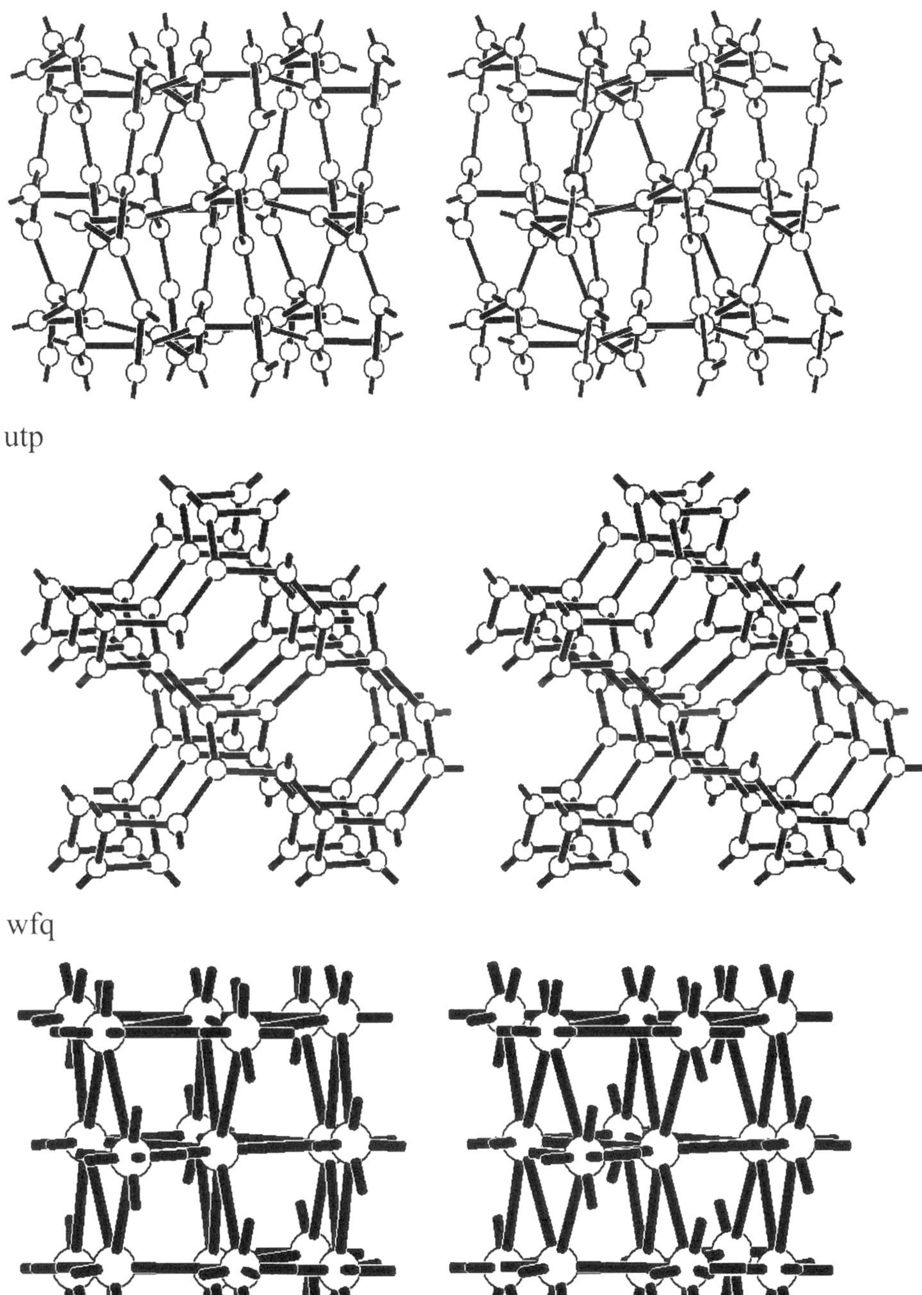

zni

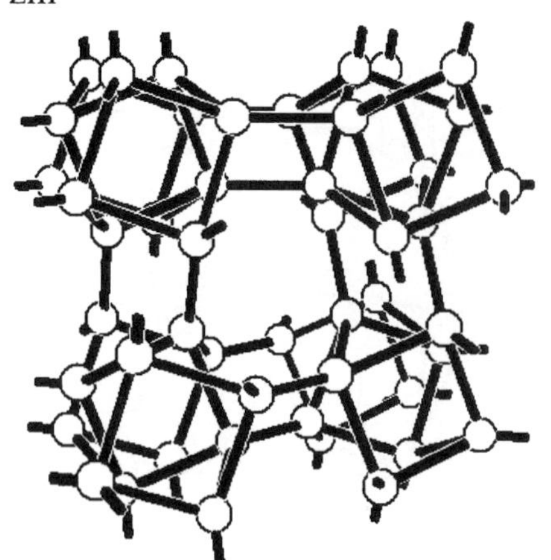 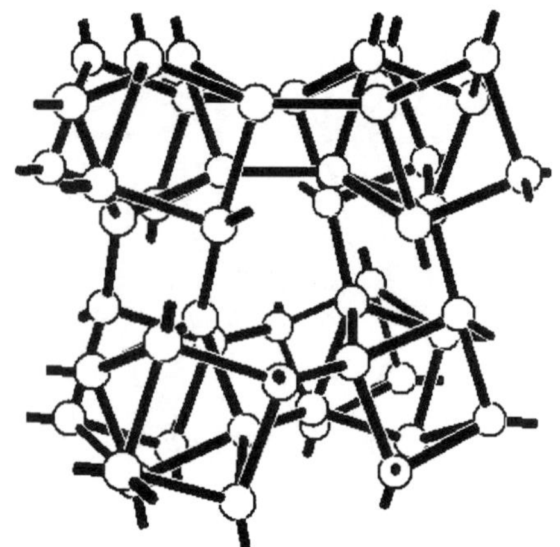

Index